T0215585

THE GEOMETRICAL LANGUAGE OF CONTINUUM MECHANICS

This book presents the fundamental concepts of modern Differential Geometry within the framework of Continuum Mechanics. It is divided into three parts of roughly equal length. The book opens with a motivational chapter to impress upon the reader that Differential Geometry is indeed the natural language of Continuum Mechanics or, better still, that the latter is a prime example of the application and materialization of the former. In the second part, the fundamental notions of Differential Geometry are presented with rigor using a writing style that is as informal as possible. Differentiable manifolds, tangent bundles, exterior derivatives, Lie derivatives, and Lie groups are illustrated in terms of their mechanical interpretations. The third part includes the theory of fibre bundles, G-structures, and groupoids, which are applicable to bodies with internal structure and to the description of material inhomogeneity. The abstract notions of Differential Geometry are thus illuminated by practical and intuitively meaningful engineering applications.

Marcelo Epstein is currently a Professor of Mechanical Engineering at the University of Calgary, Canada. His main research revolves around the various aspects of modern Continuum Mechanics and its applications. A secondary related area of interest is biomechanics. He is a Fellow of the American Academy of Mechanics, recipient of the Cancam prize, and University Professor of Rational Mechanics. He is also adjunct Professor in the Faculties of Humanities and Kinesiology at the University of Calgary.

The Geometrical Language of Continuum Mechanics

Marcelo Epstein
University of Calgary

CAMBRIDGE
UNIVERSITY PRESS

CAMBRIDGE UNIVERSITY PRESS
Cambridge, New York, Melbourne, Madrid, Cape Town,
Singapore, São Paulo, Delhi, Mexico City

Cambridge University Press
The Edinburgh Building, Cambridge CB2 8RU, UK

Published in the United States of America by Cambridge University Press, New York

www.cambridge.org
Information on this title: www.cambridge.org/9781107617032

First published 2010
First paperback edition 2013

A catalogue record for this publication is available from the British Library

Library of Congress Cataloguing in Publication Data
Epstein, M. (Marcelo)
    The geometrical language of continuum mechanics / Marcelo Epstein.
        p.   cm.
    ISBN 978-0-521-19855-4 (hardback)
    1. Continuum mechanics.   I. Title.
    QA808.2.E67 2010
    531–dc22   2010019217

ISBN 978-0-521-19855-4 Hardback
ISBN 978-1-107-61703-2 Paperback

To Silvio's memory
"Heu . . . frater adempte mihi!"

Contents

Preface

Il naufragar m'è dolce in questo mare.

GIACOMO LEOPARDI

In mathematics and the sciences, just as in the arts, there is no substitute for the masterpieces. Fortunately, both in Differential Geometry and in Continuum Mechanics, we possess a veritable treasure trove of fundamental masterpieces, classical as well as modern. Reading them may elicit a pleasure, even an emotion, comparable to that aroused by playing a Mozart piano sonata, the same sensation of perfection and beauty, the sweetness of drowning in such an ocean. This comparison is fair also in a different sense, namely, that for common mortals to achieve these spiritual heights requires a considerable amount of study and work, without which we must resign ourselves to the intuitive feelings evoked by the senses and to the assurances of the critics or the arbiters of taste. The aim of this book is to provide some familiarity with the basic ideas of Differential Geometry as they become actualized in the context of Continuum Mechanics so that the reader can feel more at home with the masters.

Differential Geometry is a rather sophisticated blend of Algebra, Topology, and Analysis. My selection of topics as a nonmathematician is rather haphazard and the depth and rigor of the treatment vary from topic to topic. As befits human nature, I am understandably forgiving of myself for this lack of consistency, but the reader may not be so kind, in which case I will not take it as a personal offence. One way or the other, there is no escape from the fact that the subject matter is extremely difficult and requires dedication and patience, particularly for the various interconnected levels of generalization and abstraction. In this respect, the presence of Continuum Mechanics as a materialization of many of the important geometric notions is of invaluable help. Learning can be regarded as a helical process, whereby each turn elevates us to the next storey of a building with an infinite number of floors. On each floor the same ideas dwell as on all others, but the perspective is wider. The speed of ascent is strongly dependent on the natural abilities and background of the visitor. Ideally, a good book should help to accelerate this process, regardless of the particular floor of departure. In some cases, a good book may help the visitor to abandon the building altogether in justifiable alarm and despair.

The plan and structure of the book emerge quite clearly from a cursory glance at the Contents, which probably reflects my own predilections and priorities. Having been trained initially in Civil Engineering, my own ascent to modern Differential Geometry was rather tortuous and slow: from beams and shells to classical geometry of curves and surfaces, from Linear Elasticity to Continuum Mechanics, General Relativity, continuous distributions of dislocations, and beyond. Along the way, I was privileged to have the help of Jędrzej Śniatycki (Calgary), Reuven Segev (Beer Sheva), Marek Elżanowski (Portland), and Manuel de León (Madrid) – to each of whom I owe more than I can express. These notes are in part the result of two graduate courses in engineering, the first at the invitation of Carlos Corona at the Universidad Politécnica de Madrid in the spring of 2006 and the second at the invitation of Ben Nadler at the University of Alberta in the spring of 2009. Both courses were helpful in providing a sounding board for the presentation and, in the second case, getting assistance from the students in spotting errors in the manuscript and providing valuable suggestions.

Calgary, July 2009

PART ONE

MOTIVATION AND BACKGROUND

1 The Case for Differential Geometry

If Mathematics is the language of Physics, then the case for the use of Differential Geometry in Mechanics needs hardly any advocacy. The very arena of mechanical phenomena is the *space-time continuum*, and a continuum is another word for a *differentiable manifold*. Roughly speaking, this foundational notion of Differential Geometry entails an entity that can support *smooth fields*, the physical nature of which is a matter of context. In Continuum Mechanics, as opposed to Classical Particle Mechanics, there is another continuum at play, namely, the *material body*. This continuous collection of particles, known also as the *body manifold*, supports fields such as temperature, velocity and stress, which interact with each other according to the physical laws governing the various phenomena of interest. Thus, we can appreciate how Differential Geometry provides us with the proper mathematical framework to describe the two fundamental entities of our discourse: the space-time manifold and the body manifold. But there is much more.

When Lagrange published his treatise on analytical mechanics, he was in fact creating, or at least laying the foundations of, a Geometrical Mechanics. A classical mechanical system, such as the plane double pendulum shown in Figure 1.4, has a finite number of degrees of freedom. In this example, because of the constraints imposed by the constancy of the lengths of the links, this number is 2. The first mass may at most describe a circumference with a fixed centre, while the second mass may at most describe a circumference whose centre is at the instantaneous position of the first mass. As a result, the surface of a torus emerges in this example as the descriptor of the *configuration space* of the system. Not surprisingly, this configuration space is, again, a differentiable manifold. This notion escaped Lagrange, who regarded the degrees of freedom as coordinates, without asking the obvious question: coordinates of what? It was only later, starting with the work of Riemann, that the answer to this question was clearly established. The torus, for example (or the surface of a sphere, for that matter), cannot be covered with a single coordinate system. Moreover, as Lagrange himself knew, the choice of coordinate systems is quite arbitrary. The underlying geometrical object, however, is always one and the same. This distinction between the underlying geometrical (and physical) entity, on the one hand, and the coordinates used to represent it, on the other hand, is one of the essential features of modern Differential Geometry. It is also an essential feature of modern Physics. The formulation of physical principles, such as the principle of virtual power, may attain a high degree of simplicity when expressed in geometrical terms. When moving into the realm of Continuum Mechanics, the situation gets complicated by the fact that the number of degrees of freedom of a continuous system is infinite. Nevertheless, at least in principle, the geometric picture is similar.

There is yet another aspect, this time without a Particle Mechanics counterpart, where Differential Geometry makes a natural appearance in Continuum Mechanics. This is the realm of *constitutive equations*. Whether because of having smeared the molecular interactions, or because of the need to agree with experimental results at a macroscopic level in a wide variety of materials, or for other epistemological reasons, the physical laws of Continuum Mechanics do not form a complete system. They need to be supplemented with descriptors of material response known as *constitutive laws* expressed in terms of constitutive equations. When seen in the context of infinite dimensional configuration spaces, as suggested above, the constitutive equations themselves can be regarded as geometric objects. Even without venturing into the infinite-dimensional domain, it is a remarkable fact that the specification of the constitutive equations of a material body results in a well-defined differential geometric structure, a sort of *material geometry*, whose study can reveal the presence of continuous distributions of material defects or other kinds of *material inhomogeneity*.

In the remainder of this motivational chapter, we will present in a very informal way some basic geometric differential concepts as they emerge in appropriate physical contexts. The concept of differentiable manifold (or just *manifold*, for short) will be assumed to be available, but we will content ourselves with the mental picture of a continuum with a definite dimension.[1] Not all the motivational lines suggested in this chapter will be pursued later in the book. It is also worth pointing out that, to this day, the program of a fully fledged geometrization of Continuum Mechanics cannot be said to have been entirely accomplished.

1.1. Classical Space-Time and Fibre Bundles

1.1.1. Aristotelian Space-Time

We may think separately of time as a 1-dimensional manifold $\mathcal{Z}$ (the time line) and of space as a 3-dimensional manifold $\mathcal{P}$. Nevertheless, as soon as we try to integrate these two entities into a single space-time manifold $\mathcal{S}$, whose points represent *events*, we realize that there are several possibilities. The first possibility that comes to mind is what we may call *Aristotelian space-time*, whereby time and space have independent and absolute meanings. Mathematically, this idea corresponds to the product:

$$\mathcal{S}_A = \mathcal{Z} \times \mathcal{P}, \tag{1.1}$$

---

[1] The rigorous definition of a manifold will be provided later in Chapter 4.

where $\times$ denotes the Cartesian product. Recall that the *Cartesian product* of two sets is the set formed by all ordered pairs such that the first element of the pair belongs to the first set and the second element belongs to the second set. Thus, the elements s of $\mathcal{S}_A$, namely, the events, are ordered pairs of the form (t,p), where $t \in \mathcal{Z}$ and $p \in \mathcal{P}$. In other words, for any given $s \in \mathcal{S}_A$, we can determine independently its corresponding temporal and spatial components. In mathematical terms, we say that the 4-dimensional (product) manifold $\mathcal{S}_A$ is endowed with two *projection maps*:

$$\pi_1 : \mathcal{S}_A \longrightarrow \mathcal{Z}, \tag{1.2}$$

and

$$\pi_2 : \mathcal{S}_A \longrightarrow \mathcal{P}, \tag{1.3}$$

defined, respectively, by:

$$\pi_1(s) = \pi_1(t,p) := t, \tag{1.4}$$

and

$$\pi_2(s) = \pi_2(t,p) := p. \tag{1.5}$$

1.1.2. Galilean Space-Time

The physical meaning of the existence of these two natural projections is that any observer can tell independently whether two events are simultaneous and whether or not (regardless of simultaneity) they have taken place at the same location in space. According to the principle of *Galilean relativity*, however, this is not the case. Two different observers agree, indeed, on the issue of simultaneity. They can tell unequivocally, for instance, whether or not two light flashes occurred at the same time and, if not, which preceded which and by how much. Nevertheless, in the case of two nonsimultaneous events, they will in general disagree on the issue of position. For example, an observer carrying a pulsating flashlight will interpret successive flashes as happening always "here," while an observer receding uniformly from the first will reckon the successive flashes as happening farther and farther away as time goes on. Mathematically, this means that we would like to get rid of the nonphysical second projection (the spatial one) while preserving the first projection.

We would like, accordingly, to construct an entity that looks like $\mathcal{S}_A$ for each observer, but which is a different version of $\mathcal{S}_A$, so to speak, for different observers. This delicate issue can be handled as follows. We define space-time as a 4-dimensional manifold $\mathcal{S}$ endowed with a projection map:

$$\pi : \mathcal{S} \longrightarrow \mathcal{Z}, \tag{1.6}$$

together with a collection of smooth and (smoothly) invertible maps:

$$\phi : \mathcal{S} \longrightarrow \mathcal{S}_A, \tag{1.7}$$

onto the naive Aristotelian space-time $\mathcal{S}_A$. Each of these maps, called a *trivialization* and potentially representing an observer, cannot be completely arbitrary, in a sense that we will now explain.

Fix a particular point of time $t \in \mathcal{Z}$ and consider the inverse image $\mathcal{S}_t = \pi^{-1}(\{t\})$. We call $\mathcal{S}_t$ the *fibre of $\mathcal{S}$ at t*. Recall that the *inverse image* of a subset of the range of a function is the collection of all the points in its domain that are mapped to points in that subset. With this definition in mind, the meaning of $\mathcal{S}_t$ is the collection of all events that may happen at time t. We clearly want this *collection* to be the same for all observers, a fact guaranteed by the existence of the projection map π. Different observers will differ only in that they will attribute possibly different locations to events in this fibre. Therefore, we want the maps ϕ to be *fibre preserving* in the sense that each fibre of $\mathcal{S}$ is mapped to one and the same fibre in $\mathcal{S}_A$. In other words, we don't want to mix in any way whatsoever the concepts of space and time. We require, therefore, that the image of each fibre in $\mathcal{S}$ by each possible ϕ be exactly equal to a fibre of $\mathcal{S}_A$. More precisely, for each $t \in \mathcal{Z}$ we insist that:

$$\phi(\mathcal{S}_t) = \pi_1^{-1}(\{t\}). \tag{1.8}$$

A manifold $\mathcal{S}$ endowed with a projection π onto another manifold $\mathcal{Z}$ (called the *base manifold*) and with a collection of smooth invertible fibre-preserving maps onto a product manifold $\mathcal{S}_A$ (of the base times another manifold $\mathcal{P}$) is known as a *fibre bundle*. Note that the fibres of $\mathcal{S}_A$ by π_1 are all exact copies of $\mathcal{P}$. We say that $\mathcal{P}$ is the *typical fibre* of $\mathcal{S}$. A suggestive pictorial representation of these concepts is given in Figure 1.1. A more comprehensive treatment of fibre bundles will be presented in Chapter 7.

EXAMPLE 1.1. **Microstructure:** A completely different application of the notion of fibre bundle is the description of bodies with internal structure, whereby the usual kinematic degrees of freedom are supplemented with extra degrees of freedom intended to describe a physically meaningful counterpart. This idea, going at least as far back as the work of the Cosserat brothers,[2] applies to diverse materials, such as liquid crystals and granular media. The base manifold represents the matrix, or *macromedium*, while the fibres represent the *micromedium* (the elongated molecules or the grains, as the case may be).

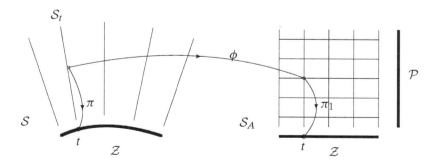

Figure 1.1. A fibre bundle

[2] Cosserat E, Cosserat F (1909), *Théorie des corps déformables*, Hermann et Fils, Paris.

Notice in Figure 1.1 how the fibres are shown hovering above (rather than touching) the base manifold. This device is used to suggest that, although each fibre is assigned to a specific point of the base manifold, the fibre and the base do not have any points in common, nor is there any preferential point in the fibre (such as a zero). Quite apart from the ability of Differential Geometry to elicit simple mental pictures to describe very complex objects, such as a fibre bundle, another important feature is that it uses the minimal amount of structure necessary. In the case of the space-time bundle, for instance, notice that we have not made any mention of the fact that there is a way to measure distances in space and a way to measure time intervals. In other words, what we have presented is what might be called a *proto-Galilean space-time*, where the notion of simultaneity has a physical (and geometrical) meaning. Beyond that, we are now in a position to impose further structure either in the base manifold, or in the typical fibre, or in both. Similarly, restrictions can be placed on the maps ϕ (governing the change of observers). In classical *Galilean space-time*, the fibre $\mathcal{P}$ has the structure of an *affine space* (roughly a vector space without an origin). Moreover, this vector space has a distinguished *dot product*, allowing to measure lengths and angles. Such an affine space is called a *Euclidean space*. The time manifold $\mathcal{Z}$ is assumed to have a Euclidean structure as well, albeit 1-dimensional. Physically, these structures mean that there is an observer-invariant way to measure distances and angles in space (at a given time) and that there is also an observer-invariant way to measure intervals of time. We say, accordingly, that Galilean space-time is an *affine bundle*. In such a fibre bundle, not only the base manifold and the typical fibre are affine spaces, but also the functions ϕ are limited to affine maps. These are maps that preserve the affine properties (for example, parallelism between two lines). In the case of Euclidean spaces, the maps may be assumed to preserve the metric structure as well.

1.1.3. Observer Transformations

Having identified an observer with a trivialization ϕ, we can consider the notion of *observer transformation*. Let $\phi_1 : \mathcal{S} \to \mathcal{S}_A$ and $\phi_2 : \mathcal{S} \to \mathcal{S}_A$ be two trivializations. Since each of these maps is, by definition, invertible and fibre preserving, the composition:

$$\phi_{1,2} = \phi_2 \circ \phi_1^{-1} : \mathcal{S}_A \to \mathcal{S}_A, \tag{1.9}$$

is a well-defined fibre-preserving map from $\mathcal{S}_A$ onto itself. It represents the transformation from observer number 1 to observer number 2. Because of fibre preservation, the map $\phi_{1,2}$ can be seen as a smooth collection of time-dependent maps $\tilde{\phi}_{1,2}^t$ of the typical fibre $\mathcal{P}$ onto itself, as shown schematically in Figure 1.2. In Galilean space-time proper, we limit these maps to affine maps that preserve the orientation and the metric (Euclidean) structure of the typical fibre $\mathcal{P}$ (which can be seen as the usual 3-dimensional Euclidean space).

Among all such maps $\tilde{\phi}_{1,2}^t : \mathcal{P} \to \mathcal{P}$, it is possible to distinguish some that not only preserve the Euclidean structure but also represent changes of observers that travel with respect to each other at a fixed inclination (i.e., without angular velocity) and at

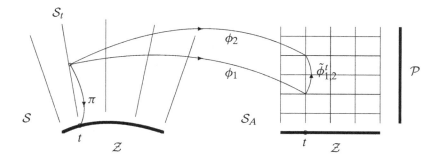

Figure 1.2. Observer transformation

a constant velocity of relative translation. Observers related in this way are said to be *inertially related*. It is possible, accordingly, to divide the collection of all observers into equivalence classes of inertially related observers. Of all these inertial classes, Isaac Newton declared one to be privileged above all others.[3] This is the class of *inertial observers*, for which the laws of Physics acquire a particularly simple form.

1.1.4. Cross Sections

A *cross section* (or simply a *section*) of a fibre bundle $\mathcal{S}$ is a map Γ of the base manifold $\mathcal{Z}$ to the fibre bundle itself:

$$\Gamma : \mathcal{Z} \to \mathcal{S}, \qquad (1.10)$$

with the property:

$$\pi \circ \Gamma = id_{\mathcal{Z}}, \qquad (1.11)$$

where $id_{\mathcal{Z}}$ is the identity map of the base manifold and where "$\circ$" denotes the composition of maps. This property expresses the fact that the image of each element of the base manifold is actually in the fibre attached to that element. A convenient way to express this fact is by means of the following commutative diagram:

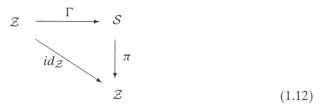

$$(1.12)$$

Pictorially, as shown in Figure 1.3, the image of a section looks like a line cutting through the fibres, hence its name. For general fibre bundles, there is no guarantee that a smooth (or even continuous) cross section exists.

In the case of Galilean space-time, a section represents a *world line* or, more classically, a *trajectory* of a particle.

---

[3] This appears to be the meaning of Newton's first law of motion.

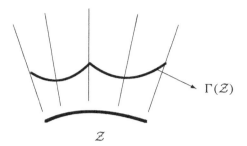

Figure 1.3. A (not necessarily smooth) section

EXERCISE 1.1. Explain on physical grounds why a world-line must be a cross section, that is, it cannot be anywhere tangent to a (space) fibre of space-time.

EXERCISE 1.2. How does the world-line of a particle at rest look in Aristotelian space-time? Let $\phi : \mathcal{S} \to \mathcal{S}_A$ be a trivialization of Galilean space-time, and let $\sigma : \mathcal{Z} \to \mathcal{S}_A$ be a constant section of Aristotelian space-time. Write an expression for the world-line of a particle at rest with respect to the observer defined by ϕ. Do constant sections exist in an arbitrary fibre bundle?

EXERCISE 1.3. Draw schematically a world diagram for a collision of two billiard balls. Comment on smoothness.

EXERCISE 1.4. How does the motion of a cloud of particles look in space-time? Justify the term "world-tube" for the trajectory of a continuous medium of finite extent.

1.1.5. Relativistic Space-time

The revolution brought about by the theory of Relativity (both in its special and general varieties) can be said to have destroyed the bundle structure altogether. In doing so, it in fact simplified the geometry of space-time, which becomes just a 4-dimensional manifold $\mathcal{S}_R$. On the other hand, instead of having two separate metric structures, one for space and one for time, Relativity assumes the existence of a space-time metric structure that involves both types of variables into a single construct. This type of metric structure is what Riemann had already considered in his pioneering work on the subject, except that Relativity (so as to be consistent with the Lorentz transformations) required a metric structure that could lead both to positive and to negative squared distances between events, according to whether or not they are reachable by a ray of light. In other words, the metric structure of Relativity is not positive definite. By removing the bundle structure of space-time, Relativity was able to formulate a geometrically simpler picture of space-time, although the notion of simplicity is in the eyes of the beholder.

1.2. Configuration Manifolds and Their Tangent and Cotangent Spaces

1.2.1. The Configuration Space

We have already introduced the notion of configuration space of a classical mechanical system and illustrated it by the example of a plane double pendulum. Figure 1.4 is a graphical representation of what we had in mind.

The coordinates θ_1 and θ_2, an example of so-called *generalized coordinates*, cannot be used globally (i.e., over the entire torus) since two configurations that differ by 2π in either coordinate must be declared identical. Moreover, other coordinate choices are also possible (for example, the horizontal deviations from the vertical line at the point of suspension). Each point on the surface of the torus corresponds to one configuration, and vice versa. The metric properties of the torus are not important. What matters is its *topology* (the fact that it looks like a doughnut, with a hole in the middle). The torus is, in this case, the *configuration space* (or *configuration manifold*) Q of the dynamical system at hand.

EXERCISE 1.5. Describe the configuration space of each of the following systems:

(1) A free particle in space.
(2) A rigid bar in the plane.
(3) An inextensible pendulum in the plane whose point of suspension can move along a rail.
(4) A pendulum in space attached to a fixed support by means of an inextensible string that can wrinkle.

1.2.2. Virtual Displacements and Tangent Vectors

We are now interested in looking at the concept of *virtual displacement*. Given a configuration $q \in Q$, we consider a small perturbation to arrive at another, neighbouring, configuration, always moving over the surface of the torus (since the system cannot

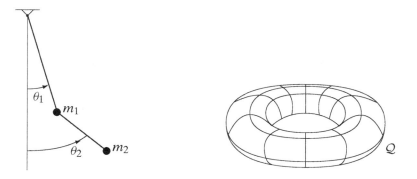

Figure 1.4. The plane double pendulum and its configuration manifold

escape the trap of its own configuration space). Intuitively, what we have is a small piece of a curve in Q, which we can approximate by a *tangent vector*.

To make this notion more precise, imagine that we have an initially unstretched thin elastic ruler on which equally spaced markers have been drawn, including a zero mark. If we now stretch or contract this ruler, bend it and then apply it to the surface of the torus at some point q, in such a way that the zero mark falls on q, we obtain an intuitive representation of a *parametrized curve* γ on the configuration manifold. Let us now repeat this procedure ad infinitum with all possible amounts of bending and stretching, always applying the deformed ruler with its zero mark at the same point q. Among all the possible curves obtained in this way, there will be a subcollection that shares the same tangent and the same stretch with γ. We call this whole collection (technically known as an *equivalence class* of parametrized curves) a *tangent vector to the configuration manifold* at q. Notice that, although when we draw this tangent vector **v** in the conventional way as an arrow, it seems to contradict the fact that we are supposed to stay on the surface, the definition as an equivalence class of curves (or, less precisely, a small piece of a curve) removes this apparent contradiction. Any of the curves in the equivalence (e.g., the curve γ of departure) can be used as the *representative* of the vector. The vector can also be regarded as a *derivation* with respect to the curve parameter (the equally spaced markers).

The collection of all tangent vectors at a point $q \in Q$ is called the *tangent space of Q at q* and is denoted by $T_q Q$. In the case of the torus, the interpretation of $T_q Q$ is the tangent plane to the torus at q, as shown in Figure 1.5. The tangent space at a point q of the configuration space is the carrier of all the possible virtual displacements away from the configuration represented by q. A physically appealing way to look at virtual displacements is as *virtual velocities* multiplied by a small time increment.

1.2.3. The Tangent Bundle

We now venture into a further level of abstraction. Assume that we attach to each point its tangent space (just like one would attach a small paper sticker at each point of a globe). We obtain a collection denoted by TQ and called the *tangent bundle* of Q. An element of this contraption consists of a point plus a tangent vector attached to it

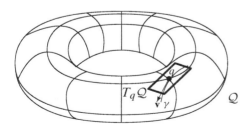

Figure 1.5. A tangent space

(i.e., a configuration and a virtual displacement away from it). It requires, therefore, four parameters: the two coordinates of the point and the two components of the vector with respect to some basis of the tangent space. Since everything is smooth, it is not difficult to imagine that we are now in the presence of a 4-dimensional manifold TQ. Moreover, given a tangent vector (a sticker, say), we can unambiguously say to which point it is attached. In other words, we a have a natural *projection map* τ:

$$\tau : TQ \to Q, \tag{1.13}$$

Thus, the tangent bundle TQ has the natural structure of a fibre bundle. Its dimension is always equal to twice the dimension of its base manifold Q. Its typical fibre is an n-dimensional vector space. Since all finite dimensional vector spaces of the same dimension are equivalent to each other, we may say that the typical fibre of the tangent bundle of an n-dimensional manifold is $\mathbb{R}^n = \underbrace{\mathbb{R} \times \cdots \times \mathbb{R}}_{n \text{ times}}$.

A section of the tangent bundle is a map $\Gamma : Q \to TQ$ satisfying condition (1.11), namely:

$$\tau \circ \Gamma = id_Q. \tag{1.14}$$

The physical meaning of a section of the tangent bundle is a *vector field*, since at each point q of Q the map Γ appoints a tangent vector $\Gamma(q)$.

1.2.4. The Cotangent Bundle

We move one step further into the physics and the geometry of the situation. Consider the notion of a linear operator on virtual displacements. This entity would be a black box that produces real numbers whenever it receives as an input a virtual displacement, just as a slot machine produces a pack of gum whenever you insert a coin. The fact that we are looking at a *linear* operator means that multiplying the input by a number results in the output being multiplied by the same number. Moreover, an input equal to the sum of two other inputs elicits an output exactly equal to the sum of the outputs elicited by the inputs acting separately. If the inputs are virtual displacements, the physical entity which behaves just like that is nothing but a *force*! And the corresponding output of our linear machine is nothing but the *virtual work* of the force over the given virtual displacement. Thus, forces are not vectors that cohabit with the virtual displacements upon which they act. They are vectors, indeed, in the sense that they also can be added and multiplied by numbers. They, however, live in their own vector space of linear operators and manifest themselves by the result of their (linear) action on the virtual displacement vectors. The collection of linear operators over the vectors of a vector space V is known as the *dual space* of V, denoted by V^*. Its elements are sometimes called *covectors*, to make a distinction from the vectors of V. Thus, the collection of all forces (covectors) that may act upon the virtual displacements issuing from the configuration $q \in T_qQ$ is the dual space T_q^*Q, called the *cotangent space of Q at q*.

The dimension of the dual space of a vector space can be shown to be equal to the dimension of the original vector space. Thus, if we glue at each point of the n-dimensional configuration space $\mathcal{Q}$ its cotangent space $T_q^*\mathcal{Q}$, we obtain a new differentiable manifold $T^*\mathcal{Q}$ of dimension $2n$, called the *cotangent bundle* of $\mathcal{Q}$. It is the carrier of all possible forces that may act on the mechanical system represented by $\mathcal{Q}$. It is endowed with a natural projection π (assigning to each force the configuration on whose virtual displacements it acts). It can be shown that it is in fact a fibre bundle whose typical fibre is $\mathbb{R}^n$. Such fibre bundles are known as *vector bundles*, because their typical fibre is a vector space (rather than a more general differentiable manifold).

A section $\Omega : \mathcal{Q} \to T^*\mathcal{Q}$ assigns to each point q of the configuration manifold a covector $\Omega(q)$. This field of covectors is also known as a 1-*form*. In the case of the configuration manifold, a 1-form represents a *force field*. Since, according to Newton's second law of motion, forces are proportional to changes of momentum, it follows that momenta are also elements of the cotangent bundle $T^*\mathcal{Q}$. For this reason, $T^*\mathcal{Q}$, whose coordinates can be seen as positions and momenta, is also known as the *phase space* of the mechanical system. It is the ground upon which Hamiltonian Mechanics is built.

We remark that the notions of tangent and cotangent bundle are automatically available as soon as a differentiable manifold has been defined. The fact that these bundles have a clear physical meaning (whenever the manifold of departure has a physical meaning, too) is one of the nice features that Differential Geometry offers. In other words, it spontaneously points the way toward the physical quantities that matter in a properly formulated physical model.

EXERCISE 1.6. For the case of a rigid bar in the plane, discuss the meaning of a generalized force. Adopt a local coordinate system that brings into evidence the notion of force couple.

1.3. The Infinite-dimensional Case

We have already anticipated that a crucial difference between a classical (particle) mechanical system and a continuous deformable system is that the configuration space of the latter is infinite-dimensional. Rather than dealing with the most general case, we will try to illustrate the setting of the continuous case by means of what is possibly the simplest continuous mechanical system that one may think of, namely, the deformable bar. It consists of a long prismatic member that deforms in such a way that its cross sections move parallel to themselves, by translation along the axis of the bar. A rubber band with a weight attached to one of its ends, while suspended from the other, is a good example. There are many structural, mechanical and bio-mechanical engineering components that can be idealized in this way. It is not difficult to produce many significant applications in various engineering contexts, just as the following examples show.

EXAMPLE 1.2. In the oil industry, oil needs to be pumped from the bottom of the well to the surface. To this effect, a down-hole pump is activated by means of a very long metal string (called a sucker rod) connected to an oscillating pump-jack placed on the surface. The sucker rod is an example of a deformable bar. An important technological problem, for instance, is the propagation of stress waves along the sucker rod. Because of its extreme thinness, the rod can be idealized as a segment of the real line $\mathbb{R}$. Moreover, because of its confinement within the narrow tubing of the well, all its displacements are such that the points of the bar stay in $\mathbb{R}$ throughout the process of deformation.

EXAMPLE 1.3. In the construction industry, some soil conditions require that the foundation of a building be secured in place by means of long piles driven into the ground by a hydraulic hammer. These piles constitute another example of a deformable bar in one dimension.

EXAMPLE 1.4. A somewhat different example is provided by the analysis of the so-called water hammer. Here, a fluid is confined to the interior of a rigid cylindrical pipe and, in first approximation, it is assumed that the waves generated, for example, by the sudden closing of a valve, propagate by displacing each fluid cross section parallel to itself along the axis of the pipe.

EXAMPLE 1.5. The fundamental force-producing unit in skeletal muscle is called a *sarcomere*. It consists of a thin filament sliding past a thick counterpart with the ability of engaging (or disengaging) connections between equidistantly placed molecular sites at either filament. The connecting units are called *cross-bridges*. A sarcomere and, more generally, a muscle fibre (consisting of a large number of sarcomeres arranged unidirectionally) is another example of a deformable bar in one dimension.

1.3.1. How Many Degrees of Freedom Does a Bar Have?

We want to treat the deformable bar just as we did with classical discrete dynamical systems, such as the double pendulum. We have seen that the collection of all possible configurations of a classical (discrete) system is a differentiable manifold whose dimension can be interpreted physically as the number of degrees of freedom of the system. A legitimate question, therefore, is: how many degrees of freedom does a deformable bar have?

A heuristic, albeit not very creative, way to arrive at the answer to this question might be the following. We may think of a deformable bar as the limit of a finite collection of material particles interconnected by means of springs or some other massless connecting elements (dashpots, for instance). The continuous bar is obtained when we let the number of particles go to infinity, while reducing the size of the springs accordingly. This frame of mind (already considered by the pioneers of the theory of elasticity in the 18th and 19th centuries) can indeed be perfected to provide a more or less rigorous a-posteriori justification of Continuum Mechanics in terms of Molecular Dynamics. It is clear from this picture that the number of degrees of freedom of a deformable bar is infinite, but is it a countable infinity or one with the power of the continuum?

In the sequel, we want to provide a more consistent answer to our question, namely, an answer that is independent of any underlying molecular model or any passage to a limit. To this end, we need only to agree on the answer to the following question: what is a configuration of a deformable bar? Once this question is answered one way or another, all we need to do is to count how many parameters are needed to completely specify any such configuration, and we are done.

1.3.2. What Is a Configuration of a Deformable Bar?

Having established that a bar can be identified with a segment $[a,b]$ of the real line $\mathbb{R}$, and having decided that our analysis will confine its attention to deformations along $\mathbb{R}$, we need to decide whether any further physically reasonable restrictions are to be imposed upon these deformations. We will settle, at least provisionally, on the following three restrictions:

(1) Continuity;
(2) Impenetrability of matter;
(3) Preservation of orientation.

In order to give a more precise geometrical interpretation of these restrictions, we start by working with two copies of $\mathbb{R}$, which we will call the *referential* and the *spatial* manifolds, denoted by $\mathbb{R}_B$ and $\mathbb{R}_P$, respectively. The body itself is identified with the closed subset $\mathcal{B} = [a,b]$ of $\mathbb{R}_B$. Before the imposition of the three restrictions listed above, a *configuration* is defined as an arbitrary map:

$$\kappa : \mathcal{B} \longrightarrow \mathbb{R}_P. \tag{1.15}$$

The image $\kappa(\mathcal{B})$ (sometimes also called a configuration, by abuse of terminology) can so far be any subset of $\mathbb{R}_P$. We would prefer this image to be a connected set, so that effectively we are not allowing the body to break into disjoint pieces. This we achieve by imposing the continuity condition (1) upon the function κ, as shown in Figure 1.6.

Although the continuity condition may seem to be quite satisfactory to develop at least a rudimentary form of Continuum Mechanics (by having prevented the breakup of the bar into separate pieces), a moment's reflection reveals that we may run into a different kind of trouble. Indeed, we would like to avoid a situation whereby two different material particles end up occupying one and the same position in space. To prevent this from happening, we impose condition (2) above. In mathematical terms, we want the function to be *injective* (or *one-to-one*). Figure 1.7 shows examples of noninjective and injective functions. An injective continuous real function of a single real variable is necessarily strictly increasing or decreasing.

Finally, without loss of generality, we may assume that the configurations under consideration are only those that preserve the orientation of the body (condition (3) above), thus effectively ruling out the strictly decreasing functions.

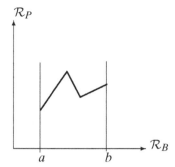

Figure 1.6. The continuity condition

We are now in a position of counting the number of degrees of freedom of our system. It is clear that the graph of a real function of one real variable $X \in [a,b]$ is completely determined by specifying for each point the value of the function. In principle, therefore, the number of degrees of freedom of our system is equal to the number of points in the interval. The condition of continuity, however, severely restricts this number. Indeed, let us assume that we have specified the values of the function just at the rational values of X. Then, since the rational numbers are dense in the real line,[4] the values of the function at all points (rational and irrational) will be automatically available by the continuity condition. Since, on the other hand, it is clear that no finite number of coordinates will be enough to attain that knowledge, it follows that the number of degrees of freedom of the deformable bar is the same as that of the rational numbers. In other words, the number of degrees of freedom is countably infinite.

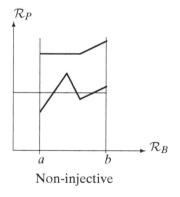

Non-injective

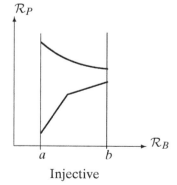

Injective

Figure 1.7. The impenetrability condition

---

[4] Namely, every neighbourhood of a real number $X \in \mathbb{R}$ contains rational numbers.

1.3.3. Is Continuity Sufficient?

While the condition of impenetrability of matter seems to be beyond reproach, a question of some importance is whether or not the continuity condition just imposed is a reflection of all the geometrical properties that we wish to assign to configurations. We have justified it in terms of the physically based idea that continuity is necessary and sufficient to prevent the body from breaking up into separate pieces. What this means is that, as long as we explicitly exclude such phenomena as formation of cracks, the continuity condition is sufficient to describe the intended behavior. On the other hand, for certain problems where the loading and the material properties can be assumed to vary smoothly throughout the bar, we may wish to impose further restrictions upon the possible configurations. We may wish, for instance, to require that configurations be not only continuous but also differentiable, with continuous derivatives up to order r, where r is an integer greater than 0. A function with such property is called a function of class C^r. The symbol C^0 is reserved for continuous functions, while C^∞ describes the class of *smooth* (infinitely differentiable) functions. Finally, C^ω denotes the *analytic* functions. There exist also many other classes of functions worth considering. The choice of the class of admissible configurations is an important one, both physically and mathematically. For example, the study of shock waves requires that the configurations be as general as C^0, while for many problems in linear elastostatics analyticity will do.

Those functions of class C^r which happen to be injective (that is, which satisfy the impenetrability condition) have the following property: they are invertible, provided the domain of the inverse is restricted to the image of the original domain. If this inverse is of class C^r, we say that the function is a *diffeomorphism* of degree r. We will denote the collection of such diffeomorphisms by D^r. The diffeomorphisms of degree 0 are called homeomorphisms. They consist of continuous functions with a continuous inverse. From now on, we will use the symbol D^r to denote orientation-preserving diffeomorphisms.

1.3.4. The Configuration Space

Just as in the case of systems with a finite number of degrees of freedom, we will now define the configuration space $\mathcal{Q}$ as the collection of all possible configurations, namely:

$$\mathcal{Q} = \{\kappa \mid \kappa : \mathcal{B} \to \mathbb{R}, \kappa \in D^r\}. \tag{1.16}$$

In other words, we identify the configuration space with the space D^r of orientation-preserving diffeomorphisms of some order r. A point of this space, namely, a configuration of the system, is now a whole function κ. Pictorially, we may say that the configuration space consists of all strictly increasing curves defined over the interval $[a,b]$ (with the desired degree of differentiability).

In the space $\mathcal{Q}$ just introduced we can define the addition of two functions of class D^r in point-wise fashion, and similarly with the multiplication by a scalar. Nevertheless, $\mathcal{Q}$ is not a vector space, since it is not closed under these operations. For example, multiplying an increasing function by -1 we obtain a decreasing function. Instead, Q is an open subset of a vector space. Notice that, even in the case of the double pendulum, the configuration space was not a vector space. More generally, therefore, we would like to be able to assert the following principle: The configuration space Q of a material body is an infinite-dimensional differentiable manifold. More precisely, Q is a *manifold of maps* between the body manifold and the space manifold.

There are many technical questions that need to be addressed to substantiate this statement. An n-dimensional differentiable manifold looks locally like $\mathbb{R}^n$, as can be seen by choosing a system of generalized coordinates. We say that a finite dimensional manifold is *modelled after* $\mathbb{R}^n$. In the infinite-dimensional case, therefore, one of the fundamental issues is the precise definition of the *modelling space*. The most tractable case is that in which the infinite-dimensional manifold is modelled after a Banach space.[5] At this point, however, we ask the reader to suspend disbelief and to accompany us in the intellectual exercise of the remainder of this section as if the above principle made sense. After all, one of the most useful and beautiful features of the geometric approach is precisely its ability to provide a visual framework as the basis for correct intuitive reasoning.

1.3.5. The Tangent Bundle and Its Physical Meaning

From the position of temporary suspension of disbelief just adopted, we can afford the luxury of exploiting our just-earned experience with the configuration space of classical (discrete) systems, as discussed in the previous section. And what have we learned? Briefly, we have found that the geometric formulation of mechanics is best achieved by enlarging the arena of discussion from the configuration space to its tangent and cotangent bundles (and possibly beyond, to their respective iterated tangent and cotangent bundles). The tangent bundle TQ, a natural geometric construct erected upon the original manifold Q, turned out (in the finite-dimensional case, at least) to be the repository of all possible virtual displacements, or velocities, of the system. The concept of (generalized) force arose then naturally as a linear real-valued operator on virtual displacements at a given configuration. In other words, the cotangent bundle T^*Q turned out to be the repository of all possible generalized forces. Our hope is, therefore, that if we just emulate these ideas (leaving the fine technical details for specialized mathematical treatment), we will end up with the most natural, and at the same time the most general, concept of *generalized force* in Continuum Mechanics. The golden rule of the geometric approach is, in fact, that the choice of configuration space (a choice that is essentially geometric, as we have seen) brings about also, as a geometrical necessity, the emergence of the nature of

---

[5] It can be shown that this is the case when the body is a closed interval of $\mathbb{R}$. An open interval will fail to lead to a Banach configuration manifold. For a somewhat more general statement, see Section 4.17 of Chapter 4.

the whole mechanical apparatus: virtual displacements, forces, stresses, equilibrium, and so on.

We start with the notion of a tangent vector at a point $\kappa \in Q$. Recall that a tangent vector at a point of a differentiable manifold is an equivalence class of (parametrized) smooth curves passing through that point and sharing the same derivative (with respect to the parameter). To avoid the abstractness of this idea, however, in the infinite-dimensional case we will content ourselves (always in the spirit of suspension of disbelief) with replacing the notion of equivalence class of curves with the more intuitive idea of a very small (first-order) piece of a curve, which is, after all, what the equivalence class is intended to represent. And what is a parametrized curve γ through a point κ of Q? It is nothing but a map:

$$\gamma : (-\epsilon, \epsilon) \to Q, \tag{1.17}$$

such that:

$$\gamma(0) = \kappa, \tag{1.18}$$

where ϵ is a positive real number. The interval $(-\epsilon, \epsilon)$ is precisely our marked ruler as described in Section 1.2.2.

Let us attempt a geometric representation of this idea. We must assign to each value of a real parameter s in the interval $(-\epsilon, \epsilon)$ a configuration $\kappa_s = \kappa(s)$, that is, a point of Q. Figure 1.8 suggests such an assignment:

Notice that for $s = 0$ we have the configuration κ of departure, while for other values of s we obtain nearby configurations. In this picture, it is not easy to convey the idea that the assignment has been done smoothly. While a precise definition of differentiability can be given, it is intuitively clear that, in the case of our deformable bar, it must mean that, if we regard the family (or curve) of configurations (as we certainly can) as a real function $\kappa(s, X)$ of two variables (namely, $s \in (-\epsilon, \epsilon)$ and $X \in [a, b]$), then this function should be differentiable as many times as needed with respect to s, and at least r times with respect to X. If we now fix a point X in the body manifold, as s varies this point will be seen as describing a smooth trajectory in $\mathbb{R}_P$. In other words, our curve of configurations can also be seen as a family of smooth

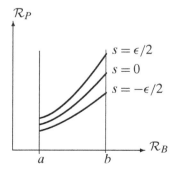

Figure 1.8. Three points on a parametrized curve in Q

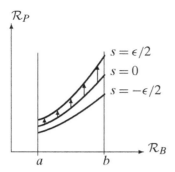

Figure 1.9. A tangent vector to a curve in $\mathcal{Q}$ can be regarded as a vector field on κ

trajectories described by each particle of the body! When we consider the meaning of an equivalence class of such trajectories sharing their first derivative (with respect to s), we conclude that the meaning of a tangent vector to a configuration is nothing but a *vector field* defined over the configuration itself. In other words, the meaning of a tangent vector is now a *virtual displacement field*, in the ordinary sense of the term. The concepts just described are illustrated in Figure 1.9:

EXAMPLE 1.6. Just in case you want to get a more concrete picture of the meaning of a tangent vector to $\mathcal{Q}$ at a given configuration κ, consider the following numerical example. A deformable bar is identified with the closed interval $[1, 1.6]$ in $\mathbb{R}$. The configuration κ is given by the equation:

$$x = 0.1\left(1 + e^{x^2}\right) + 0.5 + X, \tag{1.19}$$

where X and x denote, respectively, the running variables in $\mathbb{R}_B$ and $\mathbb{R}_P$. You should check that this function is indeed an orientation preserving diffeomorphism between the interval $[1, 1.6]$ and its image in $\mathbb{R}_P$. A curve in $\mathcal{Q}$ passing through the configuration κ is, by definition, a one-parameter family of configurations such that, by convention, when the parameter s vanishes we obtain the configuration κ of departure. It is enough to define this family for values of s varying over any open interval containing the origin. For our purposes, we will adopt the open interval $(-0.25, 0.25)$. We consider, as an example, the following family:

$$x = x(X, s) = 0.1\left(e^{5sX^2} + e^{X^2}\right) + 0.5 + 2s + X. \tag{1.20}$$

For each value of the parameter s lying within the interval $(-0.25, 0.25)$ we obtain a configuration (you should check that this is indeed true). The following Mathematica code will plot the configurations corresponding to the five values $s = -0.2, -0.1, 0.0, 0.1, 0.2$:

```
Do[f[i] = 0.1 * (Exp[5 * (-0.2 + 0.1 * i) * X^2] + Exp[X^2]) + 0.5 - 0.4 + X, {i, 0, 4, 1}]
Plot[f[0], f[1], f[2], f[3], f[4], X, 1.0, 1.6, AxesLabel -> {"Body", "Space"}]
```

On the other hand, Equation (1.20) can legitimately be seen as describing a function of 2 variables, X and s. The following Mathematica code will plot a graph of this function within the domain of interest.

$$Plot3D[0.1 * (Exp[5 * s * X^2] + Exp[X^2]) + 0.5 + 2 * s + X,$$
$$\{s, -0.2, 0.2\}, \{X, 1.0, 1.6\}, AxesLabel -> \{"s", "Body'", "Space"\}].$$

But there is still a third way of looking at this formula. We may fix a material particle (namely, set X to a fixed value) and let s vary, thus obtaining a trajectory of this particle in space. The following Mathematica code will plot five such trajectories, for $X = 1.00, 1.15, 1.30, 1.45, 1.60$:

$$Do[g[i] = 0.1 * (Exp[5 * (s) * (1 + 0.15 * i)^2] + Exp[(1 + 0.15 * i)^2]) + 1.5 + 2 * s + 0.15 * i, \{i, 0, 4, 1\}]$$
$$Plot[g[0], g[1], g[2], g[3], g[4], \{s, -0.2, 0.2\}, AxesLabel -> \{"s", "Space"\}]$$

Notice that, apart from the condition of differentiability, these trajectories (unlike the configurations themselves) can be completely arbitrary (increasing, decreasing, constant, etc.), although in our example they turn out to be increasing.

Each of these trajectories has a definite tangent vector at $s = 0$. It is the collection of the tangent vectors to these trajectories at $s = 0$ that provides us with the tangent vector to the original curve of configurations in the infinite-dimensional manifold $\mathcal{Q}$. In our particular example, therefore, the tangent vector $\delta\kappa$ is given by the formula:

$$\delta x = 0.5X^2 + 2. \tag{1.21}$$

EXERCISE 1.7. Do virtual displacements (or velocities) need to satisfy some form of some principle of impenetrability of matter? Why, or why not?

EXERCISE 1.8. Draw schematically a one-parameter family of configurations for a 2-dimensional body as a succession of nearby configurations (leaving the parameter s out of the picture). Draw the corresponding trajectories for a few selected points and the corresponding tangent vectors at the initial configuration. Interpret the result as a virtual displacement (or velocity) field. Why is this picture clearer than the 1-dimensional counterpart?

The collection of all virtual displacement fields at a point $\kappa \in \mathcal{Q}$ can be shown to constitute a vector space $T_\kappa \mathcal{Q}$, namely, the tangent space to $\mathcal{Q}$ at κ. As in the finite-dimensional case, one may be able to show (after dealing with difficult technical issues) that the union of all these tangent spaces can be given the structure of an (infinite-dimensional) differentiable manifold $T\mathcal{Q}$, namely, the tangent bundle of $\mathcal{Q}$.

1.3.6. The Cotangent Bundle

Having introduced and thoroughly discussed the notion of tangent vectors at a configuration $\kappa \in \mathcal{Q}$, we find ourselves, as it were, in automatic pilot. The vector space

$T_\kappa Q$ formed by all these tangent vectors, called the tangent space at κ, has a well-defined dual space $T_\kappa^* Q$, called the cotangent space at κ. It consists, by definition, of all (continuous) real-valued linear functions on $T_\kappa Q$. From the physical point of view the cotangent space is a veritable bonanza: it contains all those physical entities which produce linearly a real number δW, called virtual work, when acting on virtual displacements. And what can these entities be but generalized forces acting on the configuration κ? There can be no other meaning of a generalized force. This being so, we have, by merely following the natural geometric constructions arising from the very definition of the configuration space Q, stumbled upon the most general possible notion of generalized force in Continuum Mechanics. This concept is so general that it must surely encompass any notion of force that you may think of: external body forces, surface tractions of all kinds, internal stresses, and so on.

EXERCISE 1.9. Let the configuration space Q of a bar be the space of D^1 functions over the interval $[a,b]$, and let $\delta\kappa \in T_\kappa Q$ be a virtual displacement away from some given configuration $\kappa \in Q$. Verify that each of the following equations expresses a generalized force by means of the formula providing the value of the virtual work δW over $\delta\kappa$:

$$\delta W(\delta\kappa) = \int_a^b f(X)\delta\kappa(X)dX, \tag{1.22}$$

$$\delta W(\delta\kappa) = F\delta\kappa(b), \tag{1.23}$$

$$\delta W(\delta\kappa) = \int_a^b \sigma(X)\frac{d(\delta\kappa(X))}{dX}dX, \tag{1.24}$$

where F is a constant and f and σ are real functions of X. Explain the physical meaning of each of these forces.

1.4. Elasticity

Abandoning the 1-dimensional example of a bar, it is clear that for a more realistic 3-dimensional material body manifold B we can repeat the same ideas developed for the case of the bar. Fixing, for simplicity, a particular observer and ignoring the time variable, a *configuration* κ is a smooth map of B into the affine Euclidean physical space $\mathbb{E}^3$ of Classical Mechanics:

$$\kappa : B \longrightarrow \mathbb{E}^3 . \tag{1.25}$$

The *configuration space* Q of the body B is the set of all such maps.

The tangent bundle TQ has the following physical meaning: An element $\delta\kappa \in T_\kappa Q$ can be regarded as a vector field over the spatial image $\kappa(B)$ and is, therefore, a *virtual displacement field* at the configuration κ. Similarly, the cotangent bundle T^*Q can be interpreted physically as the collection of all possible *generalized forces* that may act on the material body. These forces are general enough to encompass any possible notion of external or internal force (stress). The evaluation of a covector on a vector can, correspondingly, be interpreted as *virtual work*. A rigorous distinction between external and internal forces can be established by imposing for the latter a condition of invariance under the Galilean group, a task that we will not pursue at this point.

A *global elastic constitutive law* consists of the assignment of an internal force σ to every configuration κ. In other words, such a constitutive equation is nothing but a 1-form on $\mathcal{Q}$. A material body whose material response is completely characterized by such a 1-form is said to be *globally elastic*.

EXERCISE 1.10. Give an example of a genuinely global elastic constitutive law for a bar.

1.5. Material or Configurational Forces

It should be abundantly clear by now that Differential Geometry suggests, of its own accord, physically meaningful entities by the mere fact that they have an intrinsic geometric meaning. Another example of this power is displayed when we consider the collection of diffeomorphisms of the material body manifold $\mathcal{B}$ onto itself, without reference to the physical space altogether. Assuming that the collection of such maps has the structure of an infinite-dimensional differentiable manifold (just as the configuration space), its tangent bundle can be interpreted as the repository of all *material virtual displacements*, where the word "displacement" is to be interpreted in the appropriate physical context as some small *material* change that may take place in the body itself. An element of the cotangent bundle will, accordingly, be interpreted as a *material or configurational force*. These forces play an important role in the description of phenomena of anelastic material evolution and growth.

2 Vector and Affine Spaces

Vector spaces are essentially linear structures. They are so ubiquitous in pure and applied mathematics that their review in a book such as this is fully warranted. In addition to their obvious service as the repository of quantities, such as forces and velocities, that historically gave birth to the notion of a vector, vector spaces appear in many other, sometimes unexpected, contexts. Affine spaces are often described as vector spaces without a zero, an imprecise description which, nevertheless, conveys the idea that, once an origin is chosen arbitrarily, the affine space can be regarded as a vector space. The most important example for us is the Galilean space of Classical Mechanics and, if nothing else, this example would be justification enough to devote some attention to affine spaces.

2.1. Vector Spaces: Definition and Examples

One of the most creative ideas of modern mathematics (starting from the end of the nineteenth century) has probably been to consider particularly important examples and, divesting them of as much of their particularity as permitted without loss of essence, to abstract the remaining structure and elevate it to a general category in its own right.[1] Once created and named, the essential commonality between particular instances of this structure, within what may appear to be completely unrelated fields, comes to light and can be used to great advantage. One such creation was the very idea of a *set* as a collection of objects (satisfying certain consistency conditions). The creation of other categories usually involves the notion of one or more sets and one or more *operations* between elements of those sets. The notion of a *group*, involving just one set and one binary operation, was rightly predicted more than one hundred years ago to become the dominant structure underlying most mathematical and physical disciplines during the twentieth century. It is in this spirit that we will present the

---

[1] It is interesting to remark that the proponent of the philosophical school of Phenomenology, Edmund Husserl (1859–1938), was a mathematician turned philosopher. His doctoral thesis dealt with certain aspects of the calculus of variations.

notion of a *vector space* in an axiomatic fashion, so that we will recognize its structure whenever it appears under any guise.

In this book, we will deal only with *real vector spaces*, namely, vector spaces that presuppose the existence of the set $\mathbb{R}$ of real numbers, which will play a role in the definition. There exist other possibilities, most notably *complex vector spaces*, which presuppose the complex plane $\mathbb{C}$ or another auxiliary set. This auxiliary entity is technically called a *field* and its elements are called *scalars*. A real vector space consists of a set V (whose elements are called *vectors*) endowed with two operations:

(1) Vector addition:

$$V \times V \longrightarrow V$$
$$(\mathbf{u}, \mathbf{v}) \mapsto \mathbf{u} + \mathbf{v}, \tag{2.1}$$

and
(2) Multiplication by a scalar:

$$\mathbb{R} \times V \longrightarrow V$$
$$(\alpha, \mathbf{v}) \mapsto \alpha \mathbf{v}, \tag{2.2}$$

satisfying the following properties:

(1) Associativity of vector addition:

$$(\mathbf{u} + \mathbf{v}) + \mathbf{w} = \mathbf{u} + (\mathbf{v} + \mathbf{w}), \quad \forall \mathbf{u}, \mathbf{v}, \mathbf{w} \in V. \tag{2.3}$$

(2) Existence of a zero vector:

$$\exists \mathbf{0} \in V \text{ such that } \mathbf{0} + \mathbf{v} = \mathbf{v} + \mathbf{0} = \mathbf{v}, \quad \forall \mathbf{v} \in V. \tag{2.4}$$

(3) Existence of an inverse:

$$\text{For each } \mathbf{v} \in V, \exists (-\mathbf{v}) \in V \text{ such that } \mathbf{v} + (-\mathbf{v}) = (-\mathbf{v}) + \mathbf{v} = \mathbf{0}. \tag{2.5}$$

(4) Commutativity of the addition:

$$\mathbf{u} + \mathbf{v} = \mathbf{v} + \mathbf{u}, \quad \forall \mathbf{u}, \mathbf{v} \in V. \tag{2.6}$$

(5) Distributivity to vector addition:

$$\alpha(\mathbf{u} + \mathbf{v}) = \alpha \mathbf{u} + \alpha \mathbf{v}, \quad \forall \alpha \in \mathbb{R}, \mathbf{u}, \mathbf{v} \in V. \tag{2.7}$$

(6) Distributivity to scalar addition:

$$(\alpha + \beta)\mathbf{u} = \alpha\mathbf{u} + \beta\mathbf{u}, \quad \forall \alpha, \beta \in \mathbb{R}, \ \mathbf{u} \in V. \tag{2.8}$$

(7) Associativity to multiplication of scalars:

$$(\alpha\beta)\mathbf{u} = \alpha(\beta\mathbf{u}), \quad \forall \alpha, \beta \in \mathbb{R}, \ \mathbf{u} \in V. \tag{2.9}$$

(8) Multiplicative unit consistency:

$$1\,\mathbf{u} = \mathbf{u}, \quad \forall\mathbf{u} \in V. \tag{2.10}$$

The first three properties (which refer exclusively to vector addition, without any mention of the field of scalars) constitute the definition of a *group*. The fourth property means that this group is *commutative* (or *Abelian*). The group operation in a commutative group is usually denoted as an addition.

REMARK 2.1. **Notation**: Notice the subtlety of the notational abuse: In (2.8) the addition on the left is not of the same type as the addition on the right, while in (2.9) the first multiplication on the left is not of the same type as the first multiplication on the right. In mathematics, though, notation is everything and this particular notation has been chosen so that one can confidently operate on vectors (open brackets, etc.) as if they were numbers,

EXERCISE 2.1. **Some derived properties:** Show that: (a) the zero vector is unique; (b) the inverse of any given vector is unique; (c) $0\mathbf{u} = \mathbf{0}$, $\forall\mathbf{u} \in V$; (d) $(-1)\mathbf{u} = (-\mathbf{u})$, $\forall\mathbf{u} \in V$. Make sure that you use just the eight properties above and that you are not carried away by the notation. In fact, this exercise amply justifies this being carried away from now on. In particular, it justifies the subtraction of vectors: $\mathbf{u} - \mathbf{v} := \mathbf{u} + (-\mathbf{v})$.

2.2. Linear Independence and Dimension

A *linear combination* of a finite subset $\{\mathbf{v}_1, \mathbf{v}_2, \ldots, \mathbf{v}_k\} \subset V$ is an expression of the form:

$$\sum_{i=1}^{k} \alpha_i \mathbf{v}_i = \alpha_1 \mathbf{v}_1 + \ldots + \alpha_k \mathbf{v}_k, \tag{2.11}$$

where the real numbers α_i are called the *coefficients* of the linear combination. When evaluated with given coefficients, a linear combination produces a vector. By the properties of vector operations, a linear combination will always vanish if each of its coefficients is zero. Such a linear combination (namely, one all of whose coefficients are zero) is called *trivial*. On the other hand, a linear combination may vanish even if some of its coefficients are not zero. For instance, if the zero vector is a member of our subset, then we can simply choose *its* coefficient as, say, 1, and the remaining coefficients as 0. By the properties of the vector operations, we will thus obtain a vanishing nontrivial linear combination.

A finite subset $\{\mathbf{v}_1, \mathbf{v}_2, \ldots, \mathbf{v}_k\} \subset V$ is said to be *linearly independent* (equivalently, the members of this subset are said to be mutually linearly independent) if there exists no vanishing nontrivial linear combination. Put in other words, the subset is linearly independent if, and only if, the only zero linear combination is the trivial one (namely, the one with all coefficients equal to zero). It is clear that such subsets always exist, provided the vector space has at least one element different from the zero element. Indeed, in that case the subset consisting of just that one element is a linearly independent subset. A subset that is not linearly independent is said to be *linearly dependent*. A linearly independent subset is said to be *maximal* if the addition of *any* (arbitrary) vector to the subset renders it linearly dependent. Although we have demonstrated the existence of linearly independent subsets, there is no a-priori reason for the existence of maximal linearly independent subsets.[2]

BOX 2.1. **Examples of Vector Spaces**

1. Virtual velocities of a particle: In Classical Mechanics, a particle occupying instantaneously a position p in space can be assigned any *virtual* velocity. In this context, a velocity is seen as an oriented segment, or arrow, whose tail is at p and whose tip may point in any direction. At this level, we have the luxury of straight lines and the concepts of length and angle. The set of all arrows emerging from a point forms a vector space under the following two operations: (i) vector addition: by the parallelogram rule; (ii) multiplication by a scalar: by changing the length of the arrow through multiplication by the absolute value of the scalar (and changing the sense if the scalar is negative). All eight properties of the definition are satisfied by these two operations. [Question: what is the zero vector? Warning: proving the associativity of the vector addition may take some work.] Historically, the parallelogram rule was first discovered for forces. Galileo Galilei (and later Newton) used this idea to add velocities as well.

2. Real polynomials of degree n: For a fixed natural number n, consider the collection of polynomials of the form: $P(x) = a_0 + a_1 x + a_2 x^2 + \cdots + a_n x^n$, where $a_0, \ldots, a_n$ are real constants and x is a real variable. Defining the addition of two such polynomials as the polynomial resulting from adding the coefficients of equal powers of x, and multiplication of a polynomial by a scalar as the polynomial resulting from multiplying each coefficient by the given scalar, all eight properties of the definition are satisfied. In this example, no concept of length or angle was used. An equivalent example is the collection of all ordered $(n+1)$-tuples of real numbers with the appropriate operations. Thus, $\mathbb{R}^n$ has a natural vector-space structure.

---

[2] If we do not impose the restriction that the subset under consideration must be finite, then the existence of maximal linearly independent subsets is guaranteed by Zorn's lemma. An infinite subset $W \subset V$ is linearly independent if every nonempty finite subset of W is linearly independent. A maximal (infinite) linearly independent subset is called a Hamel basis.

3. Continuous functions: Given an (open or closed) interval of the real line, we look at the collection of all continuous real functions over this interval. By point-wise addition and point-wise multiplication by a scalar, it is a straightforward matter to verify that this collection forms a vector space. As in the case of the polynomials, the vector-space structure is independent of any metric notion. We will soon see that this example differs from the previous two ones in another fundamental respect.

4. The free vector space of a set: All the previous examples share the feature that the operations had a natural pre-existing meaning in the context in which they are defined (in other words, we knew how to add two polynomials already, and we only checked that this operation is a right one to confer upon the set of polynomials the structure of a vector space). We now consider a completely different type of example, which shows the startling potential of a mathematical concept once it has been properly introduced. Let A be an arbitrary set. Consider the collection of all *formal expressions* $\alpha_1 a_1 + \alpha_2 a_2 + \cdots + \alpha_k a_k$, where k is an arbitrary (finite) natural number and where $\alpha_i \in \mathbb{R}$, $a_i \in A$, $(i = 1, \ldots, k)$. All such expressions (for all k and all choices of scalars and elements of A) are part of the intended collection. If there is a repeated a_i in an expression ($a_i = a_j$, say), the expression is considered equivalent to one with one less term, namely, $\alpha_i a_i + \alpha_j a_j = (\alpha_i + \alpha_j)a_i$. Similarly, if $\alpha_i = 0$, the expression is considered equivalent to the one obtained by simply eliminating the corresponding term. Thus, all expressions all of whose coefficients α_i vanish are equivalent. This equivalence class is the zero vector. A coefficient 1 can be ignored. Vector addition and multiplication by a scalar are defined in the same way as one would for the set of polynomials of all orders.

A maximal linearly independent subset (if it exists) is called a *basis* of the vector space. The reason for this terminology is contained in the following theorem: Every vector can be expressed uniquely as a linear combination of the elements of a basis. Indeed, let $\{\mathbf{e}_1, \ldots \mathbf{e}_k\}$ be a basis and let $\mathbf{v} \in V$. Since the basis is, by definition, maximal, the augmented subset $\{\mathbf{v}, \mathbf{e}_1, \ldots \mathbf{e}_k\}$ must be linearly dependent. This means that there exists a nontrivial linear combination that vanishes. Moreover, in this nontrivial linear combination the coefficient of $\mathbf{v}$ must be different from zero. Otherwise, the linear combination would be trivial, since the elements in the basis are already linearly independent. Let, therefore, the vanishing non-trivial linear combination be given by:

$$\beta\mathbf{v} + \alpha_1\mathbf{e}_1 + \cdots + \alpha_k\mathbf{e}_k = \mathbf{0}, \tag{2.12}$$

with $\beta \neq 0$. Dividing throughout by β and using the algebraic properties of the operations, we can write:

$$\mathbf{v} = -(\alpha_1/\beta)\mathbf{e}_1 - \ldots - (\alpha_k/\beta)\mathbf{e}_k, \tag{2.13}$$

which proves that $\mathbf{v}$ is a linear combination of the basis.

EXERCISE 2.2. **Uniqueness:** Complete the proof of the theorem by showing that the linear combination just obtained is unique. [Hint: assume that there exists a different one, subtract and invoke linear independence.]

The unique coefficients of the linear combination just obtained are called the *components* of **v** in the given basis. They are denoted by v^i, $(i = 1, \ldots k)$. An important corollary of this theorem is that all bases of a vector space (if they exist) have the same number of elements.[3] This common number is called the *dimension* of the vector space. If there are no (finite) maximal linearly independent subsets, the vector space is said to be of *infinite* dimension.

EXERCISE 2.3. **Dimension:** Prove the above corollary, namely: show that all bases have the same number of elements. [Hint: assume that there exist two bases with different numbers of elements and express the extra vectors of one in terms of the other.]

EXERCISE 2.4. **Examples:** Determine the dimension of each of the vector spaces of Box 2.1.

EXERCISE 2.5. Show that if every vector is expressible as a linear combination of a finite number of linearly independent vectors, then these vectors form a basis of the vector space.

A subset of a vector space V is said to be a *subspace* if it is closed under the operations of vector addition and multiplication by a scalar. A subspace is itself a vector space. Given a subset A of a vector space V, the set obtained by constructing all possible (finite) linear combinations of elements of A is a subspace of V called the *subspace generated, or spanned, by A*. Given a subspace U of a vector space V, we can define an equivalence relation in V as follows: two vectors are equivalent if, and only if, their difference belongs to the subspace U. The *quotient space* by this equivalence relation is denoted as V/U. Recall that, by definition of quotient space, each element of V/U is precisely a whole equivalence class defined by the equivalence relation. The quotient space has automatically the structure of a vector space (by just choosing a representative of each class and applying the operations of V). If the dimension of V is n and the dimension of U is $m \leq n$, the dimension of V/U is $n - m$. The (unique) equivalence class to which an element $\mathbf{v} \in V$ belongs is also called the *coset* of **v** with respect to the subspace U.

EXERCISE 2.6. **Subspaces:** Verify all the statements made in the previous paragraph. Describe the subspace U of $\mathbb{R}^3$ generated by the vectors $(1,0,0)$ and $(0,1,0)$. What are, geometrically speaking, the elements of $\mathbb{R}^3/U$? What is the meaning of the vector addition in $\mathbb{R}^3/U$? Give an example.

---

[3] For Hamel bases the corresponding statement is that all Hamel bases have the same cardinality.

2.3. Change of Basis and the Summation Convention

We will concentrate for now on finite-dimensional vector spaces. Once a basis has been given, each vector $\mathbf{v} \in V$ is expressible uniquely as:

$$\mathbf{v} = v^1 \mathbf{e}_1 + \cdots + v^n \mathbf{e}_n = \sum_{i=1}^{n} v^i \mathbf{e}_i, \tag{2.14}$$

where n is the dimension of the vector space V and $\{\mathbf{e}_1, \ldots \mathbf{e}_n\}$ is the given basis. A convenient notational device, attributed to Einstein, is the so-called *summation convention*. Its purpose is to do away with the summation symbol Σ by establishing that: in any indexed quantity or product of indexed quantities, where an index appears diagonally repeated, the summation over that index from 1 to the dimension of the space is understood. By diagonally repeated we mean that the index appears once as a subscript and once as a superscript. Thus, we may write (2.14) as:

$$\mathbf{v} = v^i \mathbf{e}_i. \tag{2.15}$$

The expression may consist of just one symbol. For example:

$$a_i^i = a_1^1 + a_2^2 + \cdots + a_n^n. \tag{2.16}$$

EXERCISE 2.7. **Einstein's summation convention:** For $n = 3$, write the following expressions in full: (1) $c = a_i^i b_j^j - a_j^i b_i^j$, (2) $c_i = a_{ijk} v^j v^k$.

Since a diagonally repeated index is summed on, it is a *dummy index*, in the sense that it can be changed to any other index (provided this other index does not already appear in the same expression). An index repeated in any other fashion (not diagonally, or more than once) is considered to be suspect and calls for an explanation. There are, of course, deep reasons for this self-correcting power of the summation convention. To get a glimpse, consider Equation (2.15). The left-hand side (namely, the vector $\mathbf{v}$ itself) is an element of the vector space V and has, therefore, an intrinsic geometric meaning dictated by the context at hand (e.g., it can represent the velocity of a particle). The components v^i, on the other hand, are only a representation of $\mathbf{v}$ in some basis. Change the basis, and these numbers will also change. But how? In a manner to compensate for the change of basis, so that the resulting object $\mathbf{v}$ turns out to be the same! This compensatory feature (technically called *contravariance*) is indicated by placing the index as a superscript for the components v^i, while the vectors of the basis are indicated with subscripts (*covariance*).[4] To see how this works in practice, let us consider two different bases, $\{\mathbf{e}_1, \ldots \mathbf{e}_n\}$ and $\{\mathbf{f}_1, \ldots \mathbf{f}_n\}$, of the vector space V. Each vector of the new basis is expressible uniquely as a linear combination of the vectors of the old basis, viz.:

$$\mathbf{f}_i = f_i^{\,j}\, \mathbf{e}_j. \tag{2.17}$$

---

[4] When we study the theory of principal bundles, we will give a further formalization of these concepts under the heading of associated bundles.

Notice that the index i is a *free index*, as opposed to j, which acts here as a dummy index. Free indices must be balanced, on each side of the equation (and on each term, if there were more than one). In this example, the free index denotes which of the new base vectors ($\mathbf{f}_i$) is treated, The upper index in $f_i^{\ j}$ indicates a component of this vector in the old basis. For safety, we also make sure that the order in which these indices are to be read is indicated somehow (in this case, we have chosen to have the lower index first, and indicated so by placing the upper index slightly to the right). Now, the vector $\mathbf{v}$ has an expression in terms of components in either basis, namely:

$$\mathbf{v} = v^i \mathbf{e}_i = w^i \mathbf{f}_i. \tag{2.18}$$

Using (2.17), we obtain:

$$\mathbf{v} = v^i \mathbf{e}_i = w^i \mathbf{f}_i = w^i f_i^{\ j} \, \mathbf{e}_j = w^j f_j^{\ i} \, \mathbf{e}_i. \tag{2.19}$$

Notice that in the last step we have cavalierly exchanged i with j, as we are allowed to do, since they are both dummy indices. We thus obtain two different representations of the same vector in the same basis. But we know that vector components on a given basis are unique. We conclude, therefore, that:

$$v^i = w^j f_j^{\ i}. \tag{2.20}$$

We have obtained that the components of vectors change, on changing the basis, according to the same coefficients $f_i^{\ j}$ as the bases themselves. Nevertheless, we notice that, whereas these coefficients were used to obtain the new base vectors from the old, now they are used to obtain the old components from the new! This is exactly what we meant by the compensatory nature of contravariance. If one were to arrange the coefficients $f_i^{\ j}$ in matrix format, and the base vectors and the components in columns, it would follow that the base vectors change by premultiplication by this matrix, while the components of vectors change according to its inverse-transpose.

EXERCISE 2.8. **Invertibility:** Show that the matrix of coefficients $f_i^{\ j}$ is always invertible. [Hint: invoke linear independence.]

2.4. The Dual Space

Vector spaces are essentially *linear* structures. This linearity is rooted in the distributivity properties of the product of vectors by scalars. One of the interesting properties of linearity is that it permits, in a completely natural way, creation of new linear structures out of old ones ad infinitum. The first, and perhaps crucial, manifestation of such a self-replicating feature of linearity is the concept of the dual space of a vector space.

A *linear function* on a vector space is a mapping (onto the field of scalars) that respects the vector operations. More precisely, a map $f : V \to \mathbb{R}$ is said to be a linear function if

$$f(\alpha \mathbf{u} + \beta \mathbf{v}) = \alpha f(\mathbf{u}) + \beta f(\mathbf{v}), \quad \forall \alpha, \beta \in \mathbb{R}, \ \mathbf{u}, \mathbf{v} \in V. \tag{2.21}$$

REMARK 2.2. **The principle of superposition**: In elementary courses of Strength of Materials, Electric Circuits, and many other Engineering disciplines, frequent use is made of the so-called principle of superposition. There is no such principle in nature, but many successful models of important phenomena are based upon the assumption of (approximate) linearity between causes and effects. Hooke's law, Ohm's law, and Fourier's law are just three examples from different areas. The principle of superposition asserts that the effect due to the combined action of several causes acting together is equal to the sum (superposition) of the effects that would be produced by each cause acting individually.[5] Equation (2.21) can be interpreted as stating precisely this fact. Linear functions on a vector space always exist and one has the right to investigate their geometrical or physical meaning in any particular context. The result is invariably useful: it provides the dual counterpart of the original concept. Velocities and momenta in Classical Mechanics, position and momentum in Quantum Mechanics are just two examples.

EXAMPLE 2.1. **Forces:** We have introduced (in Box 2.1) the vector space V of virtual displacements of a particle at a position in space. A *force* is a linear function on V. The evaluation of a force on a virtual displacement is called *virtual work*. Similarly, the evaluation of a force on a virtual velocity is called *virtual power*. It is important to point out that, although virtual displacements or velocities are often defined in a space where lengths and angles are available, this restriction is by no means necessary, as we know from Lagrangian Mechanics. As a linear function, the concept of force (or generalized force) is always well defined, since the linearity of a function does not presuppose any concept of length. This example shows that the very nature of a force is different from that of a velocity: they belong to different worlds, and they are related only through the ability of evaluating one on the other. To draw a force and a velocity as arrows in the same diagram is, strictly speaking, wrong.[6]

One might wonder whether linear functions always exist. We shall presently give a more detailed description of the set of all linear functions on a vector space, but it is comforting at the outset to notice that this set cannot be empty since the zero function $\mathbf{0}: V \to \mathbb{R}$ (assigning to each vector the number zero) is clearly linear. In fact, the set of linear functions on V has the natural structure of a vector space. Indeed, let f and g be two linear functions on a vector space V. We can define their sum as the linear operator $f + g$ whose action on vectors is given by:

$$(f + g)(\mathbf{v}) := f(\mathbf{v}) + g(\mathbf{v}), \quad \forall\, \mathbf{v} \in V. \tag{2.22}$$

---

[5] Even when the principle of superposition does apply, one should be very careful to use it correctly. A good example of what can go wrong is provided by a linear spring: although the displacement caused by forces abides by the principle of superposition, the stored elastic energy does not. Indeed, the latter is a quadratic function of the forces.

[6] An interesting observation pertaining to the method of finite elements in Solid Mechanics can be made: to express the stresses in components on the same basis as the displacements leads to stress fields with artificial discontinuities.

Similarly, the multiplication of a scalar α by a linear function f is, by definition, the linear function αf given by:

$$(\alpha f)(\mathbf{v}) := \alpha f(\mathbf{v}), \quad \forall \mathbf{v} \in V. \tag{2.23}$$

We are indulging in the typical abuse of notation so kindly allowed by vector spaces (for instance, the $+$ sign on the left-hand side of (2.22) refers to the operation being defined, while the $+$ sign on the right-hand side is the ordinary addition of scalars).

EXERCISE 2.9. **Vector-space structure:** Verify that, with the operations just defined, the set of all linear functions on a vector space V is a vector space.

The vector space of all linear functions on a vector space V is called the *dual space of V* and is denoted by V^*. Its elements are called *covectors* or *1-forms*. Let $\{\mathbf{e}_1, \dots \mathbf{e}_n\}$ be a basis in V, and, for each $i = 1, \dots, n$, consider the real function $\mathbf{e}^i : V \to \mathbb{R}$ defined by:

$$\mathbf{e}^i(\mathbf{v}) = v^i. \tag{2.24}$$

In words, the function $\mathbf{e}^i$ assigns to each vector $\mathbf{v} \in V$ its i-th component in the given basis. It is clear that each of these functions is linear, so that we have effectively constructed n nontrivial elements of V^*.

EXERCISE 2.10. Show that:

$$\mathbf{e}^i(\mathbf{e}_j) = \delta^i_j, \tag{2.25}$$

where the *Kronecker symbol* δ^i_j is defined as:

$$\delta^i_j = \begin{cases} 1 & \text{if } i = j, \\ 0 & \text{if } i \neq j. \end{cases} \tag{2.26}$$

Assume that n scalars α_i $(i = 1, \dots, n)$ have been found such that $\alpha_i \mathbf{e}^i = \mathbf{0}$. Applying both sides of this equation to the vector $\mathbf{e}_j$, we conclude that each of the coefficients α_j must vanish. Therefore, the covectors $\mathbf{e}^i$ are linearly independent. Now let ω be a covector. By linearity we have, for all $\mathbf{v} \in V$:

$$\omega(\mathbf{v}) = \omega(v^i \mathbf{e}_i) = v^i \omega(\mathbf{e}_i). \tag{2.27}$$

Invoking (2.24) and denoting $\omega_i = \omega(\mathbf{e}_i)$, we can write:

$$\omega(\mathbf{v}) = \omega_i \mathbf{e}^i(\mathbf{v}), \quad \forall \mathbf{v} \in V, \tag{2.28}$$

which shows that every covector ω is expressible (uniquely) as a linear combination $\omega_i \mathbf{e}^i$ of the n linearly independent covectors $\mathbf{e}^i$. In other words, the covectors $\mathbf{e}^1, \dots, \mathbf{e}^n$ constitute a basis of the dual space V^*, whose dimension, therefore, is the same as that of the original vector space V. A basis of V^* obtained in this way is called the *dual basis* of the corresponding basis in V.

EXERCISE 2.11. **Recovering the original basis:** Assume that a basis $\{\mathbf{f}^1, \dots, \mathbf{f}^n\}$ is given in V^* ab initio. Find the (unique) basis $\{\mathbf{f}_1, \dots, \mathbf{f}_n\}$ in V, of which $\{\mathbf{f}^1, \dots, \mathbf{f}^n\}$ is the dual. [Hint: choose an arbitrary basis in V and use (2.25).] [Remark: observe that this procedure would not work in general for an infinite-dimensional vector space.]

An alternative, and in some contexts more suggestive, notation for the evaluation of a covector ω on a vector $\mathbf{v}$ is the following one:

$$\langle \omega, \mathbf{v} \rangle := \omega(\mathbf{v}). \tag{2.29}$$

We will use both notations interchangeably.

EXERCISE 2.12. **Components:** (i) Show that if v^i $(i = 1, \ldots, n)$ are the components of $\mathbf{v} \in V$ in a given basis and if ω_i $(i = 1, \ldots, n)$ are the components of $\omega \in V^*$ in the dual basis, then: $\langle \omega, \mathbf{v} \rangle = \omega_i v^i$. (ii) Prove the important formulas:

$$v^i = \langle \mathbf{e}^i, \mathbf{v} \rangle, \tag{2.30}$$

and

$$\omega_i = \langle \omega, \mathbf{e}_i \rangle. \tag{2.31}$$

EXERCISE 2.13. **Change of basis:** Show that under a change of basis $\mathbf{f}_i = a_i^j \mathbf{e}_j$ in V, the dual basis changes contravariantly, while the components of covectors change covariantly, thus justifying the placement of the indices.

2.5. Linear Operators and the Tensor Product

What we have called the self-replicating nature of linearity can be demonstrated once again by considering maps between vector spaces. Since the real line $\mathbb{R}$ is endowed with a natural vector-space structure, it turns out that our previous treatment of the dual space can be seen as a particular case.

A *linear operator* T between two vector spaces U and V is a map $T : U \to V$ that respects the vector-space structure. More precisely,

$$T(\alpha \mathbf{u}_1 + \beta \mathbf{u}_2) = \alpha T(\mathbf{u}_1) + \beta T(\mathbf{u}_2), \quad \forall \alpha, \beta \in \mathbb{R}, \mathbf{u}_1, \mathbf{u}_2 \in U, \tag{2.32}$$

where the operations are understood in the corresponding vector spaces. When the source and target vector spaces coincide, the linear operator is called a *tensor*. Occasionally, the terminology of *two-point tensor* (or just tensor) is also used for the general case, particularly when the dimension of both spaces is the same. We will use these terms (linear operator, linear map, tensor, and so on) liberally.

Our next step should not come as a surprise: we can consider the collection $L(U, V)$ of *all* linear operators between two given vector spaces, and endow it with the natural structure of a vector space. To do so, we define the sum of two linear operators S and T as the linear operator $S + T$ whose action on an arbitrary vector $\mathbf{u} \in U$ is given by:

$$(S + T)(\mathbf{u}) := S(\mathbf{u}) + T(\mathbf{u}). \tag{2.33}$$

Similarly, we define the product of a scalar α by a linear operator T as the linear operator αT given by:

$$(\alpha T)(\mathbf{u}) := \alpha T(\mathbf{u}). \tag{2.34}$$

It is a straightforward matter to verify that the set $L(U,V)$, with these two operations, is a vector space. Such is the power of linearity. In the case of the dual space V^* (which can be identified with $L(V,\mathbb{R})$), we were immediately able to ascertain that it was never empty, since the zero map is linear. The same is true for $L(U,V)$, whose zero element is the linear map $\mathbf{0} : U \to V$ assigning to each vector of U the zero vector of V. Inspired by the example of the dual space, we will attempt now to construct a basis of $L(U,V)$ starting from given bases at U and V. This point takes some more work, but the result is, both conceptually and notationally, extremely creative.

Let $\omega \in U^*$ and $\mathbf{v} \in V$ be, respectively, a covector of the source space and a vector of the target space of a linear operator $T : U \to V$. We define the *tensor product* of $\mathbf{v}$ with ω as the linear operator $\mathbf{v} \otimes \omega \in L(U,V)$ obtained as follows[7]:

$$(\mathbf{v} \otimes \omega)(\mathbf{u}) := \langle \omega, \mathbf{u} \rangle \, \mathbf{v}, \quad \forall \mathbf{u} \in U. \tag{2.35}$$

We emphasize that the tensor product is fundamentally noncommutative. We note, on the other hand, that the tensor product is a bilinear operation, namely, it is linear in each of the factors, viz.:

$$(\alpha \mathbf{u}_1 + \beta \mathbf{u}_2) \otimes \omega = \alpha(\mathbf{u}_1 \otimes \omega) + \beta(\mathbf{u}_2 \otimes \omega), \tag{2.36}$$

and

$$\mathbf{u} \otimes (\alpha \omega_1 + \beta \omega_2) = \alpha(\mathbf{u} \otimes \omega_1) + \beta(\mathbf{u} \otimes \omega_2) \tag{2.37}$$

for all $\alpha, \beta \in \mathbb{R}$.

One of the reasons for the conceptual novelty of the tensor product is that it does not seem to have an immediately intuitive interpretation. In fact, it is a very singular linear operator, since, fixing the first factor, it squeezes the whole vector space U^* into an image consisting of a single line of V (the line of action of $\mathbf{v}$).

Let the dimensions of U and V be, respectively, m and n, and let $\{\mathbf{e}_\alpha\}$ ($\alpha = 1, \ldots, m$) and $\{\mathbf{f}_i\}$ ($i = 1, \ldots, n$) be respective bases. It makes sense to consider the $m \times n$ tensor products $\mathbf{f}_i \otimes \mathbf{e}^\alpha \in L(U,V)$. We want to show that these linear operators (considered as vectors belonging to the vector space $L(U,V)$) are in fact linearly independent. Assume that a vanishing linear combination has been found, namely, $\rho_\alpha^i \, \mathbf{f}_i \otimes \mathbf{e}^\alpha = \mathbf{0}$, where $\rho_\alpha^i \in \mathbb{R}$ and where the summation convention is appropriately used (Greek indices ranging from 1 to m, and Latin indices ranging from 1 to n). Applying this linear combination to the base vector $\mathbf{e}_\beta \in V$, we obtain $\rho_\beta^i \, \mathbf{f}_i = \mathbf{0}$, whence $\rho_\beta^i = 0$, proving that the only vanishing linear combination is the trivial one.

Let $T \in L(U,V)$ be an arbitrary linear operator. By linearity, we may write:

$$T(\mathbf{u}) = T(u^\alpha \mathbf{e}_\alpha) = u^\alpha \, T(\mathbf{e}_\alpha). \tag{2.38}$$

Each $T(\mathbf{e}_\alpha)$, being an element of V, can be written as a unique linear combination of the basis, namely:

$$T(\mathbf{e}_\alpha) = T_\alpha^i \, \mathbf{f}_i, \quad T_\alpha^i \in \mathbb{R}. \tag{2.39}$$

---

[7] Some authors define the tensor product backwards with respect to our definition.

We form now the linear operator $T^i_\alpha \, \mathbf{f}_i \otimes \mathbf{e}^\alpha$ and apply it to the vector $\mathbf{u}$:

$$T^i_\alpha \, \mathbf{f}_i \otimes \mathbf{e}^\alpha \, (\mathbf{u}) = u^\alpha \, T^i_\alpha \, \mathbf{f}_i. \tag{2.40}$$

Comparing this result with (2.38) and (2.39) we conclude that the original operator T and the operator $T^i_\alpha \, \mathbf{f}_i \otimes \mathbf{e}^\alpha$ produce the same result when operating on an arbitrary vector $\mathbf{u} \in U$. They are, therefore, identical and we can write:

$$T = T^i_\alpha \, \mathbf{f}_i \otimes \mathbf{e}^\alpha. \tag{2.41}$$

In other words, every linear operator in $L(U,V)$ can be written as a linear combination of the $m \times n$ linearly independent operators $\mathbf{f}_i \otimes \mathbf{e}^\alpha$, showing that these operators form a basis of $L(U,V)$, whose dimension is, therefore, the product of the dimensions of U and V. For these reasons, the vector space $L(U,V)$ is also called the *tensor product space of V and U^**, and is denoted as $V \otimes U^*$. The unique coefficients T^i_α are called the *components* of the tensor T in the corresponding basis.

EXERCISE 2.14. What are the components of the tensor product of a covector and a vector? Express your result in indices and also using the column-row notation of matrix algebra. For the case $U = V$, contrast this last result with the expression for the action of the covector on the vector. [Remark: note, in passing, that not every linear operator can be expressed as the tensor product of two vectors. In fact, those that can (called *decomposable* or *reducible operators*) are the exception rather than the rule.]

We have spoken about the conceptual novelty of the tensor product. No less important is its notational convenience. In the case of a vector, say $\mathbf{v} \in V$, it is obviously advantageous to be able to express the master concept of a vector in terms of the subsidiary notion of its components in a particular basis by simply writing: $\mathbf{v} = v^i \mathbf{f}_i$. If we change basis, for instance, the fact that what we have just written is an invariant expression, with a meaning beyond the particular basis chosen, can be exploited, as we have already done. In the case of linear operators, we have now obtained, according to Equation (2.41), a similar way to express the "real thing" invariantly in terms of its components on a basis arising from having chosen arbitrary bases in both the source and the target spaces. We can now *show* a tensor itself, much in the same way as we are able to show a vector itself. So powerful is this idea that, historically, the notation was invented before the concept of tensor product had been rigorously defined. In old physics texts it was called the *diadic* notation, and it consisted of simply apposing the elements of the bases involved (usually in the same (Cartesian) vector space: $\mathbf{ii}, \mathbf{ij}$, etc.). It is also interesting to recall that in Quantum Mechanics the prevailing notation for covectors and tensors is the ingenious device introduced by Dirac in terms of "bras" and "kets."

2.6. Isomorphisms and Iterated Dual

The *kernel*, $\ker(T)$, of the linear map $T : U \to V$ is the subset of U which is mapped to the zero of V. Equivalently, the kernel is the inverse image of zero: $\ker(T) = T^{-1}(\mathbf{0})$.

EXERCISE 2.15. **The kernel as subspace:** Show that the kernel of a linear operator $T : U \to V$ is a vector subspace of U.

The kernel of a linear map is never empty, since the zero of the source space is, by linearity, always in it.

EXERCISE 2.16. **Injectivity:** Show that a linear map is injective (one-to-one) if, and only if, the kernel contains no other element but the zero.

Recall that a general map $\phi : A \to B$ between two sets, A and B, is *invertible* if, and only if, it is one-to-one and onto. Recall also that the map ϕ is *onto* or *surjective* if its range equals B. The *inverse map* is then uniquely defined, and it is also one-to-one and onto. If a linear map $T : U \to V$ is invertible, its inverse $T^{-1} : V \to U$ is also a linear map.

EXERCISE 2.17. **Invertibility and linearity:** Show that a one-to-one linear map between vector spaces of the same (finite) dimension is automatically onto and, therefore, invertible. [Hint: show that the image of a basis is a basis.] Show that if a linear map is invertible, then its inverse is also a linear map.

An invertible linear map between vector spaces is called an *isomorphism*, and the two spaces are said to be *isomorphic*. An isomorphism of a space with itself is called an *automorphism*.

EXERCISE 2.18. **Isomorphism and dimension:** Show that all n-dimensional vector spaces (for a fixed n) are isomorphic. [Hint: choose a basis in each.] In particular, a vector space is isomorphic to its dual.

An isomorphism between vector spaces is *natural* or *canonical* if it is independent of the choice of bases.

EXERCISE 2.19. **Isomorphism with the dual space:** Show that the isomorphism between a space and its dual that is obtained by choosing a basis and assigning to each vector the covector that has the same components (in the dual basis), is not a natural isomorphism. [Hint: change the basis and check what happens when you reapply the rule.]

EXAMPLE 2.2. **Forces and velocities, again:** According to what we have just learned, a statement such as "the force is aligned with the velocity" makes no sense in general (e.g., in terms of generalized forces and velocities). It is only after having chosen, once and for all, a particular isomorphism between the space of virtual velocities and its dual (where forces dwell) that we can make such a statement. One way to choose an isomorphism between a space and its dual is by means of an inner product, which we will study soon. There are other meaningful possibilities in mechanics (a so-called symplectic structure, for example).

Since the dual of a vector space V is itself a vector space V^*, it makes sense to investigate what is the nature of its dual, namely, $(V^*)^*$ (or simply V^{**}). The tongue becomes somewhat tangled when trying to say it is the space of linear functions defined on the space of linear functions of the original vector space V. Fortunately, in the finite-dimensional case at least, it is possible to prove the following remarkable result: The iterated dual V^{**} is *canonically* isomorphic to the original space V. When two vector spaces are canonically isomorphic, they can be considered as being the same, since the mutual identification of elements is dictated by the very nature of the spaces involved, without the intervention of any external factors, such as the choice of a basis. With this identification in mind, we can safely say that (in the finite-dimensional case) the operation of taking a dual is self-cancelling, just as, for example, the operation of transposing a matrix, or of changing the sign of a number. There are two possible ways to exhibit a canonical isomorphism between vector spaces. The first, and preferred, way is by explicitly showing the correspondence between elements of the two spaces, without recourse to a basis. The second way consists of establishing the correspondence by means of a particular basis, but then proving that the correspondence is independent of the choice of basis. We will follow the first route. Let $\mathbf{v} \in V$ be a vector. We are looking for the corresponding element, say $\mathbf{A}_v \in V^{**}$ of the iterated dual. Since an element of the iterated dual is an operator acting on covectors, we will have done the job if we show what is the real number that $\mathbf{A}_v$ assigns to each element $\omega \in V^*$. We define:

$$\mathbf{A}_v(\omega) := \omega(\mathbf{v}), \quad \forall \omega \in V^*, \tag{2.42}$$

or, equivalently, in the alternative notation of Equation (2.29):

$$\langle \mathbf{A}_v, \omega \rangle := \langle \omega, \mathbf{v} \rangle, \quad \forall \omega \in V^*. \tag{2.43}$$

We have thus obtained a map $I : V \to V^{**}$.

BOX 2.2. **A More Abstract View of the Tensor Product**

The tensor product can be shown to have a deeper meaning as the most general linear construct that can be formed out of vector spaces. To outline the basic ideas, we start from the Cartesian product of two vector spaces, $U \times V$, and form the corresponding free vector space Z (see Box 2.1). By definition, this vector space consists of all formal linear combinations of pairs $(\mathbf{u}, \mathbf{v})$, $\mathbf{u} \in U$, $\mathbf{v} \in V$, with real coefficients. What we want to do now is to create a new entity in which we somehow identify different linear combinations within Z so that the rules of the desired bilinearity of the tensor product of two spaces are satisfied (see Equations (2.36) and (2.37)). For instance, since the free vector space Z is defined only formally, an element such as $(\mathbf{u}_1 + \mathbf{u}_2, \mathbf{v}) - (\mathbf{u}_1, \mathbf{v}) - (\mathbf{u}_2, \mathbf{v})$ is not equivalent to the zero element, an equivalence we would like to enforce in our new entity. To this end, we consider the collection of the elements of Z of the form:

$$(\alpha \mathbf{u}_1 + \beta \mathbf{u}_2, \rho \mathbf{v}_1 + \sigma \mathbf{v}_2) - \alpha\rho(\mathbf{u}_1, \mathbf{v}_1) - \alpha\sigma(\mathbf{u}_1, \mathbf{v}_2) - \beta\rho(\mathbf{u}_2, \mathbf{v}_1) - \beta\sigma(\mathbf{u}_2, \mathbf{v}_2), \tag{2.44}$$

where $\alpha, \beta, \rho, \sigma \in \mathbb{R}$, $\mathbf{u}_1, \mathbf{u}_2 \in U$, $\mathbf{v}_1, \mathbf{v}_2 \in V$, and where we recall the rule for eliminating, in any formal linear combination, terms with a zero coefficient. If we now define the subspace of W spanned by members of this collection, it is clear that W captures exactly all those elements of Z that we want to identify with the zero element of the new entity we want to create. We now define this new entity (namely, the tensor product of the spaces of departure) as the quotient space: $U \otimes V = Z/W$. To establish the relation between this definition and the previous one, it can be proved that the tensor product as just defined enjoys the so-called *universal factorization property*. Roughly speaking, this means that any bilinear function from $U \times V$ into any vector space X, can be essentially understood as a *linear* map from the tensor product $U \otimes V$ into X. The key to the proof is in the fact that any conceivable bilinear map automatically vanishes on the subspace W defined above. This theorem, showing that the abstract definition is more general, can be used not only to establish the link between the two definitions, but also to prove many other properties of the tensor product.

EXERCISE 2.20. Show that the map just defined (assigning to each element $\mathbf{v} \in V$ the element $\mathbf{A}_v \in V^{**}$) is linear and injective.

To complete the proof, we need to establish that this map is also onto, namely, that every element of the iterated dual can be obtained in this way. It is here that the finite dimensionality of the spaces needs to be invoked (see Exercise 2.17).

Having established the natural isomorphism between a space and its iterated dual, we can now regard the operation $\langle \omega, \mathbf{v} \rangle$ as commutative. Indeed, when we write $\langle \mathbf{v}, \omega \rangle$ we can agree to interpret it as $\langle \mathbf{A}_v, \omega \rangle$, which has the same value as $\langle \omega, \mathbf{v} \rangle$. So, for example, we can say that a force acts on a virtual velocity to produce virtual power, or that a virtual displacement acts on a force to produce the same result.

We can phrase the foregoing conclusion by saying that every finite-dimensional vector space can be regarded as a dual space (namely, it is the dual of its own dual space). In terms of the tensor product, this fact allows us to work comfortably with the tensor product between any two vector spaces. Thus, for instance, $U \otimes V$ can be identified with the space $L(V^*, U)$. Since the various spaces of linear operators are themselves vector spaces, one can take multiple tensor products, such as $(U \otimes V) \otimes W$. It is not difficult to show that there is a natural isomorphism between $(U \otimes V) \otimes W$ and $U \otimes (V \otimes W)$, so that the tensor product of vector spaces can be considered associative.

Commutativity is a different matter. Let $T : U \to V$ be a linear map. We define the *transpose* of T as the map $T^T : V^* \to U^*$ obtained by the prescription:

$$\langle T^T(\omega), \mathbf{u} \rangle := \langle \omega, T(\mathbf{u}) \rangle. \tag{2.45}$$

EXERCISE 2.21. **The transpose of a linear operator:** Show that all the operations involved in Equation (2.45) make sense. Show that, if $\{\mathbf{e}_\alpha\}$ ($\alpha = 1, \ldots, m$) and $\{\mathbf{f}_i\}$ ($i = 1, \ldots, n$) are, respectively, bases of U and V, whereby T is expressed as

$$T = T^i_\alpha \, \mathbf{f}_i \otimes \mathbf{e}^\alpha, \tag{2.46}$$

then the transpose of T is expressed as:

$$T^T = T^i_{\ \alpha}\, \mathbf{e}^\alpha \otimes \mathbf{f}_i. \tag{2.47}$$

In other words, the transpose of a tensor is obtained by leaving the components unchanged and switching around the base vectors. On the other hand, we may want to express the transpose in its own right by the standard formula:

$$T^T = (T^T)_\alpha^{\ i}\, \mathbf{e}^\alpha \otimes \mathbf{f}_i, \tag{2.48}$$

applicable to the components of a tensor in terms of a basis. Comparing the last two equations, we conclude that:

$$(T^T)_\alpha^{\ i} = T^i_{\ \alpha}. \tag{2.49}$$

Notice the precise order of the indices in each case.

A linear operator T is said to be *symmetric* if $T = T^T$ and *skew-* (or *anti-*) *symmetric* if $T = -T^T$. Recall, however, that in general T and T^T operate between different spaces. This means that the notion of *symmetry* should be reserved to the very special case in which the target space is precisely the dual of the source space, namely, when the linear operator belongs to some $L(U, U^*)$ or, equivalently to $U^* \otimes U^*$. We conclude that, as expected from the very tensor-product notation, a linear operator and its transpose are of the same nature (and may, therefore, be checked for symmetry) if, and only if, they belong to a tensor product of the form $V \otimes V$. Having said this, it is clear that if some artificial (noncanonical) isomorphism is introduced between a space and its dual (by means of an inner product, for example, as we shall presently do), then the concept of symmetry can be extended.

EXAMPLE 2.3. **Impossible symmetries in Continuum Mechanics:** Two-point tensors, such as the deformation gradient $\mathbf{F}$ and the first Piola-Kirchhoff stress $\mathbf{T}$, *cannot* be symmetric, by their very nature.

EXERCISE 2.22. **Natural isomorphism:** Show that the map that assigns to each linear operator $T : U \to V$ its transpose is a natural isomorphism between $V \otimes U^*$ and $U^* \otimes V$.

The *composition* of linear operators is a particular case of the composition of functions. Let $T : U \to V$ and $S : V \to W$ be linear operators between the vector spaces U, V and W. The composition $S \circ T : U \to W$ is usually denoted as ST and is called the *product* of the operators.

EXERCISE 2.23. **Product:** Choose bases in the vector spaces U, V and W and express the operators $T : U \to V$ and $S : V \to W$ in components. Show that the composition $S \circ T$ is represented in components by the *matrix product* $[S][T]$ of the matrices of components of S and T. This is the best justification of the, at first sight odd, rule for multiplying matrices.

EXERCISE 2.24. Can a linear operator $T \in L(U, V)$ in general be multiplied by its transpose? Why or why not?

There exists a natural isomorphism between the space of linear operators $L(U,V)$ and the space $L(U,V^*;\mathbb{R})$ of *bilinear*, real-valued maps $B: U \times V^* \to \mathbb{R}$. Bilinearity means that B is linear in each argument, namely:

$$B(\alpha\mathbf{u}_1 + \beta\mathbf{u}_2, \omega) = \alpha B(\mathbf{u}_1, \omega) + \beta B(\mathbf{u}_2, \omega), \tag{2.50}$$

and

$$B(\mathbf{u}, \alpha\omega_1 + \beta\omega_2) = \alpha B(\mathbf{u}, \omega_1) + \beta B(\mathbf{u}, \omega_2), \tag{2.51}$$

for all $\alpha, \beta \in \mathbb{R}$, $\mathbf{u}, \mathbf{u}_1, \mathbf{u}_2 \in U$ and $\omega, \omega_1, \omega_2 \in V^*$. The isomorphism assigns to each operator $T: U \to V$ the bilinear operator B_T defined by the formula:

$$B_T(\mathbf{u}, \omega) := \langle \omega, T(\mathbf{u}) \rangle. \tag{2.52}$$

EXERCISE 2.25. **Linear and bilinear operators:** Prove that the above correspondence (2.52) is a natural isomorphism. Write it in components.

REMARK 2.3. **Different ways**: Notice that we have found that the collection of linear operators between the vector spaces U and V can be expressed (modulo a canonical isomorphism) in at least three different ways: $L(U,V)$, $V \otimes U^*$ and $L(U,V^*;\mathbb{R})$.

Our review of linear operators is by no means complete. Nevertheless, we pause and turn to the introduction of inner-product spaces, leaving for the next chapter a more general treatment of the algebra of linear operators.

2.7. Inner-product Spaces

2.7.1. Generalities and Definition

We have come a long way without the need to speak about metric concepts, such as the length of a vector or the angle between two vectors. That even the concept of power of a force can be introduced without any metric background may have seemed somewhat surprising, particularly to those accustomed to hear about "the magnitude of the force multiplied by the magnitude of the velocity and by the cosine of the angle they form." It is very often the case in applications to particular fields (Mechanics, Theoretical Physics, Chemistry, Engineering, and so on) that there is much more structure to go around than really needed to formulate the basic concepts. For the particular application at hand, there is nothing wrong in taking advantage of this extra structure. Quite the contrary: the extra structure may be the carrier of implicit assumptions that permit, consciously or not, the formulation of the physical laws. The most dramatic example is perhaps the adherence to Euclidean Geometry as the backbone of Newtonian Physics. On the other hand, the elucidation of the minimal (or nearly so) structure necessary for the formulation of a fundamental notion, has proven time and again to be the beginning of an enlightenment that can lead to further developments and, no less importantly, to a better insight into the old results.

We have seen how the concept of the space dual to a given vector space arises naturally from the consideration of linear functions on the original vector space.

On the other hand, we have learned that, intimately related as they are, there is no natural isomorphism between these two spaces. In other words, there is no natural way to associate a covector to a given vector, and vice versa. In Newtonian Mechanics, however, the assumption of a Euclidean metric, whereby the theorem of Pythagoras holds globally, provides such identification. In Lagrangian Mechanics, it is the kinetic energy of the system that can be shown to provide such extra structure, at least locally. In General Relativity, this extra structure (but of a somewhat different nature) becomes the main physical quantity to be found by solving Einstein's equations. In all these cases, the identification of vectors with covectors is achieved by means of the introduction of a new operation called an *inner product* (or a *dot product* or, less felicitously, a *scalar product*).

A vector space V is said to be an *inner-product space* if it is endowed with an operation (called an inner product):

$$\cdot : V \times V \longrightarrow \mathbb{R}$$
$$(\mathbf{u}, \mathbf{v}) \mapsto \mathbf{u} \cdot \mathbf{v}, \tag{2.53}$$

satisfying the following properties[8]:

(1) Commutativity:

$$\mathbf{u} \cdot \mathbf{v} = \mathbf{v} \cdot \mathbf{u}, \quad \forall \mathbf{u}, \mathbf{v} \in V; \tag{2.54}$$

(2) Bilinearity[9]:

$$(\alpha \mathbf{u}_1 + \beta \mathbf{u}_2) \cdot \mathbf{v} = \alpha(\mathbf{u}_1 \cdot \mathbf{v}) + \beta(\mathbf{u}_2 \cdot \mathbf{v}), \quad \forall \alpha, \beta \in \mathbb{R}, \mathbf{u}_1, \mathbf{u}_2, \mathbf{v} \in V; \tag{2.55}$$

(3) Positive definiteness[10]:

$$\mathbf{v} \neq \mathbf{0} \implies \mathbf{v} \cdot \mathbf{v} > 0. \tag{2.56}$$

EXERCISE 2.26. Show that $\mathbf{0} \cdot \mathbf{v} = 0$, for all $\mathbf{v}$.

The *magnitude* or *length* of a vector $\mathbf{v}$ is defined as the non-negative number $\sqrt{\mathbf{v} \cdot \mathbf{v}}$. Two vectors $\mathbf{u}, \mathbf{v} \in V$ are called *orthogonal* (or *perpendicular*) to each other if $\mathbf{u} \cdot \mathbf{v} = 0$.

EXERCISE 2.27. Show that the zero vector is perpendicular to all vectors, and that it is the only vector with this property.

---

[8] It is to be noted that in the case of a complex vector space, such as in Quantum Mechanics applications, these properties need to be altered somewhat.

[9] The term bilinearity refers to the fact that the inner product is linear in each of its two arguments. Nevertheless, given that we have already assumed commutativity, we need only to show linearity with respect to one of the arguments.

[10] In Relativity this property is removed.

2.7.2. The Isomorphism between V and V^*

We want now to show how the existence of an inner product induces an isomorphism between a space and its dual (always in the finite-dimensional case). Let $\mathbf{v} \in V$ be a fixed element of V. By the linearity of the inner product, the product $\mathbf{v} \cdot \mathbf{u}$ is linear in the second argument. Accordingly, we define the covector $\omega_v \in V^*$ corresponding to the vector $\mathbf{v} \in V$, by:

$$\langle \omega_v, \mathbf{u} \rangle := \mathbf{v} \cdot \mathbf{u}, \quad \forall \mathbf{u} \in V. \tag{2.57}$$

It is not difficult to prove that this linear map from V to V^* is one-to-one and that, therefore, it constitutes an isomorphism between V and V^*. We conclude that in an inner-product space there is no need to distinguish notation-wise between vectors and covectors.

EXERCISE 2.28. Prove that the map from V to V^* represented by Equation (2.57) is one-to-one (injective). [Hint: use the result of Exercise 2.27.]

2.7.3. The Reciprocal Basis

We call *reciprocal basis* the basis of V that corresponds to the dual basis in the isomorphism induced by the inner product. We already know that the dual basis operates on vectors in the following way:

$$\langle \mathbf{e}^i, \mathbf{v} \rangle = v^i, \quad \forall \mathbf{v} \in V, \tag{2.58}$$

where v^i is the i-th component of $\mathbf{v} \in V$ in the basis $\{\mathbf{e}_j\}$ $(j = 1, \dots n)$. The reciprocal basis, therefore, consists of *vectors* $\{\mathbf{e}^j\}$ $(j = 1, \dots n)$ such that:

$$\mathbf{e}^i \cdot \mathbf{v} = v^i, \quad \forall \mathbf{v} \in V. \tag{2.59}$$

Let the components of the reciprocal base vectors be expressed as:

$$\mathbf{e}^i = g^{ij} \mathbf{e}_j. \tag{2.60}$$

In other words, we denote by g^{ij} the j-th component of the i-th member of the reciprocal basis we are seeking. It follows from (2.59) that:

$$\mathbf{e}^i \cdot \mathbf{v} = (g^{ij} \mathbf{e}_j) \cdot (v^k \mathbf{e}_k) = g^{ij} (\mathbf{e}_j \cdot \mathbf{e}_k) v^k = v^i, \quad \forall v^k \in \mathbb{R}. \tag{2.61}$$

Looking at the very last equality, it follows that

$$g^{ij} (\mathbf{e}_j \cdot \mathbf{e}_k) = \delta_k^i. \tag{2.62}$$

Indeed, regarded as a matrix equation, (2.61) establishes that the matrix with entries $[g^{ij} (\mathbf{e}_j \cdot \mathbf{e}_k)]$ (summation convention understood), when multiplied by an arbitrary column-vector, leaves it unchanged. It follows that this matrix must be the identity. This is possible only if the matrix with entries:

$$g_{ij} = \mathbf{e}_i \cdot \mathbf{e}_j, \tag{2.63}$$

is the inverse of the matrix with entries g^{ij}. So, the procedure to find the reciprocal basis is the following: (i) Construct the (symmetric) square matrix with entries $g_{ij} = \mathbf{e}_i \cdot \mathbf{e}_j$; (ii) invert this matrix to obtain the matrix with entries g^{ij}; (iii) define $\mathbf{e}^i = g^{ij}\mathbf{e}_j$.

REMARK 2.4. The *metric matrix* $\{g_{ij}\}$ is always invertible, as it follows from the linear independence of the basis.

A basis of an inner-product space is called *orthonormal* if all its members are of unit length and mutually orthogonal.

EXERCISE 2.29. Show that the reciprocal of an orthonormal basis coincides with the original basis.

2.7.4. Consequences

Having identified an inner-product space with its dual, and having brought back the dual basis to the original space under the guise of the reciprocal basis, we have at our disposal contravariant and covariant components of vectors. Recall that before the introduction of an inner product, the choice of a basis in V condemned vectors to have contravariant components only, while the components of covectors were covariant.

REMARK 2.5. If the basis of departure is orthonormal, then the covariant and contravariant components of a vector coincide. This fact leads to the somewhat confusing habit of adopting the summation convention for repeated indices at the same level (rather than diagonally). Even in a Euclidean space, however, the distinction between covariant and contravariant components needs to be made if a non-Cartesian system of coordinates is used.

EXERCISE 2.30. **Index gymnastics:** Starting from $\mathbf{v} = v^i\mathbf{e}_i = v_i\mathbf{e}^i$ and using Equations (2.62) and (2.63), prove the following formulas:

$$v^i = g^{ij}v_j, \tag{2.64}$$

$$v_i = g_{ij}v^j, \tag{2.65}$$

$$\mathbf{e}_i = g_{ij}\mathbf{e}^j, \tag{2.66}$$

$$v^i = \mathbf{v} \cdot \mathbf{e}^i, \tag{2.67}$$

$$v_i = \mathbf{v} \cdot \mathbf{e}_i, \tag{2.68}$$

$$\mathbf{e}^i \cdot \mathbf{e}^j = g^{ij}, \tag{2.69}$$

$$\mathbf{e}^i \cdot \mathbf{e}_j = \delta^i_j. \tag{2.70}$$

EXERCISE 2.31. **Covariant and contravariant components:** Consider the space $\mathbb{R}^2$ with its natural basis $\mathbf{e}_1 = (1,0)$, $\mathbf{e}_2 = (0,1)$. Show that the vectors $\mathbf{g}_1 = 2\mathbf{e}_1 + \mathbf{e}_2$, $\mathbf{g}_2 = \mathbf{e}_1 + 3\mathbf{e}_2$ form a new (covariant) basis of $\mathbb{R}^2$. Draw these vectors in the plane. Using the ordinary inner product (dot product), construct the reciprocal (contravariant) basis $\mathbf{g}^1$, $\mathbf{g}^2$. Draw this basis, clearly indicating any orthogonality with the covariant basis.

Given the vector $\mathbf{v} = 3\mathbf{e}_1 + 3\mathbf{e}_2$, find its contravariant and covariant components, v^i and v_i, with respect to the new basis. Do this by using the parallelogram rule and then verify that Equations (2.67) and (2.68) hold true.

EXERCISE 2.32. **A space of polynomials:** Consider the vector space $\mathcal{P}_n$ of all real n-degree polynomials P_n of a real variable x. Show that the real function $\omega : \mathcal{P}_n \to \mathbb{R}$ defined by:

$$\omega(P_n(x)) = P_n(0), \tag{2.71}$$

belongs to the dual space $\mathcal{P}_n^*$. Show that the operation $\cdot : \mathcal{P}_n \times \mathcal{P}_n \to \mathbb{R}$ given by the (Riemann) integral:

$$P_n \cdot Q_n := \int_{-1}^{1} P_n(x)\, Q_n(x)\, dx, \tag{2.72}$$

is a well-defined inner product in $\mathcal{P}_n$. Consider the natural basis of $\mathcal{P}_2$, namely:

$$\mathbf{e}_1 = 1, \quad \mathbf{e}_2 = x, \quad \mathbf{e}_3 = x^2. \tag{2.73}$$

Is this an orthonormal basis, with respect to the inner product just defined? If not, construct an orthonormal basis of $\mathcal{P}_2$. Construct the reciprocal basis of the natural basis (2.73). Represent the linear operator (2.71) as a vector $\mathbf{w}$, in accordance to (2.57). Generalize this last result for $\mathcal{P}_n$ with $n > 2$. Using perhaps a computer program, observe the general features of $\mathbf{w}$ as n grows. Can you discern any general (intuitive) features? [Hint: think of the Dirac delta, or pulse.]

The definition of the tensor product of two vectors $\mathbf{u}$ and $\mathbf{v}$ chosen, respectively, from any two inner-product spaces U and V can now be phrased as follows:

$$\langle \mathbf{u} \otimes \mathbf{v}, \mathbf{w} \rangle := (\mathbf{v} \cdot \mathbf{w})\, \mathbf{u}, \quad \forall w \in V. \tag{2.74}$$

Accordingly, the transpose of a linear operator $T : U \to V$ can be regarded as the linear operator $T^T : V \to U$ with the property:

$$T(\mathbf{u}) \cdot \mathbf{v} = \mathbf{u} \cdot T^T(\mathbf{v}), \quad \forall \mathbf{u} \in U, \mathbf{v} \in V. \tag{2.75}$$

It is only now that we are in a position to define the symmetry of a tensor of the form $T : U \to U$. Let $T : U \to U$ be a linear operator of an inner-product space into itself. If $T^T = T$ the operator is called *symmetric*, and if $T = -T^T$ it is called *skew-symmetric* or *antisymmetric*.

EXERCISE 2.33. **Cauchy-Green tensors:** Show that, for an arbitrary linear operator between inner-product spaces $T : U \to V$, the operators $T^T T$ and $T T^T$ are symmetric. On which space does each of these products act? Notice that for the case of the deformation gradient $\mathbf{F}$ of Continuum Mechanics, we obtain the right and left Cauchy-Green tensors. Write these tensors in components and show precisely at which point the inner-product structure needs to be invoked.

Just as with vectors, linear operators can now have covariant, contravariant, or mixed components. Thus:

$$T = T^{i\alpha}\mathbf{e}_i \otimes \mathbf{f}_\alpha = T^i_{\ \alpha}\mathbf{e}_i \otimes \mathbf{f}^\alpha = T_i^{\ \alpha}\mathbf{e}^i \otimes \mathbf{f}_\alpha = T_{i\alpha}\mathbf{e}^i \otimes \mathbf{f}^\alpha. \tag{2.76}$$

EXERCISE 2.34. Find the relation between the different kinds of components in (2.76).

A linear map $Q : U \to V$ between inner-product spaces is called *orthogonal* if $QQ^T = id_V$ and $Q^TQ = id_U$, where *id* stands for the identity map in the subscript space. The components of an orthogonal linear map in orthonormal bases of both spaces comprise an *orthogonal matrix*.

EXERCISE 2.35. **Preservation of inner product:** Prove that a linear map T between inner-product spaces preserves the inner product if, and only if, it is an orthogonal map. By preservation of inner product we mean that: $T(\mathbf{u}) \cdot T(\mathbf{v}) = \mathbf{u} \cdot \mathbf{v}, \ \forall \mathbf{u}, \mathbf{v} \in \mathbf{U}$.

EXERCISE 2.36. **Tensors in Continuum Mechanics:** For each of the tensors $(\mathbf{F}, \mathbf{C}, \mathbf{R}, \mathbf{t}, \mathbf{T}$, etc.) introduced in our Continuum Mechanics Primer (Appendix A), describe the input and output vector spaces and the action of the tensor on the input to produce the output. [Remark: in some cases it is easier to describe the meaning of a tensor as a bilinear real-valued operator.] The idea of looking at a linear operator as some kind of machine (say, a coin-operated candy dispenser) that accepts an input of some kind (money, say) to produce an output of a different kind (candy), naive as it may be, is often illuminating.

2.8. Affine Spaces

2.8.1. Introduction

A naive view of the world tends to suggest that physical space is some kind of vector space with a fixed preferential basis. All observers know what upward or downward *really* mean, and where things are to be measured from (each civilization, of course, disagreeing with the others on this last point). An event, such as the sudden extinguishing of a candle, is described (apart from its time of occurrence) by its position vector. Moreover, this space is also an inner-product space, where lengths and angles have a definite meaning. The space (whether 2- or 3-dimensional) that Euclid presents in his Elements is rather more sophisticated than that, at least in one respect: a vector is to be associated not to single points but rather to each *pair* of points. Moreover, the correspondence between pairs of points, on the one hand, and vectors, on the other, is not one-to-one: different pairs of points can be assigned the same vector. This property is what allows Euclid to make statements such as: "now we will draw a parallel to a given line from a point that does not belong to it." Also, Euclid's space has no preferred orientations. Not a single theorem of Euclid's elements necessitates a statement as to the triangle having a horizontal base or anything of the sort. Part of the scientific revolution brought about by Galileo and Newton has been precisely the adoption of Euclidean space, rather than what we would call today $\mathbb{R}^3$, as their model of the universe. This new allegiance brought into play the notions of *observer* and of (Galilean) *relativity*, notions undreamt of by ancient physics, though (as usual) ancient mathematics had already laid down the formal apparatus needed to develop them.

The space of Euclid is a particular case of a more general entity called an *affine space*. Affine spaces are not vector spaces, but they are *modelled* after a vector space. We shall later see more general entities (differentiable manifolds) modelled after vector spaces in a more sophisticated way. The particularity of a Euclidean space, as compared to a general affine space, is that the former is modelled after an inner-product space, whereas the latter can be modelled after a general vector space. In the next section we will give a more precise meaning to the expression "modelled after."

2.8.2. Definition

An affine space consists of a set $\mathcal{A}$ (the total space of *points*), a vector space V (the *supporting* or modelling vector space) and a map:

$$\mathsf{F} : \mathcal{A} \times \mathcal{A} \longrightarrow V$$
$$(p,q) \;\mapsto\; \mathsf{F}(p,q) = \vec{pq} = q - p, \tag{2.77}$$

with the following properties:

(1) Anticommutativity:

$$\mathsf{F}(q,p) = -\mathsf{F}(p,q), \quad \forall p,q \in \mathcal{A}, \tag{2.78}$$

(2) Triangle rule
$$\mathsf{F}(p,q) = \mathsf{F}(p,r) + \mathsf{F}(r,q), \quad \forall p,q,r \in \mathcal{A}, \tag{2.79}$$

(3) Arbitrary choice of origin

For each $p_0 \in \mathcal{A}$, $\mathbf{v} \in V$, $\exists$ a unique $q \in \mathcal{A}$ such that $\mathsf{F}(p_0,q) = \mathbf{v}$. $\tag{2.80}$

BOX 2.3. An Alternative Presentation

We want to explore the potential of the point-difference notation already introduced in Equation (2.77), namely, $q - p = \mathsf{F}(p,q)$, to reformulate the very notion of affine space. Let $\mathbf{v} = q - p$ be the vector pointing from p to q. We can exploit the notation to write:

$$q = p + \mathbf{v}, \tag{2.81}$$

and say that the point p is translated by the vector $\mathbf{v}$ to the point q. This hybrid addition operation satisfies the following rules:

(1) Vector associativity

$$(p + \mathbf{u}) + \mathbf{v} = p + (\mathbf{u} + \mathbf{v}), \quad \forall p \in \mathcal{A}, \, \mathbf{u}, \mathbf{v} \in V. \tag{2.82}$$

(2) Invariance

$$\text{For any } p \in \mathcal{A}: \ p + \mathbf{u} = p \ \Leftrightarrow \ \mathbf{u} = \mathbf{0}. \tag{2.83}$$

(3) Vector supply

$$p, q \in \mathcal{A} \ \Rightarrow \ \exists \text{ a unique } \mathbf{u} \in V \text{ such that } p + \mathbf{u} = q. \tag{2.84}$$

To prove (2.82) we start from the notational tautologies: $\mathbf{u} = \mathsf{F}(p, p + \mathbf{u})$, $\mathbf{v} = \mathsf{F}((p + \mathbf{u}), (p + \mathbf{u}) + \mathbf{v})$ and $\mathbf{u} + \mathbf{v} = \mathsf{F}(p, p + (\mathbf{u} + \mathbf{v}))$. Adding the first two and equating to the third, (2.82) follows immediately by the triangle rule (2.79). To prove (2.83), assume that for some $p \in \mathcal{A}$ we have $p = p + \mathbf{u}$. Since $\mathbf{u} = \mathsf{F}(p, p + \mathbf{u})$ we can write $\mathbf{u} = \mathsf{F}(p, p) = 0$, by anticommutativity. Conversely, let $p + \mathbf{0} = q$. Then, $\mathsf{F}(p, q) = q - p = \mathbf{0}$. But we also know that $\mathsf{F}(p, q) = 0$. By the uniqueness implied in (2.80) we conclude that $q = p$. Finally, (2.84) follows by setting $\mathbf{u} = \mathsf{F}(p, q)$ and invoking (2.80).

Properties (2.82), (2.83), and (2.84) can be used ab initio to define affine spaces. In that case one can prove properties (2.78), (2.79), and (2.80), showing that the two definitions are equivalent. The notations of addition and subtraction can be mixed, resulting in a very flexible notational scheme that can be conveniently abused. Thus, for instance, we may open brackets as follows:

$$\alpha(p - q) + \beta(q - r) = \alpha p + (\beta - \alpha)q - \beta r, \tag{2.85}$$

where $\alpha, \beta \in \mathbb{R}$ and $p, q, r \in \mathcal{A}$. We are thus multiplying scalars by points and constructing linear combinations thereof! Some caution must be exercised when doing this. More precisely, the linear combination of points given by $\alpha_0 p_0 + \alpha_1 p_1 + \cdots + \alpha_k p_k$ represents a vector if, and only if, the sum of the coefficients vanishes. It represents a point if, and only if, the sum of the coefficients is equal to 1.

EXERCISE 2.37. **Linear combinations of points:** Prove the above two assertions.

The concept of linear combination of points is particularly useful for the definition and manipulation of *affine simplexes*.

Before proceeding, some intuitive comments are called for. Thinking of the prototypical affine space, the anticommutative property simply asserts that the segment $\overline{pq}$ joining two points has exactly two possible senses and that the vector that goes from p to q is exactly equal to minus the vector that goes from q to p. As a consequence of this property, we obtain that $\mathsf{F}(p, p) = \overrightarrow{pp} = \mathbf{0}$. The triangle rule permits the application of the parallelogram rule in the large. Finally, the property of arbitrary choice of origin reveals the nature of an affine space in the sometimes used imprecise words: "an affine space is a vector space without an origin." In other words, once an origin is arbitrarily chosen, to every point of $\mathcal{A}$ there corresponds a unique vector of V, and vice versa.

EXERCISE 2.38. **Notation:** Write Equations (2.78), (2.79) and (2.80) using the alternative notations shown in Equation (2.77) and draw your own conclusions on their relative merits. See also Box 2.3

The *dimension* of an affine space $\mathcal{A}$ is, by definition, the dimension of its supporting vector space V. A subset $\mathcal{B}$ of $\mathcal{A}$ is an m-dimensional *affine subspace* if $\mathsf{F}(p,q) \in W$, for every pair $(p,q) \in \mathcal{B} \times \mathcal{B}$, where W is an m-dimensional subspace of V. A 1-dimensional affine subspace is called a *line*; an $(n-1)$-dimensional affine subspace is called a *hyperplane*. If not empty, the intersection of two affine subspaces is itself an affine subspace whose supporting space is equal to the intersection of the supporting subspaces. Two affine subspaces with the same supporting subspace W are said to be *parallel*. Using the additive notation of Box 2.3, an affine subspace is always of the form $p + W$, where p is any point in the subspace. Two parallel subspaces, $p + W$ and $q + W$, are different if, and only if, $q - p \notin W$. Thus, two parallel subspaces have no common points, unless they coincide. By a forgivable abuse of terminology, we say that a basis of the vector subspace W is also a basis of the affine subspace $\mathcal{B}$ it supports. Thus, a basis of an affine subspace is also a basis of any of its parallel subspaces.

EXERCISE 2.39. **The affine space generated by a vector space:** Show that a vector space can be canonically considered as an affine space modelled on itself. [Hint: call vectors "points" and assign to two points their vector difference.]

We have already alluded to the fact that in general the map F is not one-to-one. The inverse images $\mathsf{F}^{-1}(\mathbf{v})$ divide the set $\mathcal{A} \times \mathcal{A}$ into equivalence classes. Given, for the same total space $\mathcal{A}$ and vector space V, a different map, G, satisfying the above properties, it may happen that it respects the equivalence classes. Accordingly, all possible maps abiding by the above properties may themselves be divided into equivalence classes according to the above criterion. Any such equivalence class of maps is called an *affine structure*. From now on, when we speak of an affine space we will actually mean an affine structure.

EXERCISE 2.40. **Affine structures:** Show that two maps F and G satisfying properties (2.78), (2.79), and (2.80) belong to the same affine structure if, and only if, they are related by:

$$\mathsf{G} = T \circ \mathsf{F}, \tag{2.86}$$

where $T \in L(V, V)$ is an invertible map (i.e., an automorphism of V).

Within a fixed affine structure, according to the basic properties of affine spaces, a bijection (one-to-one and onto map) $\mathcal{F} : \mathcal{A} \to V$ between the total space and the supporting vector space is achieved by choosing an origin p_0 and a map F within the structure. The invertible map $\mathcal{F}$ is also called a *framing* or a *frame* of the affine space. More explicitly, the framing is given by:

$$\mathcal{F}(p) = \mathsf{F}(p_0, p). \tag{2.87}$$

Given two frames, $\mathcal{F}$ and $\mathcal{G}$, with origins p_0 and p_1 and maps F and G = TF, respectively, we obtain that:

$$\mathcal{G}(p) = \mathsf{G}(p_1, p) = \mathsf{G}(p_1, p_0) + \mathsf{G}(p_0, p) = \mathcal{G}(p_0) + T\mathcal{F}(p). \tag{2.88}$$

In other words, the most general *change of frame* is given by:

$$\mathcal{G}(p) = \mathbf{c} + T\mathcal{F}(p), \tag{2.89}$$

where $\mathbf{c}$ is an element of V and T is an automorphism of V, both arbitrary. Recall that an automorphism is an isomorphism of a vector space with itself.

DEFINITION 2.1. An *affine map* $f : \mathcal{A} \to \mathcal{B}$ between the affine spaces $\mathcal{A}$ and $\mathcal{B}$ is a map such that

$$f(q) - f(p) = L(q - p) \quad \forall p, q \in \mathcal{A}, \tag{2.90}$$

where $L : U \to V$ is a fixed linear map between the supporting vector spaces U and V of $\mathcal{A}$ and $\mathcal{B}$, respectively.

2.8.3. Affine Simplexes

The concept of *affine simplex* plays an important role in the theory of integration. It also provides a link between basic notions of point-set and combinatorial topology. A mundane example is provided by the finite element method in Continuum Mechanics: any material body can be approximated by a triangulation, that is, by an array of small tetrahedral elements, each of which is a simplex. For the formalization of the basic ideas, the notation of Box 2.3 is particularly useful.

A finite set of points $p_0, p_1, \ldots, p_r$ of an affine space $\mathcal{A}$ is said to be *linearly independent* if the vectors $p_1 - p_0, p_2 - p_0, \ldots, p_r - p_0 \in V$ are linearly independent. Linear independence is indifferent to the order of the given points: any one of the points may be assigned the zero subscript without altering the linear independence. We would like, therefore, to give an equivalent definition that does not single out any point of reference for the subtractions and that formally resembles the definition of linear independence of vectors. This objective is easily achieved: the points $p_0, p_1, \ldots, p_r$ are linearly independent if, and only if, the relations:

$$\alpha_0 p_0 + \alpha_1 p_1 + \cdots + \alpha_r p_r = 0, \tag{2.91}$$

and

$$\alpha_0 + \alpha_1 + \cdots + \alpha_r = 0, \tag{2.92}$$

imply that each of the scalars α_i vanishes.

EXERCISE 2.41. **Linear independence of points:** Prove the last statement.

The *affine r-simplex* associated with the linearly independent points $p_0, p_1, \ldots, p_r$ is defined as the set of points $x \in \mathcal{A}$ of the form:

$$x = \alpha_0 p_0 + \alpha_1 p_1 + \cdots + \alpha_r p_r, \tag{2.93}$$

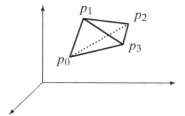

Figure 2.1. An affine 3-simplex (solid tetrahedron) in $\mathbb{R}^3$

such that

$$\alpha_0 + \alpha_1 + \cdots + \alpha_r = 1, \tag{2.94}$$

with

$$\alpha_i \geq 0, \quad i = 0, \ldots, r. \tag{2.95}$$

EXERCISE 2.42. **A simplex:** Consider three noncollinear points in ordinary 3-dimensional space and verify that the above definition of a simplex yields the (solid) triangle spanned by these points.

EXERCISE 2.43. **Dimension:** By definition, the dimension of an r-simplex is r. A 0-simplex consists of a single point. Show that an r-simplex of dimension larger than that of the supporting space V is necessarily empty.

The $r + 1$ points of departure are called the *vertices* of the r-simplex. The vertices of a (nonempty) r-simplex belong to the r-simplex. Indeed, the k-th vertex is obtained by setting $\alpha_i = \delta_{ik}$. Any numbers α_i satisfying Equations (2.94) and (2.95) are called the *barycentric coordinates* of the point x they define according to Equation (2.93).

EXERCISE 2.44. **Barycentric coordinates:** Clearly, any choice of numbers satisfying conditions (2.94, 2.95) determines a point x in the r-simplex. Prove the converse: a point x in an r-simplex uniquely determines a set of scalars satisfying those conditions. This one-to-one correspondence justifies the designation of coordinates. [Hint: invoke the linear independence of the vertices.]

Since every subset of a linearly independent set of points is again linearly independent, the collection of any number $s + 1$ ($0 \leq s \leq r$) of vertices determines an s-simplex, called an *s-face* of the original r-simplex. An s-face is said to be *proper* if $s < r$. The vertices of an r-simplex coincide with its 0-faces.

2.8.4. Euclidean or Inner-product Affine Structures

If the vector space V underlying an affine space $\mathcal{A}$ is an inner-product space, the affine space is called a *Euclidean* (or inner-product) affine space. The only difference between Euclidean and general affine spaces is that in the case of Euclidean spaces one demands that any permissible framing within a structure must preserve the inner product. In other words, to consider two maps, F and G, as belonging to the same Euclidean structure, we demand that, in addition to preserving the equivalence classes of pairs, they satisfy the condition:

$$F(p,q) \cdot F(r,s) = G(p,q) \cdot G(r,s), \quad \forall p,q,r,s \in \mathcal{A}. \tag{2.96}$$

EXERCISE 2.45. **Euclidean structures:** Show that two maps F and G satisfying properties (2.78),(2.79) and (2.80) belong to the same Euclidean structure if, and only if, they are related by:

$$G = Q \circ F, \tag{2.97}$$

where Q is an orthogonal transformation of V.

EXAMPLE 2.4. **Observer transformations:** If we identify a 3-dimensional Euclidean space with our physical space, then a frame is called an *instantaneous observer*. The most general *instantaneous observer transformation* is, therefore, given by:

$$\mathcal{G}(p) = \mathbf{c} + Q\mathcal{F}(p), \tag{2.98}$$

where Equation (2.89) has been used. The prescription (2.98) is precisely the formula for observer transformation used in Continuum Mechanics to establish the principle of material frame indifference (see Equations (A.6) and (A.87) of Appendix A).

2.9. Banach Spaces

The purpose of this section is to review a few basic notions that pertain to infinite-dimensional vector spaces, such as spaces of functions and spaces of chains, without entering into the more intricate details of functional analysis.

2.9.1. Basic Definitions

DEFINITION 2.2. Let V be a (finite- or infinite-dimensional) vector space over the reals. A *seminorm* on V is a function $p : V \to \mathbb{R}$ that satisfies the following conditions:

(1) $p(\mathbf{u} + \mathbf{v}) \leq p(\mathbf{u}) + p(\mathbf{v}) \quad \forall \mathbf{u}, \mathbf{v} \in V.$
(2) $p(\alpha\mathbf{u}) = |\alpha|\, p(\mathbf{u}) \quad \forall \mathbf{u} \in V, \alpha \in \mathbb{R}.$

EXERCISE 2.46. Given a linear function $g : V \to \mathbb{R}$ define the function p_g by $p_g(\mathbf{u}) := |g(\mathbf{u})|$. Show that $p_g : V \to \mathbb{R}$ is a seminorm on V.

LEMMA 2.1. *Let p be a seminorm on V. Then:*

$$p(\mathbf{0}) = 0, \tag{2.99}$$

where $\mathbf{0}$ is the zero vector.

PROOF. For any $\mathbf{u} \in V$, we have:

$$p(\mathbf{0}) = p(0\mathbf{u}) = 0p(\mathbf{u}) = 0, \tag{2.100}$$

where property (2) has been used. □

LEMMA 2.2. *Let p be a seminorm on V. Then:*

$$p(\mathbf{u}) \geq 0 \quad \forall \mathbf{u} \in V. \tag{2.101}$$

PROOF. By property (2), we have:

$$p(\mathbf{u}) = p(-\mathbf{u}) \quad \forall \mathbf{u} \in V. \tag{2.102}$$

Thus:

$$2p(\mathbf{u}) = p(\mathbf{u}) + p(\mathbf{u}) = p(\mathbf{u}) + p(-\mathbf{u}) \geq p(\mathbf{u} - \mathbf{u}) = p(\mathbf{0}) = 0, \tag{2.103}$$

where we have used property (1) and Lemma 2.1. □

A seminorm has some of the desirable properties of the concept of length of a vector without the need for the stricter notion of an inner product. Indeed, the seminorm is always non-negative and satisfies a triangle inequality (property (1)). Moreover, magnifying a vector by some factor has the desired effect (property (2)). What a seminorm does not guarantee is that if the value of the seminorm is zero then the vector is necessarily the zero vector. In other words, there may be nonzero vectors with a zero norm. This observation leads to the following definition:

DEFINITION 2.3. A *norm* on a vector space V is a seminorm with the additional property:

(3) $p(\mathbf{u}) = 0 \iff \mathbf{u} = \mathbf{0}$.

The usual notation for a norm is the following:

$$\|\mathbf{u}\| = p(\mathbf{u}), \tag{2.104}$$

a notation that we shall adopt. A vector space endowed with a norm is called a *normed space*.

EXERCISE 2.47. Show that an inner-product space is automatically a normed space. In other words, an inner product induces a norm.

The *distance* $d(\mathbf{u}, \mathbf{v})$ between two vectors $\mathbf{u}$ and $\mathbf{v}$ of a normed space V is defined as the norm of their difference, namely:

$$d(\mathbf{u}, \mathbf{v}) := \|\mathbf{u} - \mathbf{v}\|. \tag{2.105}$$

Thus, every normed space is automatically a *metric space*, that is, a set endowed with a distance function which is positive, definite, and symmetric and which satisfies the triangle inequality.

DEFINITION 2.4. Given a sequence $\mathbf{u}_1, \mathbf{u}_2, \ldots$ of vectors in a normed space V, we say that it converges to a vector $\mathbf{u} \in V$ if, for every real number $\epsilon > 0$, there exists some integer $N = N(\epsilon) > 0$ such that $\|\mathbf{u} - \mathbf{u}_n\| < \epsilon$ for every $n > N$. We say that $\mathbf{u}$ is the *limit* of the sequence and we write:

$$\mathbf{u} = \lim_{i \to \infty} \mathbf{u}_i. \tag{2.106}$$

If a sequence converges to a limit, it is said to be *convergent*. A direct application of the triangle inequality shows that every convergent sequence satisfies the condition that for every $\epsilon > 0$ there exists $N = N(\epsilon) > 0$ such that $\|\mathbf{u}_m - \mathbf{u}_n\| < \epsilon$, whenever $m, n > N$. A more elegant way to state this condition is:

$$\lim_{i,j \to \infty} \|\mathbf{u}_i - \mathbf{u}_j\| = 0. \tag{2.107}$$

DEFINITION 2.5. A sequence, whether convergent or not, satisfying condition (2.107) is said to be a *Cauchy sequence*.

Thus, every convergent sequence is a Cauchy sequence, but the converse is not necessarily true: Cauchy sequences may exist that are not convergent to a vector in V. A normed space V, in which every Cauchy sequence converges to a limit (in V), is called a *complete normed space* or, more commonly, a *Banach space*.

It can be shown that every finite-dimensional normed vector space is complete. For this reason, the use of the terminology "Banach space" usually implies that the underlying vector space is infinite-dimensional.

EXAMPLE 2.5. **The sup norm:** Consider the (infinite-dimensional) vector space $B(\mathcal{D})$ of *bounded* real-valued functions defined over a domain $\mathcal{D}$. The *sup norm* $\|f\|_\infty$ of a function $f \in B(\mathcal{D})$ is defined[11] as:

$$\|f\|_\infty := \sup\left(|f(x)|, x \in \mathcal{D}\right). \tag{2.108}$$

A particular case is the vector space $C[a, b]$ of continuous functions on a closed interval $[a, b] \in \mathbb{R}$. In this case the supremum may be replaced with the maximum. It can be shown that convergence of a sequence of functions under this norm is tantamount to *uniform convergence*. A well-known result of analysis is that the limit of a uniformly convergent sequence of continuous functions is itself continuous. Thus, the space $C[a, b]$ is a Banach space under the sup norm. On the other hand, the subspaces $P[a, b]$ of all polynomial functions and $C^n[a, b]$ of n-times continuously differentiable functions (with $n \geq 1$), are not complete under the sup norm.

---

[11] The notation $\|.\|_\infty$ is adopted for consistency with the family of norms $\|.\|_p$, which we do not consider in this example.

2.9.2. The Dual Space of a Normed Space

DEFINITION 2.6. A linear function $\omega : V \to \mathbb{R}$ on a normed space V is said to be *bounded* if:

$$\|\omega\| := \sup \left\{ \frac{|\omega(\mathbf{u})|}{\|\mathbf{u}\|}, \mathbf{u} \neq \mathbf{0} \right\} < \infty. \tag{2.109}$$

By linearity, (2.109) can also be written as:

$$\|\omega\| : \sup \{\omega(\mathbf{u}), \|\mathbf{u}\| = 1\} < \infty. \tag{2.110}$$

One can show that every bounded linear function is continuous and vice versa. The notation $\|\omega\|$ in Equation (2.109) is justified a posteriori, since one can show that this operation constitutes a norm in the space of bounded linear functions on V. An equivalent definition of this norm is:

$$\|\omega\| := \inf \{c \in \mathbb{R} : |\omega(\mathbf{u})| \leq c\|\mathbf{u}\|, \ \forall \mathbf{u} \in V\}. \tag{2.111}$$

Note that in this definition the infimum can be replaced with the minimum.

DEFINITION 2.7. With the norm (2.109), the normed space V^* of bounded linear functions on V is called the *dual space* of V.

Note that for a finite-dimensional vector space, this definition coincides with the usual one.

LEMMA 2.3. *The dual space V^* of a normed space (whether complete or not) is complete. In other words, V^* is always a Banach space.*

PROOF. Let $\omega_1, \omega_2, \ldots$ be a Cauchy sequence in V^*. For any fixed $\mathbf{u} \in V$, consider the sequence of real numbers $\omega_1(\mathbf{u}), \omega_2(\mathbf{u}), \ldots$. By (2.111) we have:

$$|\omega_i(\mathbf{u}) - \omega_j(\mathbf{u})| \leq \|\omega_i - \omega_j\| \, \|\mathbf{u}\|, \tag{2.112}$$

whence the sequence $\omega_1(\mathbf{u}), \omega_2(\mathbf{u}), \ldots$ is a Cauchy sequence of real numbers. The real line being complete (in the usual norm), this sequence has a (finite) limit, which we call $\omega(\mathbf{u})$. Letting $\mathbf{u}$ vary over V, we obtain a well-defined bounded[12] linear function ω. But, given any $\epsilon > 0$, we can find $N = N(\epsilon)$ such that, whenever $n > N$:

$$\|\omega - \omega_n\| = \sup\{|\omega(\mathbf{u}) - \omega_n(\mathbf{u})|, \|\mathbf{u}\| = 1\} < \epsilon, \tag{2.113}$$

which means that the given sequence $\omega_1, \omega_2, \ldots$ converges to ω. $\square$

2.9.3. Completion of a Normed Space

Since a normed space V is not necessarily complete, the question arises as to whether there is a unique way to add elements to V so as to render it complete. Informally, this can be accomplished by incorporating all the Cauchy sequences or, somewhat

---

[12] We skip the proof that ω is actually bounded.

more accurately, by constructing the space $\bar{V}$ of (equivalence classes of) all Cauchy sequences of V. A slightly more technical discussion follows.

DEFINITION 2.8. Two normed spaces U and V are *equivalent* if there exists a norm-preserving linear transformation $\phi : U \to V$, namely, if:

$$\|\phi(\mathbf{u})\|_V = \|\mathbf{u}\|_U \quad \forall \mathbf{u} \in U, \tag{2.114}$$

where $\|.\|_U$ and $\|.\|_V$ denote the norms of U and V, respectively.

DEFINITION 2.9. A normed subspace U of a normed space V is said to be *dense* in V if for every $\epsilon > 0$ and for every $\mathbf{v} \in V$, the ball $B_\epsilon(\mathbf{v}) := \{\mathbf{v}' \in V \ : \ \|\mathbf{v}' - \mathbf{v}\|_V < \epsilon\}$ contains an element $\mathbf{u}$ of U.

In simple words, there are elements of U as close as desired to any element of V.

DEFINITION 2.10. A Banach space $\bar{V}$ is said to be a *completion* of a normed space V if V is equivalent to a dense subspace of $\bar{V}$.

We have shown that the dual V^* of a normed space V is always a Banach space. Consider now the dual of V^*, denoted as V^{**}. Given an element $\mathbf{u} \in V$ we can, as we did in the case of finite-dimensional vector spaces, construct canonically a corresponding element $A_\mathbf{u} \in V^{**}$ by means of the prescription:

$$A_\mathbf{u}(\omega) := \omega(\mathbf{u}) \quad \forall \omega \in V^*, \tag{2.115}$$

which yields a well-defined bounded linear operator on V^*. The map $\phi : V \to V^{**}$ given by $\phi(\mathbf{u}) = A_\mathbf{u}$ is a linear map called the *canonical embedding*. It is not difficult to prove that this map is norm-preserving, so that, in fact, it is an equivalence between V and a subspace of V^{**}. Recalling that V^{**} is a Banach space (since it is the dual of a normed space), what we have achieved is the identification of V with a subspace $\phi(V)$ of a Banach space. What this means is that every Cauchy sequence in $\phi(V)$, being also a Cauchy sequence in V^{**}, is convergent in V^{**}. We can now consider the *closure* of $\phi(V)$ in V^{**}, namely, the collection of the limits of all Cauchy sequences of elements of $\phi(V)$. This is clearly a Banach subspace $\bar{V}$ of V^{**} and, by definition of limit, it is clear that $\phi(V)$ is dense in $\bar{V}$. We have thus obtained a completion of V. It can be shown that the completion is unique to within an equivalence.

3 Tensor Algebras and Multivectors

In Chapter 2 we have insisted on the vector-space nature of various spaces of tensors. We have also defined an associative operation between arbitrary vector spaces, namely the tensor product, thus opening the possibility of defining the structure of an *algebra*. In this chapter, we will introduce a number of different algebras that arise naturally out of a given vector space. Of particular interest for applications is the so-called *exterior algebra* of multivectors, providing a mathematical representation of the ability of a portion of space to become the carrier of physical quantities of various kinds.

3.1. The Algebra of Tensors on a Vector Space

3.1.1. The Direct Sum of Vector Spaces

Let U and V be vector spaces. The Cartesian product $U \times V$ is automatically endowed with the structure of a vector space called the *direct sum* of U and V and denoted as $U + V$. If $(\mathbf{u}_1, \mathbf{v}_1), (\mathbf{u}_1, \mathbf{v}_1) \in U \times V$, we define:

$$\alpha(\mathbf{u}_1, \mathbf{v}_1) + \beta(\mathbf{u}_2, \mathbf{v}_2) := (\alpha\mathbf{u}_1 + \beta\mathbf{u}_2, \, \alpha\mathbf{v}_1 + \beta\mathbf{v}_2), \quad \forall \alpha, \beta \in \mathbb{R}. \tag{3.1}$$

The direct sum is associative via the associativity of the Cartesian product. One may consider the (weak) direct sum of a countably infinite number of vector spaces, in which case only a finite number of entries are allowed to be nonzero. In other words, a typical element of the direct sum $\sum_{i=1}^{\infty} U_i$, where each U_i is a vector space, is of the form $(\mathbf{u}_1, \mathbf{u}_2, \ldots, \mathbf{u}_n, \mathbf{0}, \ldots, \mathbf{0}, \ldots)$, for some n (which may vary from element to element). Because of this property, it is clear that a (finite) linear combination of vectors in a direct sum (even if countably infinite) involves only a finite number of operations.

The tensor product is distributive with respect to the direct sum:

$$(U_1 + U_2) \otimes V = (U_1 \otimes V) + (U_2 \otimes V), \tag{3.2}$$

and

$$U \otimes (V_1 + V_2) = (U \otimes V_1) + (U \otimes V_2), \tag{3.3}$$

for arbitrary vector spaces.

3.1.2. Tensors on a Vector Space

Although a more general situation may be envisioned, we now consider the collection of all possible tensor products involving any finite number of factors, each factor being equal to a given vector space V or its dual V^*. The order of the factors, of course, matters, but it is customary to say that a tensor product is *of type (r,s)* if it is obtained by multiplying r copies of V and s copies of V^*, regardless of the order in which these copies appear in the product. An element of such a tensor product is also called a tensor of type (r,s). Another common terminology is to refer to r and s, respectively, as the contravariant and covariant degrees of the tensor. Thus, a vector is a tensor of type $(1,0)$, while a covector is of type $(0,1)$. By convention, a tensor of type $(0,0)$ is identified with a scalar.

EXERCISE 3.1. **The scalar field as a vector space:** Since the field of scalars ($\mathbb{R}$ in our case) has the natural structure of a vector space (whose elements are tensors of type $(0,0)$), it makes sense to take its tensor product with a vector space. Show that $\mathbb{R} \otimes V = V$.

The tensor product of a tensor of type (r_1, s_1) with a tensor of type (r_2, s_2) is a tensor of type $(r_1 + r_2, s_1 + s_2)$.

EXERCISE 3.2. **Components:** Let S be a tensor of type $(3,2)$ and T a tensor of type $(1,2)$. Assuming that in both cases all the V-factors have been placed first, express the tensor product $S \otimes T$ in terms of components induced by a base in V.

EXERCISE 3.3. **Multilinearity:** A map from a Cartesian product of vector spaces into a vector space is said to be *multilinear* if it is linear in each of the arguments. Show that a tensor T of type (r,s) can be considered as a multilinear map such that $T(\omega_1, \ldots, \omega_r, \mathbf{v}_1, \ldots, \mathbf{v}_s) \in \mathbb{R}$, where $\mathbf{v}_i$ and ω_j belong, respectively, to V and V^*, for each $i = 1, \ldots, r$ and each $j = 1, \ldots, s$. Express this action in coordinates.

3.1.3. The Tensor Algebra

The direct sum of the tensor spaces of all types (in all possible combinations within each type) looks as follows:

$$\mathcal{T}(V) = \mathbb{R} + V + V^* + V \otimes V + V \otimes V^* + V^* \otimes V +$$
$$V^* \otimes V^* + V \otimes V \otimes V + V \otimes V \otimes V^* + \cdots \tag{3.4}$$

EXERCISE 3.4. **Dimension:** Show that $\mathcal{T}(V)$ is an infinite-dimensional vector space.

We have already agreed that an element of the direct sum (3.4) has only a finite number of nonvanishing entries. Given two such elements, we already defined their sum by adding the entries slot by slot. This needs to be done, obviously, only on a finite number of slots, since beyond a large enough order, all slots in both elements will vanish. Similar remarks apply to multiplication by scalars. In other words, linear combinations are closed operations within the direct sum (3.4), as we expect in a vector space.

Consider next the tensor product. By the distributivity properties (3.2, 3.3), it is clear that the tensor product is also a closed operation.

EXERCISE 3.5. **The tensor product of two elements:** Let $\alpha \in \mathbb{R}$, $\mathbf{a}, \mathbf{b}, \mathbf{c} \in V$ and $\omega \in V^*$. Express α, $\mathbf{a} \otimes \mathbf{b} + \mathbf{b} \otimes \mathbf{c}$ and $\mathbf{a} \otimes (\alpha \omega)$ as elements (respectively, A_1, A_2, A_3) of $\mathcal{T}(V)$, as per Equation (3.4). Calculate the element $A_4 = (A_1 + A_2) \otimes (A_2 + A_3)$.

The vector space $\mathcal{T}(V)$, when supplemented with the operation of tensor multiplication, is called the *tensor algebra* of the vector space V.

3.1.4. The Operation of Contraction

We have seen that the tensor product of two tensors can never result in the decrease of either the contravariant or the covariant degrees of the factors. We will now define the operation of *contraction*. It acts on a single tensor of type (r, s), (with $rs \geq 1$) and maps it into a tensor of type $(r-1, s-1)$. To perform a contraction on T, one has first to specify, in the tensor-product space to which T belongs, one particular factor (say the i-th), which happens to be V, and a counterpart (say the j-th), which happens to be V^*. This pairing is essential to the result, which consists of a kind of mutual cancellation of the chosen factors. More precisely, if we consider first a *reducible* (or *decomposable*) *tensor* of type (r, s), namely, a tensor expressible as the tensor product of r vectors and s covectors, since the j-th factor is a covector, it can be made to act on the i-th factor, which is a vector, to produce a real number. After contraction, this number is all that remains of these two factors, which are otherwise eliminated from the product.

EXAMPLE 3.1. **Contraction of a reducible tensor:** Let a reducible tensor of type $(3, 2)$ be given by:

$$T = \omega \otimes \mathbf{u} \otimes \mathbf{v} \otimes \rho \otimes \mathbf{w}, \tag{3.5}$$

where $\mathbf{u}, \mathbf{v}, \mathbf{w} \in V$ and $\omega, \rho \in V^*$. The contraction operation C with respect to the second and fourth factors is:

$$C(T) = \langle \rho, \mathbf{u} \rangle \, \omega \otimes \mathbf{v} \otimes \mathbf{w}, \tag{3.6}$$

which is a reducible tensor of type $(2, 1)$.

This definition can now be extended by linearity to all tensors, whether reducible or not. In particular, if we choose a basis $\{\mathbf{e}_k\}$ $(k = 1,\dots,n)$ in V, every tensor of type (r,s) is a linear combination of n^{r+s} reducible tensors formed by products of vectors of the basis and the dual basis. For example, a tensor T of type $(3,2)$ may look like:

$$T = T_{a\ b}^{\ cdf}\, \mathbf{e}^a \otimes \mathbf{e}_c \otimes \mathbf{e}_d \otimes \mathbf{e}^b \otimes \mathbf{e}_f. \tag{3.7}$$

The contraction of the second and fourth indices yields the tensor:

$$C(T) = T_{a\ b}^{\ bdf}\, \mathbf{e}^a \otimes \mathbf{e}_d \otimes \mathbf{e}_f, \tag{3.8}$$

where the summation convention is at play. In other words, in components the contraction simply consists of renaming the selected indices with the same letter.

EXERCISE 3.6. **Definition through components:** Show that if one adopts the last statement (i.e., the renaming of indices) as an easy definition of contraction, the result is independent of the basis adopted.

3.2. The Contravariant and Covariant Subalgebras

The *contravariant tensor algebra* of a vector space V is the subalgebra of $\mathcal{T}(V)$ obtained by considering only tensors of covariant degree zero, namely, tensors of type $(r,0)$. The underlying vector space of the contravariant tensor algebra is, therefore:

$$\mathcal{C}(V) = \mathbb{R} + V + V \otimes V + V \otimes V \otimes V + \cdots \tag{3.9}$$

When written in components, all indices of tensors in this algebra are superscripts. A convenient notation consists of denoting by $\mathcal{C}^k(V)$ the k-fold product of V. Identifying $\mathcal{C}^0(V)$ with the scalar field $\mathbb{R}$, we may write (3.9) as:

$$\mathcal{C}(V) = \mathcal{C}^0(V) + \mathcal{C}^1(V) + \mathcal{C}^2(V) + \mathcal{C}^3(V) + \cdots = \sum_k \mathcal{C}^k(V). \tag{3.10}$$

In a similar way, one can define the *covariant tensor algebra* by considering tensors of type $(0,s)$. On the other hand, considering V^* as a vector space in its own right, we could form its contravariant tensor algebra $\mathcal{C}(V^*)$, and these two objects are obviously the same. The underlying vector space is now:

$$\mathcal{C}(V^*) = \mathbb{R} + V^* + V^* \otimes V^* + V^* \otimes V^* \otimes V^* + \cdots \tag{3.11}$$

The contravariant and covariant algebras can be considered dual to each other in the sense that there exists a canonical way to evaluate an element of one over an element of the other to produce a real number linearly. To see how this works, consider first two arbitrary reducible tensors: a tensor T of type $(k,0)$ and a tensor S of type $(0,k)$. By definition, there exist vectors $\mathbf{v}_1,\dots,\mathbf{v}_k \in V$ and covectors $\omega_1,\dots,\omega_k \in V^*$ such that $T = \mathbf{v}_1 \otimes \cdots \otimes \mathbf{v}_k$ and $S = \omega_1 \otimes \cdots \otimes \omega_k$. For these tensors, we define the evaluation

$\langle S, T \rangle$ as the real number:

$$\langle S, T \rangle = \langle \omega_1, \mathbf{v}_1 \rangle \ldots \langle \omega_k, \mathbf{v}_k \rangle = \prod_{i=1}^{k} \langle \omega_i, \mathbf{v}_i \rangle. \tag{3.12}$$

For two arbitrary, not necessarily reducible, tensors T and S of types $(k,0)$ and $(0,k)$, respectively, we extend this definition by linearity. If the tensors are of different orders (that is, a tensor of type $(r,0)$ and a tensor of type $(0,s)$ with $s \neq r$), we define the evaluation as zero. With this operation, we can identify $\mathcal{C}(V^*)$ with $(\mathcal{C}(V))^*$.

EXERCISE 3.7. **Algebra of the dual and dual of the algebra:** Show that the above definition of the evaluation of an arbitrary element of $\mathcal{C}^k(V^*)$ on an arbitrary element of $\mathcal{C}^k(V)$ is expressed in components by contracting corresponding indices (first covariant with first contravariant, and so on):

$$\langle S, T \rangle = S_{i_1 \ldots i_k} \, T^{i_1 \ldots i_k}. \tag{3.13}$$

Show how every linear function on the contravariant algebra $\mathcal{C}(V)$ can be identified with an element in the covariant algebra $\mathcal{C}(V^*)$, and vice versa. You may do this by working on a particular basis and then proving that your identification is independent of the choice of basis.

EXERCISE 3.8. **Duality and multilinearity:** A tensor T of type $(r,0)$ can be seen, as we know (Exercise 3.3), as a multilinear map:

$$T : V^*, \ldots, V^* \longrightarrow \mathbb{R}$$

$$(\omega_1, \ldots, \omega_r) \mapsto T(\omega_1, \ldots, \omega_r), \quad \omega_1, \ldots, \omega_r \in V^*. \tag{3.14}$$

Show that the evaluation of an arbitrary element T of $\mathcal{C}^k(V^*)$ (considered as a linear function) on a reducible element of $\mathcal{C}^k(V)$ gives the same result as the evaluation of T as a multilinear map operating on the vectors making up the reducible element.

For tensors in the contravariant or covariant algebras it makes sense to speak about *symmetry* and *skew-symmetry*. A tensor of type $(r,0)$ is said to be *(completely) symmetric* if the result of the operation (3.14) is independent of the order of the arguments. Put in other words, exchanging any two arguments with each other produces no effect in the result of the multilinear operator T. A similar criterion applies for completely symmetric tensors of order $(0,s)$, except that the arguments are vectors rather than covectors.

EXERCISE 3.9. **Symmetry:** Show that, choosing a basis in V, symmetry boils down to indifference to index swapping.

Analogously, a tensor of type $(r,0)$ is *(completely) skew-symmetric* if every mutual exchange of two arguments alters the sign of the result, leaving the absolute value unchanged.[1] By convention, all tensors of type $(0,0)$ (scalars), $(1,0)$ (vectors) and $(0,1)$ (covectors) are considered to be both symmetric and skew-symmetric.

---

[1] Naturally, this definition makes sense, since the parity of the number of exchanges depends only on the initial and final order of the arguments (see Box 3.1).

The collections of all symmetric or skew-symmetric tensors (whether contravariant or covariant) do not constitute a subalgebra of the tensor algebra, for the simple reason that the tensor multiplication of two symmetric (or skew-symmetric) tensors is not symmetric (skew-symmetric) in general. Nevertheless, it is possible, and convenient, to define algebras of symmetric and skew-symmetric tensors by modifying the multiplicative operation so that the results stay within the algebra. The case of skew-symmetric tensors is the most fruitful. It gives rise to the so-called *exterior algebra* of a vector space. We will explore this algebra in the next section. It will permit us to answer many intriguing questions such as: Is there anything analogous to the cross-product of vectors in dimensions other than 3? What is an area and what is the meaning of flux?

3.3. Exterior Algebra

3.3.1. Introduction

Having reached this point and gained a reasonable familiarity with the concepts of vector space, dual space, tensor product, and tensor algebra, the notion of exterior algebra can, and will, be introduced as a rather natural extension of the above concepts to the restricted class of skew-symmetric tensors, by substituting for the tensor product the operation of *exterior product*. The interpretation of the result, however, turns out to be surprisingly richer than expected and one might wonder whether it might not have been possible to develop the idea of exterior algebra on its own. A proof that this is indeed the case is provided by the historical fact that exterior algebra was developed not only before tensor algebra but also before the formalization of vector algebra!

As an intuitive motivation of the subject, consider the case of a line, that is, a 1-dimensional affine space modelled after a vector space V. Choosing an origin and an orientation, any nonzero vector $\mathbf{v}$ generates the whole line. In the absence of a distinguished inner product, it would appear that we cannot speak at all of the length attributes of $\mathbf{v}$. Nevertheless, given any other vector $\mathbf{w}$, there exists a *unique* real number α such that $\mathbf{w} = \alpha\mathbf{v}$. From the properties of the inner product it follows that, regardless of the particular inner product that may be introduced eventually, the length of $\mathbf{w}$ will always be equal to $|\alpha|$ times the length of $\mathbf{v}$. Moreover, if $\alpha < 0$, we may say that $\mathbf{w}$ points in the opposite direction of $\mathbf{v}$. In other words, *relative* magnitude and directionality are intrinsic quantities, independent of the choice of inner product.

The dual space V^* consists of all real linear functions acting on V. Given $\rho \in V^*$ and $\mathbf{v} \in V$, the evaluation $\rho(\mathbf{v}) = \langle \rho, \mathbf{v} \rangle$ can be conceived as the *content* of the physical quantity represented by ρ associated with (or contained within) the vector $\mathbf{v}$. In any particular physical application, the terminology is designed to reflect these facts. For example, ρ might be called a force and $\mathbf{v}$ a velocity, in which case the content $\langle \rho, \mathbf{v} \rangle$ is called power. In this way, vectors in V can be regarded as *potential carriers* of a physical quantity, whose *intensity* is represented by a covector. If we double the vector, by doubling the capacity of the carrier we will double the content. Or, if keeping the vector constant we should double the covector, by doubling the intensity of the physical quantity we will also double the content.[2] Changing the orientation of **v** results in a change of the sign of the content.

The aforementioned ideas are clear enough in the 1-dimensional case. We seek, therefore, a notational and conceptual means to generalize them for higher dimensions. In other words, we want to formalize the definition of an object that, belonging to a k-dimensional subspace of an n-dimensional affine space, will represent the ability to carry some physical content. The mathematical ideas involved are best expressed in terms of multivectors and forms. These entities are dual of each other and they each form an algebra, whose product operation is the *exterior product*.

3.3.2. The Exterior Product

The space of skew-symmetric contravariant tensors of type $(r,0)$ will be denoted by $\Lambda^r(V)$. The elements of $\Lambda^r(V)$ will also be called *r-vectors* and, more generally, *multivectors*. The number r is the *order* of the multivector. As before, the space $\Lambda^0(V)$ coincides with the scalar field $\mathbb{R}$, while $\Lambda^1(V)$ coincides with the vector space V. The exterior algebra of the vector space V will be defined on the direct sum:

$$\Lambda(V) = \Lambda^0(V) + \Lambda^1(V) + \Lambda^2(V) + \cdots = \sum_k \Lambda^k(V). \qquad (3.15)$$

The space $\Lambda^r(V)$ is obviously a vector subspace of $\mathcal{C}^r(V)$, since it is closed under tensor addition and multiplication by scalars. Given a subspace A of a vector space B, a *projection* $\rho : B \to A$ is a (surjective) linear map satisfying the condition $\rho \circ \rho = \rho$. This condition means that it is only the first application of the map that matters. In other words, vectors already on the subspace are not affected by the projection.

EXERCISE 3.10. **A projection map:** Let B be an n-dimensional vector space and let $\omega \in B^*$ be a nonvanishing covector. Prove that the *annihilator* A of ω (namely, the collection of all vectors **u** of B such that $\langle \omega, \mathbf{v} \rangle = 0$) is a proper subspace of B. Let $\mathbf{b} \in B$ be a vector such that $\langle \omega, \mathbf{b} \rangle = 1$ (show that such vectors must always exist). Prove that the map $P_b = id_B - \mathbf{b} \otimes \omega$ is a projection of B onto A. (You must prove that the map is linear, that its range is A and that it is equal to its square.) Prove that every vector in B can be expressed uniquely as a sum of a vector in A and a vector proportional to **b**. Conclude that the dimension of A is $n - 1$. Repeat the exercise assuming that B is an inner-product space. In this case, you may (but don't have to) identify both ω and **b** with the unit vector perpendicular to the subspace. Draw a picture for the case $n = 3$.

The only reason for our digression on the topic of projections is that we want to show that there exists a canonical projection $\mathcal{A}^r$ of $\mathcal{C}^r(V)$ onto $\Lambda^r(V)$, namely, a way

---

[2] There is an echo here of Newton's own verbal struggle in his *Principia Mathematica*: "Aer duplo densior in duplo spatio quadruplus est."

BOX 3.1. **Permutations and Determinants**

Consider a finite set $\mathcal{X}$. Its elements can, and will, be identified with the numbers $1,\ldots,n$. A *permutation* of $\mathcal{X}$ is, by definition, a bijection $\pi : \mathcal{X} \to \mathcal{X}$. It is useful (although sometimes confusing) to think of a permutation as a reordering of the elements. Permutations can be composed and it is not difficult to show that the set of all permutations of $\mathcal{X}$ under the operation of composition forms a group, known as the *symmetric group* of $\mathcal{X}$. This group, denoted by S_n, contains exactly $n!$ elements (namely, the different ways in which a set of n objects can be ordered). A *transposition* is a permutation that leaves all elements of $\mathcal{X}$ fixed except two. In the reordering picture, a transposition is simply an *exchange* of the position of two elements. It can be rigorously proven that every permutation can be expressed as a composition of transpositions. Moreover, although the representation of a given permutation as a composition of transpositions is not unique, the *parity* is. By this we mean that the number of transpositions required to express a given permutation is either even or odd, regardless of the representation chosen. Accordingly, a given permutation can be called *even* or *odd*, respectively. It is this feature that allows us to define a function $sign(\pi)$ (the *signature* of a permutation π) with a value of either $+1$ or -1, respectively, for π even or odd. The signature of a composition of permutations is equal to the product of the signatures. It follows that the even permutations form a subgroup (the so-called *alternating group*), but the odd permutations don't.

Let A be a square matrix of order n, with entries $[a_{ij}]$. Its determinant is defined as:

$$\det[a_{ij}] := \sum_{\pi} sign(\pi)\, a_{1,\pi(1)} a_{2,\pi(2)} \ldots a_{n,\pi(n)}, \qquad (3.16)$$

where the summation extends over all $n!$ different permutations π in S_n. This formula corresponds exactly to the following description in words: Form all possible products of n elements, one taken from each row in such a way that no two belong to the same column. If the column selection follows an odd permutation, change the sign of the product. The determinant is the sum of all the products thus obtained. Another useful formula is obtained by letting the row index follow an arbitrary permutation $\sigma \in S_n$ (rather than keeping the rows in the natural order):

$$\sum_{\pi} sign(\pi)\, a_{\sigma(1),\pi(1)} a_{\sigma(2),\pi(2)} \ldots a_{\sigma(n),\pi(n)} = sign(\sigma) \det[a_{ij}]. \qquad (3.17)$$

to assign to a contravariant tensor its skew-symmetric part.[3] To see that this is the case, we regard a tensor $T \in \mathcal{C}^r(V)$ as a multilinear map on the r-fold product of V^*, as we have done before. There exist exactly $r!$ permutations of the r arguments of this

---

[3] Similarly, we can assign to a tensor its symmetric part. Notice, however, that it is only in the case $r = 2$ that the tensor can be reconstructed as the sum of these two parts, as we all know from matrix algebra.

function. For any given set of arguments (that is, any ordered r-tuple of covectors) we will evaluate the function on each of these permutations and add the results, remembering to multiply a result by -1 if it corresponds to an odd permutation.[4] The multilinear function obtained in this way obviously belongs to $\Lambda^r(V)$. However, the (linear) map that assigns to each element of $C^r(V)$ the element of $\Lambda^r(V)$ constructed in this way fails to be a projection. To render it a projection ($\mathcal{A}^r$) we simply adjust it by dividing the final result by $r!$. Otherwise, an already skew-symmetric tensor would not be mapped to itself, but to a multiple of itself.

EXERCISE 3.11. **The skew-symmetric projection of a reducible tensor:** Show that for a reducible tensor $T = \mathbf{u}_1 \otimes \cdots \otimes \mathbf{u}_r$ (where each $\mathbf{u}_i \in V$) we have:

$$\mathcal{A}^r(T) = \frac{1}{r!}(\mathbf{u}_1 \otimes \mathbf{u}_2 \otimes \cdots \otimes \mathbf{u}_r - \mathbf{u}_2 \otimes \mathbf{u}_1 \otimes \cdots \otimes \mathbf{u}_r + \cdots)$$

$$= \frac{1}{r!}\sum_{\pi} sign(\pi)\, \pi\,(\mathbf{u}_1 \otimes \mathbf{u}_2 \otimes \cdots \otimes \mathbf{u}_r), \tag{3.18}$$

where π denotes a permutation of the arguments and where *sign* denotes the signature function (which assigns to a permutation π a value of 1 or -1 according to whether π is an even or an odd permutation, as described in Box 3.1). Conclude that, when working in components, the skew-symmetric part of a tensor is obtained by applying the above procedure to the indices of the array of components. If this component-wise criterion were used as a definition of skew-symmetry, prove that the criterion would be independent of the basis chosen.

EXERCISE 3.12. **The case $r = 3$:** For the case of a contravariant tensor of degree 3, namely, $T = T^{ijk}\mathbf{e}_i \otimes \mathbf{e}_j \otimes \mathbf{e}_k$, where $\mathbf{e}_h$ ($h = 1, \ldots, n \geq r$) is a basis of V, show that:

$$\mathcal{A}^3(T) = \frac{1}{6}\left(T^{ijk} + T^{jki} + T^{kij} - T^{ikj} - T^{jik} - T^{kji}\right)\mathbf{e}_i \otimes \mathbf{e}_j \otimes \mathbf{e}_k. \tag{3.19}$$

Verify that a repeated value of an index leads to a vanishing component.

EXERCISE 3.13. **Practice:** Consider a 3-dimensional V. Is the tensor $T = \mathbf{e}_1 \otimes \mathbf{e}_2 \otimes \mathbf{e}_3 - \mathbf{e}_1 \otimes \mathbf{e}_3 \otimes \mathbf{e}_2$ skew-symmetric? Calculate $\mathcal{A}^3(T)$. Apply $\mathcal{A}^3$ to the result and verify that if it were not for the factor $1/6$, the above formula (3.19) would not leave the (already skew-symmetric) tensor unchanged.

Already in possession of a projection $\mathcal{A}^r$ for each $C^r(V)$, we have in fact obtained a projection $\mathcal{A}$ over the whole $C(V)$. It is clear now that the appropriate operation to convert $\Lambda(V)$ into an algebra, called the *exterior algebra of V*, is the projection of the tensor product. This projected operation is called the *exterior product* or the *wedge product* of multivectors. It is denoted by the wedge symbol "$\wedge$." By definition:

$$a \wedge b = \mathcal{A}(a \otimes b), \quad \forall a, b \in \Lambda(V), \tag{3.20}$$

---

[4] Recall that an odd permutation of an ordered set of objects is one obtained by an odd number of exchanges.

What this definition in effect is saying is that in order to multiply two skew-symmetric tensors and obtain a skew-symmetric result, all we have to do is take their tensor product and then project back into the algebra (i.e., skew-symmetrize the result).[5] Since the projection $\mathcal{A}$ is, by definition, a linear operator, the wedge product is linear in each of the factors.

We have seen that the tensor product is not commutative. But, in the case of the exterior product, exchanging the order of the factors can at most affect the sign. To see what happens, it is enough to consider the case of the product $a \wedge b$ of two reducible multivectors a and b of degrees, say, r and s, respectively. (By reducible we now mean that each of the factors is expressible as the *wedge* product of vectors.) The result is, clearly, a multivector of degree $r+s$, as with the tensor product. To exchange the order of the factors a and b, we may think of sliding b to the left so that it becomes the first factor. Recall that we are looking at reducible multivectors, so that to slide b to the left we can slide each of its s factors, one at a time, leftward through the r factors of a. Each factor of b, therefore, produces exactly r transpositions. The total number of transpositions obtained in this way is exactly rs. If this product of integers is even, the product $a \wedge b$ commutes, and if it is odd it changes the sign upon reversal. We have operated on reducible multivectors, but (e.g., thinking in components) it is clear that the same result follows in general. We can therefore, write, that:

$$b \wedge a = (-1)^{rs} a \wedge b, \ a \in \Lambda^r V, b \in \Lambda^s(V). \tag{3.21}$$

Thus, for example, the wedge product with itself of a multivector of odd order must necessarily vanish. With some work, it is possible to show that the wedge product is associative, namely, $(a \wedge b) \wedge c = a \wedge (b \wedge c)$.

A fundamental difference between the algebras $\mathcal{C}(V)$ and $\Lambda(V)$ is their dimension. The algebra $\mathcal{C}(V)$ is clearly infinite-dimensional, since there is no obstruction to the multiplication of contravariant tensors of any degrees. In the case of the algebra of multivectors, however, the dimension of the underlying vector space V imposes a severe restriction on the availability of nonvanishing skew-symmetric multilinear maps. Indeed, a multilinear skew-symmetric map automatically vanishes if its arguments are linearly dependent. So, if the dimension of V is n, then a multilinear skew-symmetric map with more than n arguments must necessarily vanish. In other words, in an n-dimensional space V, all multivectors of order greater than n vanish automatically. Effectively, therefore, $\Lambda(V)$ can be reduced to the direct sum of a *finite* number of terms, the last one being $\Lambda^n(V)$.

To calculate the dimension of $\Lambda(V)$, we need to calculate the dimension of each of the terms $\Lambda^k(V)$, $k = 0,\ldots,n$, and add the results. Every element in $\Lambda^k(V)$ is also an element in $\mathcal{C}^k(V)$ and is, therefore, expressible as a linear combination of the n^k tensor products $\mathbf{e}_{i_1} \otimes \cdots \otimes \mathbf{e}_{i_k}$, where $\mathbf{e}_i$, $i = 1,\ldots,n$, is a basis of V. Because

---

[5] In spite of the natural character of this definition of the wedge product, many authors adopt a definition that includes a combinatorial factor. Thus, the two definitions lead to proportional results. Each definition has some advantages, but both are essentially equivalent. Our presentation of exterior algebra follows closely that of Sternberg S (1983), *Lectures on Differential Geometry*, 2nd ed., Chelsea.

of the skew-symmetry, however, we need to consider only products of the form $\mathbf{e}_{i_1} \wedge \cdots \wedge \mathbf{e}_{i_k}$. Two such products involving the same factors in any order are either equal or differ in sign, and a product with a repeated factor vanishes. This means that we need only count all possible combinations of n symbols taken k at a time without repetition. The number of such combinations is $\frac{n!}{(n-k)!k!}$. One way to keep track of all these combinations is to place the indices $i_1, \ldots, i_k$ in strictly increasing order. These combinations are linearly independent (see Exercise 3.14 below), thus constituting a basis. Therefore, the dimension of $\Lambda^k(V)$ is $\frac{n!}{(n-k)!k!}$. The dimension of the algebra is, accordingly, 2^n.

We note that the spaces of k-vectors and $(n-k)$-vectors have the same dimension. There is a kind of fusiform dimensional symmetry around the middle, the dimension starting at 1 for $k = 0$, increasing to a maximum toward $k = n/2$ (say, if n is even) and then going back down to 1 for $k = n$. Since spaces of equal dimension are isomorphic (albeit not always canonically so), this observation plays an important role in the identification (and sometimes confusion) of physical quantities, of the type we have encountered already for vectors and covectors. For example, an n-vector functions very much like a scalar, but with a subtle difference that we will revisit later.

EXERCISE 3.14. **Linear independence of the wedge products:** Consider the $\frac{n!}{(n-k)!k!}$ wedge products $\mathbf{e}_{i_1} \wedge \cdots \wedge \mathbf{e}_{i_k}$, where the indices are placed in strictly increasing order. We want to show that they are linearly independent. For the case $k = n$ we are left with just one product, namely, $\mathbf{e}_1 \wedge \cdots \wedge \mathbf{e}_n$, which obviously cannot vanish (as can be seen by applying it as a multilinear map to the dual basis). If $k < n$, assuming that a vanishing linear combination has been found, we simply wedge-multiply it by the wedge product of one choice of $n - k$ different base vectors. Thus, every term now has n factors. But only one of these can be nonzero (namely, the one that before the multiplication contained exactly the k terms not selected), since in all the others there must be at least one repeated index. Since the linear combination vanishes, the coefficient of this term must vanish.

As a result of Exercise 3.14, we conclude that the products $\mathbf{e}_{i_1} \wedge \cdots \wedge \mathbf{e}_{i_k}$ with strictly increasing order of the indices constitutes a basis of $\Lambda^k(V)$.

REMARK 3.1. Let a skew-symmetric contravariant tensor $a \in \Lambda^r(V)$ be given by means of its components on the basis of $\mathcal{C}^r(V)$ inherited from a basis $\mathbf{e}_1, \ldots, \mathbf{e}_n$ of V as:

$$a = a^{i_1 \ldots i_r} \, \mathbf{e}_{i_1} \otimes \cdots \otimes \mathbf{e}_{i_r}. \tag{3.22}$$

Recalling that the projection operator $\mathcal{A}^r$ is linear, we obtain:

$$a = \mathcal{A}^r(a) = a^{i_1 \ldots i_r} \, \mathcal{A}^r \left(\mathbf{e}_{i_1} \otimes \cdots \otimes \mathbf{e}_{i_r} \right) = a^{i_1 \ldots i_r} \, \mathbf{e}_{i_1} \wedge \cdots \wedge \mathbf{e}_{i_r}. \tag{3.23}$$

In these expressions, the summation convention is implied. We have obtained the result that, given a skew-symmetric tensor in components, we can substitute the wedge products for the tensor products of the base vectors. On the other hand, if we would like to express the r-vector a in terms of its components on the basis of $\Lambda^r(V)$

given by the wedge products of the base vectors of V taken in strictly increasing order of the indices, a coefficient of $r!$ will have to be included, namely:

$$a^{i_1,\dots i_r}\, \mathbf{e}_{i_1} \wedge \cdots \wedge \mathbf{e}_{i_r} = r!\sum_{i_1 < \cdots < i_r} a^{i_1,\dots i_r}\, \mathbf{e}_{i_1} \wedge \cdots \wedge \mathbf{e}_{i_r}. \qquad (3.24)$$

This means that the components on the basis (with strictly increasing indices) of the skew-symmetric part of a contravariant tensor of type $(k,0)$ are obtained without dividing by the factorial $k!$ in the projection algorithm. This, of course, is a small advantage to be gained at the expense of the summation convention.

EXAMPLE 3.2. **The factorial issue:** Let us consider a very simple example of a contravariant tensor of type $(2,0)$ in a 3-dimensional space: $T = 3\mathbf{e}_1 \otimes \mathbf{e}_2 + \mathbf{e}_2 \otimes \mathbf{e}_1$. Its skew-symmetric part is obtained by projection (dividing by 2!) as: $\mathcal{A}(T) = \mathbf{e}_1 \otimes \mathbf{e}_2 - \mathbf{e}_2 \otimes \mathbf{e}_1$. This is the same as $2\mathbf{e}_1 \wedge \mathbf{e}_2$. The procedure to obtain the final result directly can be described as follows: (i) replace the tensor products in the original expression by wedge products; (ii) collect terms so as to get rid of those that are not in a strictly increasing order. In our example, these two steps are: (i) $\mathcal{A}(T) = 3\mathbf{e}_1 \wedge \mathbf{e}_2 + \mathbf{e}_2 \wedge \mathbf{e}_1$; (ii) $\mathcal{A}(T) = 3\mathbf{e}_1 \wedge \mathbf{e}_2 - \mathbf{e}_1 \wedge \mathbf{e}_2 = 2\mathbf{e}_1 \wedge \mathbf{e}_2$. In short, the division by the factorial is avoided because the summation runs over exactly that many fewer terms.

EXAMPLE 3.3. **Determinants and volumes:** Consider the n-fold wedge product $a = \mathbf{v}_1 \wedge \mathbf{v}_2 \wedge \ldots \wedge \mathbf{v}_n$, where the $\mathbf{v}$'s are elements of an n-dimensional vector space V. Let $\{\mathbf{e}_1, \mathbf{e}_2, \ldots, \mathbf{e}_n\}$ be a basis of V. Since each of the $\mathbf{v}$'s is expressible uniquely in this basis, we may write:

$$a = (v_1^{i_1}\mathbf{e}_{i_1}) \wedge (v_2^{i_2}\mathbf{e}_{i_2}) \wedge \cdots \wedge (v_n^{i_n}\mathbf{e}_{i_n}) = v_1^{i_1} v_2^{i_2} \ldots v_n^{i_n}\, \mathbf{e}_{i_1} \wedge \mathbf{e}_{i_2} \wedge \cdots \wedge \mathbf{e}_{i_n}, \qquad (3.25)$$

where the summation convention is in full swing. Out of the possible n^n terms in this sum, there are exactly $n!$ that can survive, since each of the indices can attain n values, but repeated indices in a term kill it. However, since each of the surviving terms consists of a scalar coefficient times the exterior product of all the n elements of the basis, we can collect them all into a single scalar coefficient A multiplied by the exterior product of the base vectors arranged in a strictly increasing ordering of the indices, namely, we must have that $a = A\mathbf{e}_1 \wedge \mathbf{e}_2 \wedge \cdots \wedge \mathbf{e}_n$. This scalar coefficient consists of the sum of all the products $v_1^{i_1} v_2^{i_2} \ldots v_n^{i_n}$ with no repeated indices and with a minus sign if the superscripts form an odd permutation of $1, 2, \ldots, n$. This is precisely the definition of the determinant of the matrix whose entries are v_i^j. We conclude that, using in $\Lambda^n(V)$ the basis induced by a basis in V, the component of the exterior product of n vectors in an n-dimensional space is equal to the determinant of the matrix of the components of the individual vectors. Apart from providing a neat justification for the notion of determinant, this formula correctly suggests that the geometrical meaning of an n-vector is some measure of the ability of the (n-dimensional) parallelepiped subtended by the vectors to contain a volume. Since we have not introduced an inner product in our space, we cannot associate a number to this volume. Notice on the other hand that, although we cannot say how large a volume is, we can certainly tell (independently of any inner product) that a given n-parallelepiped is, say, twice as large as another. Notice, finally, that changing the order of two factors, or reversing the sense of one factor, changes the sign of the multivector. So, n-vectors represent *oriented* n-dimensional parallelepipeds.

3.4. Multivectors and Oriented Affine Simplexes

An affine r-simplex is said to be *oriented* if each possible way of ordering its vertices is assigned a $+$ or $-$ sign. Two orderings differing by an odd permutation are assigned opposite signs. It follows from this condition that there are exactly two possible orientations of an r-simplex. Unless stated otherwise, the ordering $p_0, p_1, \ldots, p_r$ used to name the r-simplex is assigned the positive sign. We can associate to this r-simplex the reducible r-vector $(p_1 - p_0) \wedge (p_2 - p_0) \wedge \cdots \wedge (p_r - p_0)$. As we have already suggested, the r-vector associated with an oriented r-simplex is a manifestation of the capacity[6] of the portion of space enclosed by the r-simplex to carry extensive physical quantities (such as volume, mass, energy). Several remarks are in order.

The first remark has to do with uniqueness: An oriented r-simplex determines its associated r-vector uniquely. For, if keeping p_0 fixed we make an even permutation of the remaining vertices, we clearly obtain the same r-vector. We need, therefore, only consider any particular even permutation that involves p_0, for example, one rotating the first three vertices[7] while leaving the rest fixed, namely, $p_1, p_2, p_0, \ldots, p_r$. The new r-vector is given by:

$$(p_2 - p_1) \wedge (p_0 - p_1) \wedge \cdots \wedge (p_r - p_1) =$$
$$((p_2 - p_0) - (p_1 - p_0)) \wedge (-(p_1 - p_0)) \wedge \cdots \wedge ((p_r - p_0) - (p_1 - p_0)) =$$
$$(p_1 - p_0) \wedge (p_2 - p_0) \wedge \cdots \wedge (p_r - p_0), \quad (3.26)$$

where we have used all the basic properties of the wedge product.

The second remark is simply a reminder that the uniqueness does not apply the other way around: given a reducible r-vector, it represents more than just one oriented r-simplex. This non-uniqueness has two aspects. If we change all the vertices of the simplex by a constant translation $\mathbf{u} \in V$ (i.e., we define new vertices by $p_i' = p_i + \mathbf{u}$), it is clear that we obtain the same r-vector. But there are less trivial ways in which we can change the r-simplex without changing its r-vector. For example, if we add to one of the vertices a linear combination of differences between the other vertices, we obtain the same r-vector. And so we should, as we know from the fact that sliding the sheets in a ream of paper relative to each other does not change the total volume (Cavalieri's theorem). The non-uniqueness is thus limited to r-simplexes belonging to parallel affine subspaces, as shown in Figure 3.1. (Two triangles in space may have the same scalar area, but regarded as 2-vectors

---

[6] This terminology, as well as the general tone and content of this section, have been suggested by R. Segev.

[7] For a simplex with less than 3 vertices, the assertion is obvious.

BOX 3.2. **An Independent Formulation of Exterior Algebra**

It is possible (and some would say desirable) to introduce the exterior algebra of an n-dimensional vector space V without direct reference to multilinear skew-symmetric operators on V^*. This objective can be achieved by introducing the rules of the game in a symbolic fashion. We start by defining the space $\Lambda^k(V)$ as made up of arbitrary linear combinations of symbols of the form $\mathbf{v}_1 \wedge \cdots \wedge \mathbf{v}_k$. These particular elements are called *monomials* and each of the participating vectors is called a *factor*. To complete the definition of the vector space $\Lambda^k(V)$ we need to introduce just two rules: (i) The exchange of any two factors in a monomial is equivalent to multiplication by the coefficient -1; (ii) monomials are linear in each factor: $(\alpha\mathbf{u} + \beta\mathbf{v}) \wedge \mathbf{v}_2 \wedge \cdots \wedge \mathbf{v}_k = \alpha\mathbf{u} \wedge \mathbf{v}_2 \wedge \cdots \wedge \mathbf{v}_k + \beta\mathbf{v} \wedge \mathbf{v}_2 \wedge \cdots \wedge \mathbf{v}_k$. It is clear that the set $\Lambda^k(V)$ of linear combinations of monomials (with the usual rules for linear combinations) is a vector space. It is also clear that for $k > n$ the dimension of this space must be zero. Indeed, if we have more than n factors in a monomial, one of the factors is expressible as a linear combination of the others. By property (ii) it can be resolved as a linear combination of monomials with at least one repeated factor. By property (i) it must vanish. For $k = n$ it is easy to see that the dimension of $\Lambda^n(V)$ is 1, since (again using properties (i) and (ii)) all monomials end up being proportional to each other. The dimensionality for $k < n$ can be calculated (by combinatorics) as $\frac{n!}{k!(n-k)!}$, and one can show that the monomials $\mathbf{e}_{i_1} \wedge \cdots \wedge \mathbf{e}_{i_k}$ with $i_1 < i_2 < \cdots < i_k$ are a basis of $\Lambda^k(V)$ whenever $\{\mathbf{e}_1, \ldots, \mathbf{e}_n\}$ is a basis of V. The elements of $\Lambda^k(V)$ are called k-vectors (or multivectors of degree k). We agree to the identifications: $\Lambda^0(V) = \mathbb{R}$ and $\Lambda^1(V) = V$.

To create an algebra, we introduce a new operation: the exterior (or wedge) product. It will be denoted with a wedge, because the notation will turn out to be compatible with that used for monomials. In other words, monomials will now be interpretable as wedge products. The new operation can be used to multiply multivectors of any degree. The result is a multivector whose degree is the sum of the degrees of the factors. The wedge product is: (iii) linear in each factor; (iv) associative; and (v) satisfies $b \wedge a = (-1)^{rs} a \wedge b$, for $a \in \Lambda^r(V)$, $b \in \Lambda^s(V)$. If one of the factors is of degree 0, the wedge product is equivalent to multiplication by a scalar.

they cannot be equal unless they reside in parallel planes. In the usual metric, we are talking about the area obtained as one-half the cross-product of two sides of the triangle. In fact, we see from this simple example how the exterior product is a very nontrivial generalization of the cross-product of 3-dimensional Euclidean space.)

A final remark concerns the fact that, although we have not obtained a scalar measure of an r-volume (length, area, volume, and so on), we can certainly compare the r-volumes of *all* r-simplexes belonging to *parallel* affine subspaces. Indeed, since the vectors generating the r-vector are a basis of the subspace to which the r-simplex

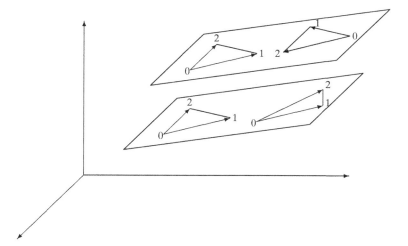

Figure 3.1. Various oriented 2-simplexes with identical multivectors

belongs (and, therefore, of all parallel subspaces), it turns out that all r-vectors in parallel r-dimensional subspaces are proportional to each other. If the constant of proportionality is negative, they have different orientations. The absolute value of the constant or proportionality is, therefore, the ratio between the scalar r-volumes. This scalar ratio will be automatically preserved by *any* inner product.

3.5. The Faces of an Oriented Affine Simplex

Recall that an s-face of an affine r-simplex is defined as the s-simplex obtained from a collection of $s+1$ vertices (with $0 \leq s \leq r$). For brevity, one often refers to the case $s = r - 1$ as simply the faces of the original simplex. An r-simplex, therefore, has exactly $r + 1$ faces, each of them a distinct $(r - 1)$-simplex. If the original simplex is oriented, it makes sense to try to define an orientation of its faces that is in some precise sense compatible with the orientation of the total simplex. This objective can be achieved as described below.

We have seen that the orientation of an affine simplex is established by choosing a particular ordering of its vertices and assigning to it the positive sign. Odd permutations of this ordering will correspond to the simplex oppositely oriented and will be assigned the negative sign. Let the orientation of the original r-simplex be given by some ordering $p_0, p_1, p_2, \ldots, p_r$ of its vertices. By definition, a face is obtained by suppressing from this list one of the vertices, p_i say, and it is customary to refer to the face thus obtained as "the face opposite to the vertex p_i." Thus, for example, the face opposite to the vertex p_1 is defined as the $(r - 1)$-simplex with vertices $p_0, p_2, p_3, \ldots, p_r$. We must now decide whether this particular ordering is to be assigned the positive or the negative sign so as to attain an orientation consistent with that of the r-simplex of departure. By convention, the choice of signature is given by the formula $(-1)^i$ (when the face is obtained by suppressing the vertex p_i while leaving the order of the remaining vertices undisturbed). So, in our example of the face opposite to the vertex p_1, the ordering $p_0,p_2,\ldots,p_r$ carries a negative signature. We may say that, to be *consistent* with the orientation of the original simplex, this face has to be given the opposite orientation (e.g., that corresponding to the ordering $p_2,p_0,p_3,\ldots,p_r$). This rule for choosing the consistent orientation of the faces is, of course, indifferent to a signature-preserving change in the ordering of the vertices of the original r-simplex. For instance, the ordering $p_1,p_2,p_0,p_3,\ldots,p_r$ will give rise to exactly the same orientation of the faces as the original ordering.

EXAMPLE 3.4. **Consistently oriented faces:** Consider 3 points p_0,p_1,p_2 constituting a triangle in $\mathbb{R}^2$ or $\mathbb{R}^3$. Following our criterion for the orientation of the faces, a set of consistently oriented faces (namely, in this case, sides) of this oriented triangle is (p_1,p_2), (p_2,p_0), (p_0,p_1). Similarly, for the oriented tetrahedron (p_0,p_1,p_2,p_3) in $\mathbb{R}^3$ (Figure 2.1) a set of consistently oriented faces is (p_1,p_2,p_3), (p_0,p_3,p_2), (p_0,p_1,p_3), (p_0,p_2,p_1). Each of these faces, in turn, can be considered as an oriented 2-simplex in its own right and a consistent orientation of *its* faces (namely, edges) can be obtained. Note that an edge (namely, a 1-face) of the original parallelepiped will have opposite orientations according to which one of two faces it is considered to belong to. Thus, for example, the edge determined by the vertices p_1 and p_2 will be oriented according to (p_1,p_2) when viewed as the side opposite to p_3 in the face (p_1,p_2,p_3), but will be oriented according to (p_2,p_1) when considered as the side opposite to (p_0) in the face (p_0,p_2,p_1).

Since the dimension of $\Lambda^{r-1}(V)$ is r, it should come as no surprise that the multivectors associated with the $r+1$ oriented faces of an oriented r-simplex are linearly dependent. In fact, a stronger result holds, namely: The sum of the multivectors of consistently oriented faces of an oriented affine r-simplex is always zero. This important proposition is a generalization of the fact that the vector sides of a triangle, when connected in tip-to-tail fashion, add up to zero. In Structural Engineering, such is the basis, among other techniques, of the once famous Maxwell-Cremona method for calculating internal forces in the members of a simple truss ("the polygons of forces must close"). The generalization just enunciated, in terms of multivectors, has important repercussions for the correct formulation of laws of balance in Continuum Mechanics.[8] The proof of the proposition is an exercise in Combinatorics, which is left to the reader.

EXERCISE 3.15. **In lieu of a proof:** If you are not planning to provide a general proof of the fact that the consistently oriented faces of an oriented r-simplex add up to zero, we propose that you at least compute explicitly the sum for $r=2,3,4$. [Hint: you may have to add and subtract a vertex in some expressions.]

3.6. Multicovectors or r-Forms

Just as we have constructed the exterior algebra $\Lambda(V)$ of multivectors (r-vectors of all orders $0 \le r \le n$) of an n-dimensional vector space V, we can construct $\Lambda(V^*)$,

---

[8] It is, for instance, used in the proof of Cauchy's theorem.

namely, the exterior algebra of the dual space:

$$\Lambda(V^*) = \sum_r \Lambda^r(V^*). \tag{3.27}$$

The elements of $\Lambda^r(V^*)$ are called *r-forms* or *multicovectors*, to emphasize the fact that, in a given context, the original vector space of departure is V rather than V^*. An *n*-form is also called a *volume form* on V. Given a basis $\{\mathbf{e}_1, \ldots, \mathbf{e}_n\}$ of V, the $\frac{n!}{(n-k)!k!}$ monomials $\mathbf{e}^{i_1} \wedge \cdots \wedge \mathbf{e}^{i_k}$ (with $i_1 < \cdots < i_k$) constitute a basis of $\Lambda^k(V^*)$. By restricting the operation (3.13) to completely skew-symmetric tensors, one can verify that each space $\Lambda^k(V^*)$ can be essentially identified with the algebraic dual of $\Lambda^k(V)$ (see Exercise 3.7). This restriction, however, is not as straightforward as one might think, so we shall work it out explicitly. It will be enough to show the result for a reducible *k*-form acting on a reducible *k*-vector.

EXAMPLE 3.5. **A simple case:** Consider the 2-form $\omega = \omega_1 \wedge \omega_2$ acting on the 2-vector $\mathbf{v} = \mathbf{v}_1 \wedge \mathbf{v}_2$, where $\mathbf{v}_1, \mathbf{v}_2 \in V$ and $\omega_1, \omega_2 \in V^*$. We have, according to Equations (3.20) and (3.12):

$$\langle \omega, \mathbf{v} \rangle = \left\langle \frac{1}{2}(\omega_1 \otimes \omega_2 - \omega_2 \otimes \omega_1), \frac{1}{2}(\mathbf{v}_1 \otimes \mathbf{v}_2 - \mathbf{v}_2 \otimes \mathbf{v}_1) \right\rangle$$

$$= \frac{1}{4}\left(\langle \omega_1, \mathbf{v}_1 \rangle \langle \omega_2, \mathbf{v}_2 \rangle - \langle \omega_1, \mathbf{v}_2 \rangle \langle \omega_2, \mathbf{v}_1 \rangle - \langle \omega_1, \mathbf{v}_2 \rangle \langle \omega_2, \mathbf{v}_1 \rangle + \langle \omega_2, \mathbf{v}_2 \rangle \langle \omega_1, \mathbf{v}_1 \rangle\right)$$

$$= \frac{1}{2}\left(\langle \omega_1, \mathbf{v}_1 \rangle \langle \omega_2, \mathbf{v}_2 \rangle - \langle \omega_1, \mathbf{v}_2 \rangle \langle \omega_2, \mathbf{v}_1 \rangle\right) = \frac{1}{2}\det\left[\langle \omega_i, \mathbf{v}_j \rangle\right]. \tag{3.28}$$

$\square$

Let $\omega = \omega_1 \wedge \cdots \wedge \omega_k$ and $\mathbf{v} = \mathbf{v}_1 \wedge \cdots \wedge \mathbf{v}_k$ be a reducible *k*-form and a reducible *k*-vector, respectively. According to our definition, we can write:

$$\omega = \mathcal{A}^k(\omega_1 \otimes \cdots \otimes \omega_k) = \frac{1}{k!}\sum_\pi sign(\pi)\, \pi(\omega_1 \otimes \omega_2 \otimes \cdots \otimes \omega_k), \tag{3.29}$$

and

$$\mathbf{v} = \mathcal{A}^k(\mathbf{v}_1 \otimes \cdots \otimes \mathbf{v}_k) = \frac{1}{k!}\sum_\pi sign(\pi)\, \pi(\mathbf{v}_1 \otimes \mathbf{v}_2 \otimes \cdots \otimes \mathbf{v}_k), \tag{3.30}$$

BOX 3.3. **Generalized Kronecker Symbol**: Recall the Kronecker symbol:

$$\delta^i_j = \begin{cases} 1 & \text{if } i = j, \\ 0 & \text{if } i \neq j. \end{cases} \tag{3.31}$$

For algebraic and numerical manipulations with multivectors and forms, it is useful to introduce the *generalized Kronecker symbol of order r* defined as follows:

$$\delta^{i_1, i_2, \ldots, i_r}_{j_1, j_2, \ldots, j_r} = \begin{cases} 1 & \text{if } i_1, i_2, \ldots, i_r \text{ are distinct and } \{j_1, j_2, \ldots, j_r\} \\ & \text{is an even permutation of } \{i_1, i_2, \ldots, i_r\} \\ -1 & \text{if } i_1, i_2, \ldots, i_r \text{ are distinct and } \{j_1, j_2, \ldots, j_r\} \\ & \text{is an odd permutation of } \{i_1, i_2, \ldots, i_r\} \\ 0 & \text{otherwise} \end{cases} \tag{3.32}$$

> The dimension of the space is $n \geq r$, so that the indices may range from 1 to n. As examples of the possible applications of this symbol consider the following:
>
> (1) The skew-symmetric projection of the tensor $T = T^{i_1,\dots,i_r} \, \mathbf{e}_{i_1} \otimes \cdots \otimes \mathbf{e}_{i_r}$ is given by:
>
> $$\mathcal{A}^r(T) = \frac{1}{r!} \delta^{i_1,\dots,i_r}_{j_1,\dots,j_r} T^{j_1,\dots,j_r} \, \mathbf{e}_{i_1} \otimes \cdots \otimes \mathbf{e}_{i_r}, \tag{3.33}$$
>
> where the sums range from 1 to $n \geq r$.
>
> (2) The determinant of the $n \times n$ matrix A with entries $\{A^i_j\}$ is given by:
>
> $$\det(A) = \delta^{i_1,\dots,i_n}_{1,\dots,n} A^{1,\dots,n}_{i_1,\dots,i_n}. \tag{3.34}$$

where π denotes a permutation of the arguments and the summation ranges over all $k!$ permutations. We obtain:

$$
\begin{aligned}
\langle \omega, \mathbf{v} \rangle &= \left\langle \frac{1}{k!} \sum_\pi sign(\pi) \, \pi(\omega_1 \otimes \omega_2 \otimes \cdots \otimes \omega_k), \frac{1}{k!} \sum_\sigma sign(\sigma) \, \sigma(\mathbf{v}_1 \otimes \mathbf{v}_2 \otimes \cdots \otimes \mathbf{v}_k) \right\rangle \\
&= \frac{1}{(k!)^2} \sum_\pi \left(sign(\pi) \left(\sum_\sigma sign(\sigma) \langle \omega_{\pi(1)}, \mathbf{v}_{\sigma(1)} \rangle \dots \langle \omega_{\pi(k)}, \mathbf{v}_{\sigma(k)} \rangle \right) \right) \\
&= \frac{1}{(k!)^2} \sum_\pi \left(sign(\pi) \left(sign(\pi) \det[\langle \omega_i, \mathbf{v}_j \rangle] \right) \right) \\
&= \frac{1}{k!} \det[\langle \omega_i, \mathbf{v}_j \rangle],
\end{aligned}
\tag{3.35}
$$

where the rules for evaluating a determinant have been used. The presence of the factorial can be eliminated by defining a new, perfectly legitimate, linear evaluation as:

$$\langle . | . \rangle := k! \, \langle . , . \rangle. \tag{3.36}$$

For the reducible case we can then write:

$$\langle \omega \, | \mathbf{v} \rangle = \det[\langle \omega_i, \mathbf{v}_j \rangle]. \tag{3.37}$$

For the nonreducible case, this formula is simply extended by linearity. Lest you may feel disheartened by the apparent complexity of this result, it is comforting to verify that, when working in components, the formula for the action of a general k-form on a general k-vector becomes identical to that for general tensors, namely, a mere contraction of corresponding indices. To check that this is the case, we start from a basis $\{\mathbf{e}_1,\dots,\mathbf{e}_n\}$ in V. The wedge products $\{\mathbf{e}_{j_1} \wedge \cdots \wedge \mathbf{e}_{j_k}\}$, $j_1 < \cdots < j_k$, constitute a basis of $\Lambda^k(V)$. Similarly, the wedge products $\{\mathbf{e}^{i_1} \wedge \cdots \wedge \mathbf{e}^{i_k}\}$, $i_1 < \cdots < i_k$, formed with the dual basis, constitute a basis of $\Lambda(V^*)$. We now compute:

$$\langle \mathbf{e}_{j_1} \wedge \cdots \wedge \mathbf{e}_{j_k} \, | \, \mathbf{e}^{i_1} \wedge \cdots \wedge \mathbf{e}^{i_k} \rangle = \det[\langle \mathbf{e}_{j_m} | \mathbf{e}^{i_n} \rangle] = \det[\delta^{i_n}_{j_m}]. \tag{3.38}$$

The $k \times k$-matrix on the right-hand side of this equation will have at least one vanishing row (and column) unless $i_m = j_m$ for each $m = 1, \ldots, k$, in which case it becomes the unit matrix. We then conclude that with the operation $\langle . | . \rangle$ the bases $\{\mathbf{e}_{j_1} \wedge \cdots \wedge \mathbf{e}_{j_k}\}$, $j_1 < \cdots < j_k$ and $\{\mathbf{e}^{i_1} \wedge \cdots \wedge \mathbf{e}^{i_k}\}$, $i_1 < \cdots < i_k$ are dual of each other.

EXERCISE 3.16. **Evaluation in components:** Show that the evaluation of a k-form:

$$\omega = \sum_{i_1 < \cdots < i_k} \omega_{i_1 \ldots i_k} \, \mathbf{e}^{i_1} \wedge \cdots \wedge \mathbf{e}^{i_k} \tag{3.39}$$

on the k-vector:

$$\mathbf{v} = \sum_{j_1 < \cdots < j_k} v^{j_1 \ldots j_k} \, \mathbf{e}_{j_1} \wedge \cdots \wedge \mathbf{e}_{j_k}, \tag{3.40}$$

is given by:

$$\langle \omega | \mathbf{v} \rangle = \sum_{i_1 < \cdots < i_k} \omega_{i_1 \ldots i_k} v^{i_1 \ldots i_k}. \tag{3.41}$$

Convince yourself that this is true by solving the next exercise.

EXERCISE 3.17. **An example:** Let $\mathbf{v_1}$ and $\mathbf{v_2}$ be vectors in a 3-dimensional space, and let ω_1 and ω_2 be covectors. Evaluate $\langle \omega_1 \wedge \omega_2 \mid \mathbf{v}_1 \wedge \mathbf{v}_2 \rangle$ according to the determinant formula (3.37). Express the given vectors and covectors symbolically in terms of components in a basis and its dual. Obtain the components of $\omega_1 \wedge \omega_2$ and of $\mathbf{v}_1 \wedge \mathbf{v}_2$ in the induced bases of $\Lambda^2(V^*)$ and $\Lambda^2(V)$, respectively. Reevaluate $\langle \omega_1 \wedge \omega_2 \mid \mathbf{v}_1 \wedge \mathbf{v}_2 \rangle$ using Equation (3.41). Verify that the two evaluations lead to identical results.

3.7. The Physical Meaning of r-Forms

We have already suggested that r-vectors convey the meaning of the ability of a portion of an affine space to contain some physical quantity. If this be so, what is then the physical meaning to be assigned to r-forms? The answer is simple: an r-form conveys the meaning of the strength of an extensive quantity which can be contained in an r-vector. These concepts are, of course, physically dual of each other, as they should, since they have just been shown to be mathematically dual. The evaluation of an r-form representing some extensive physical quantity (such as volume, mass, energy, momentum, and so on) on an r-vector (representing the ability of a portion of space to carry this quantity) is precisely the *content* of that quantity within that r-vector. Taken to the limit of a triangulation of a finite region of an affine space into infinitesimal oriented r-simplexes on each of which there is an r-form to be evaluated, we see that the collection of the r-simplexes is the *domain of integration*, while the collection of the r-forms somehow constitutes the *integrand*. This rough description can be formalized and applied not just to affine spaces but also to differentiable manifolds.

3.8. Some Useful Isomorphisms

We have already remarked that there exists a certain fusiform symmetry in the dimensionality of the various spaces of multivectors. More specifically, if n denotes the dimension of the vector space of departure V, we have:

$$\dim \Lambda^r(V) = \dim \Lambda^{n-r}(V) = \frac{n!}{r!(n-r)!} \qquad 0 \leq r \leq n. \tag{3.42}$$

Denoting by $\Lambda_r(V) = \Lambda^r(V^*)$ the space of r-forms on V, we also have:

$$\dim \Lambda_r(V) = \dim \Lambda_{n-r}(V) = \frac{n!}{r!(n-r)!} \qquad 0 \leq r \leq n. \tag{3.43}$$

Since all finite-dimensional vector spaces of the same dimension are mutually isomorphic, for each r there exist noncanonical isomorphisms between the four spaces $\Lambda^r(V)$, $\Lambda^{n-r}(V)$, $\Lambda_r(V)$ and $\Lambda_{n-r}(V)$. We are interested in exploring certain isomorphisms induced (either between pairs of these spaces or between all four of them) by potentially physically meaningful entities such as a fixed volume form or an inner-product structure on V.

Let ω be a fixed volume form on V, namely, $\omega \in \Lambda_n(V)$. For each r-vector $\mathbf{v} \in \Lambda^r(V)$ we define an $(n-r)$-form $\omega_{\mathbf{v}}$ by the formula:

$$\langle \omega_{\mathbf{v}} \mid \mathbf{w} \rangle := \langle \omega \mid \mathbf{v} \wedge \mathbf{w} \rangle \qquad \forall \, \mathbf{w} \in \Lambda^{n-r}(V). \tag{3.44}$$

EXERCISE 3.18. Show that the map from $\Lambda^r(V)$ to $\Lambda_{n-r}(V)$ defined by $\mathbf{v} \mapsto \omega_{\mathbf{v}}$ according to Equation (3.44) is indeed an isomorphism of vector spaces.

Let V be endowed with an inner product denoted by the product symbol "$\cdot$". Recall that an inner product induces an isomorphism between V and V^*, as shown explicitly in Section 2.7.2. It is not difficult to verify, on the other hand, that this inner product on V induces an inner product on each of the tensor spaces $\mathcal{C}^r(V)$ as follows:

$$(\mathbf{u}_1 \otimes \cdots \otimes \mathbf{u}_r) \cdot (\mathbf{v}_1 \otimes \cdots \otimes \mathbf{v}_r) := (\mathbf{u}_1 \cdot \mathbf{v}_1) \ldots (\mathbf{u}_r \cdot \mathbf{v}_r) \quad \mathbf{u}_i, \mathbf{v}_i \in V, \; i = 1, \ldots, r. \tag{3.45}$$

Note that, although this definition applies only to decomposable tensors, it can be extended uniquely by linearity to all tensors of the same type. Since $\Lambda^r(V)$ is a subspace of $\mathcal{C}^r(V)$, the induced inner product is automatically defined also on the former. For convenience, however, we may wish to emulate the pairing of Equation (3.36) by defining on $\Lambda^r(V)$ the modified inner product:

$$\mathbf{u} \bullet \mathbf{v} := r! \, \mathbf{u} \cdot \mathbf{v} \qquad \mathbf{u}, \mathbf{v} \in \Lambda^r(V). \tag{3.46}$$

This inner product, in turn, generates an isomorphism between $\Lambda^r(V)$ and its dual space $\Lambda_r(V)$. But an inner product on V also induces automatically a volume form on V. This follows from the fact that $\Lambda_n(V)$ is a 1-dimensional vector space. Since this vector space now has an inner product, there exist exactly two elements $\pm \omega_g$ of unit length, differing only by sign. Choosing one of them is tantamount to choosing an orientation of V (i.e., deciding on a right-hand rule for ordered n-tuples of vectors). Combining now the isomorphisms defined by (3.44) and (3.46), we obtain an isomorphism between $\Lambda^r(V)$ and $\Lambda^{n-r}(V)$, namely, between r-vectors and $(n-r)$-forms, on the one hand, and between $\Lambda_r(V)$ $\Lambda_{n-r}(V)$, on the other hand. These two isomorphisms, also called the *Hodge operator*, are denoted by the star symbol "$\star$." In conclusion, in an inner-product space V, the four members of the quadruple of the spaces indicated are mutually isomorphic.

EXAMPLE 3.6. **The case of $\mathbb{R}^3$:** The space $\mathbb{R}^3$ has a standard inner product and a standard (orthonormal) basis: $e_1 = (1,0,0)$, $e_2 = (0,1,0)$, $e_3 = (0,0,1)$. The standard dual basis coincides, so to speak, with the standard basis. Vectors and forms of the same order are identified with each other if they have the same components in the respective standard bases. The standard volume form is given by $\omega = e^1 \wedge e^2 \wedge e^3$. Let us consider the result of applying the star operator to the exterior product of two vectors $\mathbf{u} = u^i e_i$ and $\mathbf{v} = v^i e_i$. From Equation (3.44), using the volume element ω, the product $\mathbf{u} \wedge \mathbf{v}$, which is a 2-vector, can be regarded as a 1-form $\omega_{\mathbf{u} \wedge \mathbf{v}}$ whose action on a vector $\mathbf{w}$ is given by:

$$\langle \omega_{\mathbf{u} \wedge \mathbf{v}} \mid \mathbf{w} \rangle = \langle \omega \mid \mathbf{u} \wedge \mathbf{v} \wedge \mathbf{w} \rangle = \langle e^1 \wedge e^2 \wedge e^3 \mid \mathbf{u} \wedge \mathbf{v} \wedge \mathbf{w} \rangle \qquad (3.47)$$

According to Equation (3.37), therefore, we may write:

$$\langle \omega_{\mathbf{u} \wedge \mathbf{v}} \mid \mathbf{w} \rangle = \det \begin{bmatrix} u_1 & v_1 & w_1 \\ u_2 & v_2 & w_2 \\ u_3 & v_3 & w_3 \end{bmatrix}, \qquad (3.48)$$

which shows that the vector associated by the star operator to the exterior product of two vectors is precisely their *cross-product*. More formally:

$$\star(\mathbf{u} \wedge \mathbf{v}) = \mathbf{u} \times \mathbf{v} \quad \forall \mathbf{u}, \mathbf{v} \in \mathbb{R}^3. \qquad (3.49)$$

Naturally, the choice of the opposite volume element would have led to a change in the sign of the result. Thus, when presenting the definition of the cross-product in elementary courses, it is usually stated that the cross-product of two vectors is a *pseudo-vector*.

EXERCISE 3.19. Show that the Hodge (star) operator in $\mathbb{R}^3$ assigns to the vector $\mathbf{v} = A e_1 + B e_2 + C e_3$ the 2-vector $\omega_v = \star \mathbf{v} = C e_1 \wedge e_2 - B e_1 \wedge e_3 + A e_2 \wedge e_3$. Notice that this 2-vector can also be considered as a 2-form in $\mathbb{R}^3$.

PART TWO

DIFFERENTIAL GEOMETRY

4 Differentiable Manifolds

Having already established our physical motivation, we will proceed to provide a precise mathematical counterpart of the idea of a continuum. Each of the physical concepts of Continuum Mechanics, such as those covered in Appendix A (configuration, deformation gradient, differentiable fields on the body, and so on), will find its natural geometrical setting starting with the treatment in this chapter. Eventually, new physical ideas, not covered in our Continuum Mechanics primer, will arise naturally from the geometric context and will be discussed as they arise.

4.1. Introduction

The Greek historian Herodotus, who lived and wrote in the 5th century BCE, relates that the need to reconstruct the demarcations between plots of land periodically flooded by the Nile was one of the reasons for the emergence of Geometry in ancient Egypt. He thus explains the curious name of a discipline which even in his time had already attained the status and the reputation of a pure science. For "geometry" literally means "measurement of the Earth," and in some European languages to this very day the practitioner of land surveying is designated as geometer. In the light of its Earth-bound origins, therefore, it is perhaps not unworthy of notice that when modern differential geometers were looking for a terminology that would be both accurate and suggestive to characterize the notion of a continuum, they found their inspiration in Cartography, that is, in the science of making maps of the Earth.

The great cartographers (the name of Mercator comes to mind) were well aware of a number of facts that made their job very difficult. They were dealing with a bounded and continuous entity, the surface of the Earth, which could not be mapped in its entirety in a continuous way onto a sheet of paper. This first difficulty forced them to always deal with partial maps or *charts*, representing a certain portion of the globe. They also realized that the bigger the portion represented, the more distorted some regions would end up in the chart, and so (although it was clear that two charts would suffice to cover the whole Earth) they adopted the policy of mapping smaller regions and then collecting them into a book called an *atlas*. The second issue that became apparent was that, in order to make sense of the atlas, different charts needed to have some overlap. Unfortunately, the common regions in the overlap of two charts couldn't look exactly the same, but were distorted in different ways. So the need arose to establish a correspondence or *transition* device to identify clearly the common spots. This correspondence was in part achieved by labelling significant common landmarks (cities, mountain peaks, rivers) and in part by drawing (and numbering) a network of meridians and parallels. This device could serve as well to compare different atlases drawn by different cartographers. It is not necessary that the charts of two atlases cover the same domain. The only thing that matters is that in the overlapping regions the transitions be absolutely clear, so that the correspondence of places can be determined without ambiguity. The science of Cartography does not end here, but these two issues (the division into charts forming an atlas, and the unequivocal transition from chart to overlapping chart) are the main ingredients to understand not just the description of the surface of the Earth, but also the concept of the most general entity that is commonly known in Physics as a continuum.

The name "continuum" is somewhat misleading, because it tends to suggest that all that matters is some underlying continuity of the entity. In fact, what one wants is usually quite a bit more. One wants to have the possibility of defining *fields* on this entity, and one wants to be able to take derivatives (or gradients) of these fields. To give another example from the Earth, when looking at a weather report we see isotherms and isobars, and when describing the topography of a terrain in detail we see contour lines of equal height, and we expect these lines to be smooth, without corners or, if they do have corners, we want to be able to detect their presence. If we only dealt with continuity, then it wouldn't matter much whether a contour is a square or a circle. The technical name for a continuum with a notion of differentiability is *differentiable manifold*, and the particular case where only continuity matters is known as *topological manifold*. Both use the cartographic notions of charts, atlases and transition functions. There are, however, certain concepts that we don't want to import from Cartography as fundamental notions. The most important ones are the following two. The first feature that we want to leave out of the definition of differentiable manifold is that of *surrounding space*. The surface of the Earth (as we know all too well these days) can be seen from outside. It is *embedded* into a higher dimensional space (in this case, the 3-dimensional space of Classical Celestial Mechanics, the Heavens). But we want to be able to talk of a differentiable manifold in its own right. The Earth surface is a 2-dimensional entity, and so are the charts. We want the atlas of charts and the transition functions to be *all* that is involved in the definition, without mention of a surrounding space. A clear and dramatic example of this need comes from the theory of General Relativity, whereby the whole universe of our experience is the object of study.[1]

The second characteristic of Cartography that we'd rather leave aside, at least from the general definition, has to do with the fact that in the case of the Earth

---

[1] There are, however, theorems, notably Whitney's embedding theorem, that show that every manifold can be embedded into a Euclidean space of a high enough dimension.

surface (due precisely to its being embedded in a space where lengths and angles can be measured) we are able not only to trace curves but also to measure their length. Given two points, we can find a line of shortest length (*geodesic line*). If two curves intersect, we can measure the angle at which they do. Nor is it necessary, as in the case of the Earth, to have inherited these features from the surrounding space. In the case of General Relativity, for example, it is assumed that there is an underlying *metric structure* accessible to experiments. Indeed, a clock carried by an observer along a time-like path will keep measuring, as it ticks, the *proper time* along this path. Differentiable manifolds on which there is a well-defined notion of length will in general be called *Riemannian manifolds* in memory of the great mathematician Riemann who in the mid nineteenth century (1854) first proposed[2] the idea of a general geometrical entity of this kind. It is interesting, though, that years earlier[3] Lagrange had hit upon the more general notion of configuration space, but failed to realize its potential while working entirely in particular charts.

In short, we will emulate the cartographic paradigm only inasmuch as it provides us with an entity, called a *differentiable manifold*, upon which differentiable fields can be defined. This entity will have a definite integer dimension, but will not necessarily be immersed within another such entity of higher dimension. Finally, the notion of length of a curve in the manifold will not be necessarily included in the definition.

4.2. Some Topological Notions

Before arriving at the formal definition of a differentiable manifold, we need to establish or refresh a few basic notions of *Topology*. A *topological space* is a set T in which a collection of subsets (called *open sets*) is singled out with the following properties:

(1) The empty set ($\emptyset$) and the total space (T) are open sets.
(2) The union of an arbitrary collection of open sets is an open set.
(3) The intersection of any *finite* collection of open sets is an open set.

When the open sets have been specified satisfying these properties, one says that the set T has been given a *topology*. These three properties alone (inspired, naturally, by the properties of open intervals of $\mathbb{R}$), are all that is needed to introduce the notion of *continuity*. An *open neighbourhood* of a point $p \in T$ is an open set containing p. We will use the term "neighbourhood" liberally to refer to an open neighbourhood. Notice that by property (1) above, all points have at least one neighbourhood.

EXAMPLE 4.1. **The standard topology of** $\mathbb{R}^n$: An open ball in $\mathbb{R}^n$ with centre $(c^1, \ldots, c^n) \in \mathbb{R}^n$ and radius $r \in \mathbb{R}$ is the set $\{(x^1, \ldots, x^n) \in \mathbb{R}^n \mid (x^1 - c^1)^2 + \cdots + (x^n - c^n)^2 < r\}$. The collection of all open balls forms a *base* for the *standard topology of* $\mathbb{R}^n$ in the sense that every open set in $\mathbb{R}^n$ can be obtained as an arbitrary union of open balls. [Question: How is the empty set included?]

---

[2] In his epoch-making essay, *Über die Hypothesen, welcher der Geometrie zugrunde liegen.*
[3] In his not less epoch-making book *Mécanique Analytique* (1788).

DEFINITION 4.1. A *base* $\mathcal{B}$ of a topological space $\mathcal{T}$ is a collection of open sets such that every element (i.e., open set) of $\mathcal{T}$ is the union of elements of $\mathcal{B}$.

DEFINITION 4.2. A topology is said to be *second countable* if it has a countable basis.[4]

Let $\mathcal{T}$ and $\mathcal{S}$ be topological spaces, and let $f : \mathcal{T} \to \mathcal{S}$ be a function from one to the other. The function f is *continuous at the point* $p \in \mathcal{T}$ if given any neighbourhood $\mathcal{V}$ of $f(p) \in \mathcal{S}$ there exists a neighbourhood $\mathcal{U}$ of $p \in \mathcal{T}$ such that $f(\mathcal{U}) \subset \mathcal{V}$. This definition, of course, reduces to the classical ϵ-δ definition for real functions, but doesn't use any concept of radius, since the neighbourhoods are pre-defined and available without reference to any metric property. A function which is continuous at every point of its domain $\mathcal{T}$ is said to be *continuous*. It can be shown that a function is continuous if, and only if, the inverse image of every open set (in $\mathcal{S}$) is open (in $\mathcal{T}$). A continuous function $f : \mathcal{T} \to \mathcal{S}$ is a *homeomorphism* if it is a bijection (namely, one-to-one and onto) and if its inverse f^{-1} is continuous.

A subset $\mathcal{A} \subset \mathcal{T}$ is *closed* if its complement $\mathcal{T} \setminus \mathcal{A}$ is open. Notice that the notions of open and closed subsets are neither comprehensive nor mutually exclusive. A subset may be neither open nor closed, or it may be both open and closed (e.g., the empty set and the total space are both open and closed). A topological space is *connected* if the only subsets which are both open and closed are the empty set and the space itself.

A subset $\mathcal{A}$ of a topological space $\mathcal{T}$ inherits the topology of this space as follows: Its open sets are the intersections of the open sets of $\mathcal{T}$ with $\mathcal{A}$. A topology obtained in this way is called a *relative or induced topology*.

A topological space is *Hausdorff* (or a *Hausdorff space*) if, given any two different points, there exist respective neighbourhoods which are disjoint. This property makes a topological space behave more normally than it otherwise would. For example, the *coarsest* topology on a set $\mathcal{T}$ (namely, the one with the least open sets) consists of just two open sets: $\emptyset$ and $\mathcal{T}$. It is called the *trivial topology*. With the trivial topology a set is not Hausdorff. At the other extreme we have the *discrete topology* whereby every subset is open. The discrete topology is Hausdorff. A topology on a set $\mathcal{T}$ is *finer* than another topology if every open set in the second topology is also an open set in the first. Correspondingly, the second topology is then said to be *coarser* than the first. Notice that not all topologies on the same set are comparable in this way. The discrete topology is the finest possible topology on a set. A topology finer than a Hausdorff topology is necessarily Hausdorff.

A collection of open sets $\{\mathcal{U}_\alpha\}$ is a *covering* of $\mathcal{T}$ if the union of the collection equals $\mathcal{T}$. Equivalently, each point of $\mathcal{T}$ belongs to at least one set of the covering. A *subcovering* of a given covering $\{\mathcal{U}_\alpha\}$ is a subcollection of $\{\mathcal{U}_\alpha\}$ which is itself a covering. A subset $\mathcal{C} \subset \mathcal{T}$ is *compact* if it is Hausdorff and if every covering has a

---

[4] Recall that a set is *countable* if it is either finite or, if infinite, it can be put into a one-to-one correspondence with the set $\mathbb{N}$ of natural numbers $\{1,2,3,\ldots\}$.

finite subcovering (namely, one with a finite number of open sets). A compact set is necessarily closed.

Given two topological spaces, $\mathcal{T}$ and $\mathcal{S}$, the Cartesian product $\mathcal{T} \times \mathcal{S}$ is a topological space whose open sets are obtained by taking all the possible Cartesian products of open sets (one from $\mathcal{T}$ and one from $\mathcal{S}$), and then forming all arbitrary unions of these products. The resulting topology is called the *product topology*.

This is all the Topology we need to be able to proceed. Other topological notions will be introduced as the need arises. From the intuitive point of view, Topology can be described, at least in part, as the study of properties that are preserved under continuous transformations (homeomorphisms). For example, if we have a rubber membrane with a hole, the presence of the hole will be preserved no matter how we deform the membrane, as long as we don't tear it or produce any other kind of discontinuity.

4.3. Topological Manifolds

We have introduced the notion of topological space and shown that it can serve as the arena for the definition of continuous functions. Might this be the answer to the mathematical definition of a continuum? Not quite. From the introduction to this chapter we gather that one of the aspects to be learned from Cartography is that, while the surface of the Earth is not mappable *in toto* into the plane $\mathbb{R}^2$, it nevertheless looks like $\mathbb{R}^2$ piece by piece. (No wonder that a person with limited travel opportunities might still believe that the Earth is flat!) A topological space, on the other hand, in and of itself does not convey the notion of local equivalence to a flat entity such as $\mathbb{R}^n$. We, therefore, introduce the following definition:

DEFINITION 4.3. An *n-dimensional topological manifold* is a Hausdorff second-countable[5] topological space $\mathcal{T}$ such that each of its points has a neighbourhood homeomorphic to an open set of $\mathbb{R}^n$. Equivalently, we can say that there exists a covering $\mathcal{U}_\alpha$ of $\mathcal{T}$ and corresponding homeomorphisms $\phi_\alpha : \mathcal{U}_\alpha \to \phi_\alpha(\mathcal{U}_\alpha) \subset \mathbb{R}^n$. Each pair $(\mathcal{U}_\alpha, \phi_\alpha)$ is called a *chart* or a *coordinate chart*. The collection of all the given charts is called an *atlas*.

A chart $(\mathcal{U}_\alpha, \phi_\alpha)$ introduces a *coordinate system* on $\mathcal{U}_\alpha$. Indeed, the homeomorphism ϕ_α assigns to each point $p \in \mathcal{U}_\alpha$ an element of $\mathbb{R}^n$, that is, an ordered n-tuple of real numbers $x^i(p) = (\phi_\alpha(p))^i$, $i = 1, \ldots, n$, called the local coordinates of p in the given chart. Whenever two charts, $(\mathcal{U}_\alpha, \phi_\alpha)$ and $(\mathcal{U}_\beta, \phi_\beta)$, have a nonvanishing intersection, we can define the *transition function*:

$$\phi_{\alpha,\beta} = \phi_\beta \circ \phi_\alpha^{-1} : \phi_\alpha(\mathcal{U}_\alpha \cap \mathcal{U}_\beta) \longrightarrow \phi_\beta(\mathcal{U}_\alpha \cap \mathcal{U}_\beta), \tag{4.1}$$

which is a homeomorphism between open sets in $\mathbb{R}^n$. If we call the local coordinates in $\mathcal{U}_\alpha$ and $\mathcal{U}_\beta$, respectively, x^i and y^i, this homeomorphism consists simply of n

---

[5] Notice, as a technicality, that not all authors include second countability in the definition.

continuous (and continuously invertible) functions of the form:

$$y^i = y^i(x^1, \ldots, x^n), \quad i = 1, \ldots, n. \tag{4.2}$$

We have only defined finite-dimensional topological manifolds modelled locally after $\mathbb{R}^n$. It is also possible to model a topological manifold locally after other spaces, including Banach spaces, which are infinite dimensional.

4.4. Differentiable Manifolds

Our treatment has led us very close to our initial objective, except for one aspect: the issue of differentiability. We know how to define continuous functions on the manifold, but we still do not have any means of testing for differentiability. To achieve this final goal, we start by declaring two charts $(\mathcal{U}_\alpha, \phi_\alpha)$ and $(\mathcal{U}_\beta, \phi_\beta)$ as being C^k-*compatible* if their transition maps $\phi_{\alpha,\beta}$ and $\phi_{\beta,\alpha} = \phi_{\alpha,\beta}^{-1}$ are both of class C^k, namely, all the partial derivatives of (4.2) (and of its inverse) up to and including the k-th order exist and are continuous. If the domains of the two charts do not intersect, we still say that the charts are C^k-compatible. One should not get lost in the formality of the definition. What is being done here is a very clever trick of "passing the buck," an immensely creative device used often in Mathematics. Observe that, since the set $\mathcal{T}$ has been assumed to have a topology, there was no problem whatsoever in characterizing a coordinate map, ϕ_α say, as being continuous. But differentiable? The way this problem was tackled consists of deflecting attention from the maps themselves to the transition functions! We cannot say whether ϕ_α is differentiable, because there is not enough structure in a topological space to even define differentiability. But we *can* tell whether or not the transition between two such maps is differentiable, because the transition functions are defined between open subsets of $\mathbb{R}^n$.

EXAMPLE 4.2. **A body and its configurations:** Notice how creatively the same reasoning could be used in our intended physical context if we so wished: We may never see or touch the material body $\mathcal{B}$, but we may *define* it as the totality of its possible configurations.

An *atlas of class C^k* of a topological manifold $\mathcal{T}$ is an atlas whose charts are C^k-compatible. Two atlases of class C^k are compatible if their union is an atlas of class C^k. In other words, two atlases are compatible if each chart of one is C^k-compatible with each chart of the other. Given a C^k atlas one can define the compatible *maximal atlas of class C^k* as the union of all atlases that are compatible with the given one. A maximal atlas, thus, contains all its compatible atlases.

DEFINITION 4.4. An *n-dimensional differentiable manifold of class C^k* is an n-dimensional topological manifold $\mathcal{T}$ endowed with a maximal atlas of class C^k. For the particular case of class C^∞ one reserves the names *smooth manifold, C^∞-manifold,* or simply *manifold,* when there is no room for confusion. Unless otherwise stated, we will assume that all manifolds are of class C^∞. A maximal atlas on a manifold is said to define a *differentiable structure*.

EXAMPLE 4.3. **Bodies and other things:** The prime example of a differentiable manifold in the standard treatment of Continuum Mechanics is a material body $\mathcal{B}$. It is, by definition, a connected smooth manifold of dimension 3 that can be covered with just one chart. Such manifolds are sometimes called *trivial manifolds*. It is interesting to think of what this definition excludes. Examples: Two disjoint open balls in $\mathbb{R}^3$ (not connected); a thin membrane (wrong dimension); the sphere $\{x \in \mathbb{R}^4 |\ (x^1)^2 + (x^2)^2 + (x^3)^2 + (x^4)^2 = 1\}$ (right dimension, but not trivial); a closed ball in $\mathbb{R}^3$ (sorry, no boundary permitted at this stage: maybe later).

For the particular case of the space $\mathbb{R}^n$, we can choose as a (global) chart the space itself and the identity map. This particular canonical atlas induces a maximal atlas called the *standard differentiable structure of* $\mathbb{R}^n$.

A differentiable manifold is *orientable* if it admits an atlas such that all the transition functions have a positive Jacobian determinant. Such an atlas is called an *oriented atlas*. Two oriented atlases are either compatible or every transition function between two charts, one from each atlas, has an everywhere negative Jacobian determinant. Thus, there exist exactly two maximal oriented atlases for an orientable manifold. Each one of them represents a differentiable structure. Choosing one of them we obtain an *oriented manifold*. In an oriented manifold, only those coordinate changes that preserve orientation (namely, with positive Jacobian determinants) are allowed.

EXERCISE 4.1. **Vector spaces as trivial manifolds:** Show that a finite dimensional vector space can always be given the structure of a smooth trivial manifold. [Hint: Choose a basis.]

REMARK 4.1. **Different structures**: Given a topological space $\mathcal{T}$, there may exist different C^∞ structures, that is, noncompatible maximal atlases of class C^∞. Sometimes, two such atlases may nevertheless be isomorphic, in a precise sense that we will define later. On the other hand, there exist examples of nonisomorphic smooth structures on the same topological space (the most celebrated are Milnor's exotic spheres and Donaldson's strange structures of $\mathbb{R}^4$).

Given two differentiable manifolds, $\mathcal{M}$ and $\mathcal{N}$ of dimensions m and n, respectively, we can endow the Cartesian product $\mathcal{M} \times \mathcal{N}$ with a differentiable structure by adopting an atlas made up of all the Cartesian products of charts of an atlas of $\mathcal{M}$ and an atlas of $\mathcal{N}$. The resulting $(m + n)$-dimensional manifold is called a *product manifold*. Its underlying topological space is the product space with the product topology.

4.5. Differentiability

We need to show now that it was all worthwhile: that the rather complicated definition of a differentiable manifold provides us with an object on which it makes sense to say that a function is differentiable. Let $\mathcal{M}$ and $\mathcal{N}$ be two smooth manifolds of dimension m and n, respectively, and let $f : \mathcal{M} \to \mathcal{N}$ be a continuous map. If we

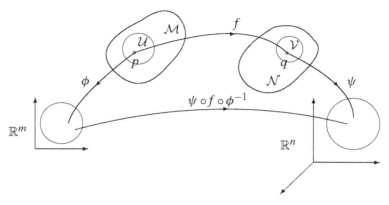

Figure 4.1. Representation of a map in two charts

choose a point $p \in \mathcal{M}$, whose image is a point $q = f(p) \in \mathcal{N}$, these points will belong to some charts $(\mathcal{U}, \phi)$ and $(\mathcal{V}, \psi)$, respectively. By continuity, we can always choose a neighbourhood of p, contained in $\mathcal{U}$, that is mapped by f into a neighbourhood of q completely contained in $\mathcal{V}$. Let us rename those neighbourhoods $\mathcal{U}$ and $\mathcal{V}$, since the restrictions of the respective charts are themselves in the maximal atlases. Each of these charts induces a local coordinate system and we ask how the function f looks in these coordinates. The answer is:

$$\psi \circ f \circ \phi^{-1} : \phi(\mathcal{U}) \longrightarrow \psi(\mathcal{V}), \tag{4.3}$$

that is, a map from an open set in $\mathbb{R}^m$ to an open set in $\mathbb{R}^n$ (see Figure 4.1). We call this map a *local coordinate representation of f*.

The map f is said to be *differentiable of class C^k* at a point $p \in \mathcal{M}$ if its local coordinate representation is of class C^k at p. It is not difficult to show that this property is independent of the particular charts chosen. A map of class C^∞ at p is said to be *smooth at p*. A map that is smooth at each point of its domain is said to be *smooth*. Notice that we have again used the technique of deflecting the idea of smoothness from the manifolds themselves to the charts.

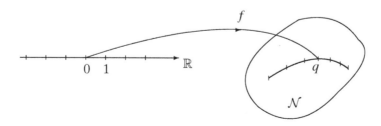

Figure 4.2. A (parametrized) curve in a manifold

In the special case $\mathcal{N} = \mathbb{R}$, the map $f : \mathcal{M} \to \mathbb{R}$ is called a (real) *function*. When, on the other hand, $\mathcal{M} = (a,b) \subset \mathbb{R}$, namely, an open interval of the line, the map $f : (a,b) \to \mathcal{N}$ is called a *parametrized curve in $\mathcal{N}$*, or just a *curve in $\mathcal{N}$* (see Figure 4.2). When $m = n$ and $f : \mathcal{M} \to \mathcal{N}$ is a homeomorphism and both f and f^{-1} are smooth, the map f is called a *diffeomorphism*, and the manifolds are said to be *diffeomorphic*. If two spaces are diffeomorphic, we say that their differentiable structures are *isomorphic*. If we now identify $\mathcal{M} = \mathcal{N}$ it may happen that two different differentiable structures are isomorphic. As pointed out above, however, there exist examples of manifolds endowed with nonisomorphic differentiable structures.

4.6. Tangent Vectors

On the surface of a sphere in space, such as the idealized Earth, we can construct the tangent plane at a point p. Any vector with origin at p and contained in the tangent plane is a tangent vector to the sphere at p. One of the useful features of the tangent plane is that it provides a good first approximation to a small neighbourhood of a point. Indeed, for many practical purposes (building a house, tilling a field, calculating property taxes, drawing a city map, and so on) we can replace the curved surface of the Earth by the tangent plane at the point of interest, as if the Earth were flat. At first sight, it would appear that in order to be able to construct the tangent plane we would need to have access to the surrounding 3-dimensional ambient space. To put it more dramatically, it would appear that an antlike being condemned to crawl on the surface of the sphere would never be aware that tangent vectors exist, since they come out, so to speak, of the surface. In Mathematics, however, there are often unexpected and creative ways to circumvent apparent obstacles, as we have already seen in this very chapter. In the case of tangent vectors, there are several different, but ultimately equivalent, intrinsic ways to define tangent vectors at a point of a manifold. By *intrinsic* we mean that the definition of a tangent vector is achieved from inside the manifold itself *without reference to a putative ambient space*.

The first intrinsic method to define tangent vectors is based on the idea that a vector can be regarded, intuitively at least, as a small piece of a curve. This intuitive idea, which can be traced back to Leibniz, is only heuristic in nature, but it can be finessed into a rigorous definition. By curve we must surely mean a *parametrized* curve: since we do not have any notion of length, the parametrization will have to play the role of something that can somehow distinguish between a vector and, say, twice that vector. Moreover, if two curves happen to be tangent to each other at the point p in question, we will have to recognize that perhaps they represent exactly the same vector, since we are only looking at a very small piece of each curve around p. In other words, we will have to consider equivalence classes of mutually tangent curves. All this, complicated as it may seem, is something an intelligent ant can do without ever leaving the surface. We will now elevate these intuitive ideas to a more formal definition.

EXAMPLE 4.4. **Strain gauges:** A common device to obtain experimental information on the state of strain at a point on the surface of a structural member consists of gluing to the surface a short piece of copper wire and then measuring the change of electrical resistance arising from the elongation of the wire. This device, based on balancing a Wheatstone bridge, is called a *strain gauge*. Because electrical measurements are very accurate, the resolution of even a medium-quality strain gauge is of the order of 10^{-6} (notice that the strain, being a relative elongation, is nondimensional). A strain gauge is a crude materialization of a tangent vector.

Recall that a parametrized curve in an m-dimensional manifold $\mathcal{M}$ is a smooth map from some open interval I of $\mathbb{R}$ to the manifold. Without loss of generality, we may assume that the open interval contains the point $0 \in \mathbb{R}$. Given a point $p \in \mathcal{M}$ we consider the collection of all curves $\gamma : I \to \mathcal{M}$ such that $\gamma(0) = p$. Let $(\mathcal{U}, \phi)$ be a chart at p. Then the composition $\phi \circ \gamma : I \to \mathbb{R}^m$ is a curve in $\mathbb{R}^m$ called the *representation of γ in the chart*. If we denote the local coordinates by x^i, $i = 1, \ldots, m$, the parametric equations of this curve will consist of m real functions, say $x^i = \gamma^i(t)$, where $t \in I \subset \mathbb{R}$. We will say that two curves, γ_1 and γ_2, in our collection are *tangent at p* if:

$$\left(\frac{d\gamma_1^i}{dt} \right)_{t=0} = \left(\frac{d\gamma_2^i}{dt} \right)_{t=0}, \quad i = 1, \ldots, m. \tag{4.4}$$

EXERCISE 4.2. Tangent curves: Verify that the definition of tangency of two curves at a point is independent of chart.

Tangency is easily seen to be an equivalence relation. A *tangent vector* $\mathbf{v}$ at p is defined as an equivalence class of tangent (parametrized) curves at p. To justify the use of the name "vector," we must define the operations of vector addition and of multiplication of a vector by a scalar. Given two equivalence classes (namely, two vectors $\mathbf{v}_1$ and $\mathbf{v}_2$), we choose representative curves γ_1 and γ_2 from each class and express them in terms of the same coordinate chart, always (without loss of generality) imposing the condition that p is mapped to the origin of $\mathbb{R}^m$. We define the sum of the two vectors as the equivalence class to which the sum of the coordinate representations belongs. Similarly, the product of a scalar by a vector $\mathbf{v}$ is defined as the equivalence class to which the multiplication of the scalar by the representation of any curve in the equivalence class of $\mathbf{v}$ belongs. With these two operations, the collection of all tangent vectors at p acquires the structure of a vector space $T_p\mathcal{M}$, called the *tangent space to the manifold at p*.

EXERCISE 4.3. Vector addition and multiplication by a scalar: Show that the above definitions of vector addition and multiplication by a scalar are independent of the choice of curves in the equivalence class and of the chart used.

EXAMPLE 4.5. The theories of shells: The traditional theory of shells is based on the adoption of the *Kirchhoff-Love hypothesis* stating that normals to the undeformed middle surface of the shell remain straight, unstretched, and perpendicular to the deformed middle surface. Other theories relax these assumptions to permit the original normals to tilt (thus admitting transverse shear strains) and/or to stretch. It is not even necessary that the special lines assumed to remain straight be originally perpendicular to the middle surface. Be that as it may, all these various formulations of the theory of shells are based on the assumption of the preservation of straightness of some line transverse to the middle surface of the shell. In reality, however, the shell being a 3-dimensional entity, there is no reason for any straight line to remain straight under arbitrary loading conditions. These theories of shells are, therefore, based on the idea that (due to the smallness of its thickness) a shell will not feel the difference between two deformations that carry a certain transverse line into curves that are tangent to each other at the middle surface. This is another way of saying that these shell theories lump all curved shapes of the deformed line into equivalence classes according to whether or not they are tangent to each other at their intersection with the middle surface. Thus, the deformation of a shell is completely characterized by the deformation of the middle surface plus the deformation of a vector field (the original special lines) into another (the equivalence class of curves).

There is another intrinsic way to look at a tangent vector. Let $f : \mathcal{M} \to \mathbb{R}$ be a function which is differentiable at the point $p \in \mathcal{M}$, and let $\mathbf{v}_p$ be a tangent vector at p, namely, and equivalence class of curves as we have just defined. Choosing a representative curve γ of this equivalence class, the composition:

$$f \circ \gamma : I \longrightarrow \mathbb{R} \tag{4.5}$$

is a real-valued function defined on the open interval $I \subset \mathbb{R}$. The derivative:

$$\mathbf{v}_p(f) := \left(\frac{d(f \circ \gamma)}{dt} \right)_{t=0}, \tag{4.6}$$

is called the *(directional) derivative of f along* $\mathbf{v}_p$. It is independent of the particular representative curve chosen.

EXERCISE 4.4. **Derivative of a function along a vector:** Prove that the derivative of a function along a vector is indeed independent of the representative curve chosen in the equivalence class.

Tangent vectors, therefore, can be regarded as linear operators on the set of differentiable functions defined on a neighbourhood of a point. The linearity is a direct consequence of the linearity of the derivative. For arbitrary real numbers, α and β, and for any differentiable function f, we have:

$$\mathbf{v}_p(\alpha f + \beta g) = \alpha \mathbf{v}_p(f) + \beta \mathbf{v}_p(g). \tag{4.7}$$

Notice, however, that not every linear operator can be regarded as a vector because, by virtue of (4.6), tangent vectors must also satisfy the Leibniz rule for the derivative of a product:

$$\mathbf{v}_p(fg) = f(p)\mathbf{v}_p(g) + \mathbf{v}_p(f)g(p), \tag{4.8}$$

where f and g are differentiable functions at p. We have thus established a one-to-one map from tangent vectors at a point to linear operators on the algebra of differentiable functions at that point that satisfy Leibniz's rule (i.e., *derivations*). This point of view, as well as the nice notation $\mathbf{v}(f)$, will be extremely useful in future considerations.

BOX 4.1. **Tangent Vectors and Derivations**:

We have not proved that there exists a bijection between tangent vectors and derivations. In other words, we have not shown that *every derivation* on the algebra of differentiable functions is actually a tangent vector (in the sense of an equivalence class of curves). In the case of a smooth (C^∞) manifold, and considering the algebra of smooth functions, it can be shown that this is the case. The proof is based on the fact that a C^k function ψ of m variables x^i (for $k \geq 2$) can always be expressed, in a sufficiently small neighbourhood of the origin, as:

$$\psi(x^1,\ldots,x^m) = \psi(0,\ldots,0) + \left(\frac{\partial \psi}{\partial x^i}\right)_0 x^i + \psi_{ij}(x^1,\ldots,x^m)x^i x^j, \qquad (4.9)$$

where (and this is very important) ψ_{ij} are C^{k-2} functions. If we start with a linear operator acting on the algebra of C^∞ functions defined in that neighbourhood, we can apply this operator term by term to the right-hand side of this equation. This would be true too if we consider analytic functions. For finite k, however, we don't have this luxury, since the algebra of C^{k-2} functions is larger than that of C^k functions. Assuming then that we have a linear operator $\mathbf{w}$, satisfying Leibniz's rule and acting on C^∞ functions, and noting that the Leibniz rule implies that such an operator must vanish on constant functions, we obtain (using Leibniz's rule repeatedly on the right-hand side) that:

$$\mathbf{w}(\psi) = \left(\frac{\partial \psi}{\partial x^i}\right)_0 \mathbf{w}(x^i), \qquad (4.10)$$

where $\mathbf{w}(x^i)$ is the value assigned to the function x^i by the linear operator $\mathbf{w}$. If we now consider the line γ defined parametrically by:

$$x^i = t\mathbf{w}(x^i), \qquad (4.11)$$

we can write:

$$\mathbf{w}(\psi) = \left(\frac{d(\psi \circ \gamma)}{dt}\right)_0. \qquad (4.12)$$

In short, the vector represented by the curve γ has exactly the same effect, when acting on smooth functions, as the operator $\mathbf{w}$. We have worked on $\mathbb{R}^m$, but it is obvious that the proof can be lifted to the manifold by choosing a local chart. In conclusion, as far as a smooth manifold is concerned, there is a complete equivalence between tangent vectors and linear operators satisfying the Leibniz rule.

    To find the dimension of the tangent space $T_p\mathcal{M}$, we choose a local chart $(\mathcal{U},\phi)$ with coordinates x^i, $i = 1,\ldots,m$, and consider the inverse map $\phi^{-1} : (\phi(\mathcal{U}) \subset \mathbb{R}^m) \to \mathcal{U}$. The restrictions of this map to each of the natural coordinate lines ($x^i = $ constant) of $\mathbb{R}^m$ passing through $\phi(p)$ are m curves in $\mathcal{M}$ passing through p. Each of these curves, called a *coordinate line in* $\mathcal{U}$, therefore, defines a tangent vector $(\mathbf{e}_p)_i$ at p, which will also be denoted suggestively by $(\partial/\partial x^i)_p$. The reason for this notation is clear: according to the interpretation of a vector as a linear Leibnizian operator we have:

$$(\mathbf{e}_p)_i(f) = (\partial/\partial x^i)(f) = \left(\frac{\partial(f \circ \phi^{-1})}{\partial x^i}\right)_{\phi(p)}. \tag{4.13}$$

In other words, the application of each of these operators delivers the partial derivative of the representative of the function f with respect to the corresponding coordinate. We will now prove that the collection $\{\partial/\partial x^i\}$ constitutes a basis of the tangent space $T_p\mathcal{M}$, called the *natural basis associated with the given coordinate system*. Let γ be a curve through p and let f be an arbitrary differentiable function. Denoting by $\mathbf{v}_p$ the operator corresponding to γ we can write, in the given coordinate system:

$$\mathbf{v}_p(f) = \left(\frac{d(f \circ \gamma)}{dt}\right)_{t=0} = \left(\frac{\partial(f \circ \gamma)}{\partial x^i}\right)_{\phi(p)}\left(\frac{d\gamma^i}{dt}\right)_{t=0}$$

$$= \left(\frac{d\gamma^i}{dt}\right)_{t=0}(\partial/\partial x^i)_p(f). \tag{4.14}$$

This equation clearly shows that every tangent vector can be expressed as a linear combination of the vectors $(\partial/\partial x^i)_p$, namely:

$$\mathbf{v}_p = v_p^i(\partial/\partial x^i)_p, \tag{4.15}$$

with:

$$v_p^i = \left(\frac{d\gamma^i}{dt}\right)_{t=0}. \tag{4.16}$$

Here, as before, $x^i = \gamma^i(t)$ is the parametric equation in the given coordinate system of a representative curve γ associated with the vector $\mathbf{v}_p$. Naturally, the component v_p^i can also be regarded as the result of operating $\mathbf{v}_p$ on the i-th coordinate line at $\mathcal{U}$. We still need to show that to every linear combination of the natural base vectors we can associate a curve. Let $\alpha^i(\partial/\partial x^i)_p$ (with $\alpha^i \in \mathbb{R}$) be a given linear combination. We construct the curve with the following parametric equations:

$$x^i(t) = x_p^i + \alpha^i t, \tag{4.17}$$

which can easily be checked to do the job. Finally, for the m vectors $(\partial/\partial x^i)_p$ to constitute a basis, they must be linearly independent. Let $\alpha^i(\partial/\partial x^i)_p = 0$. Applying this linear combination to the k-th coordinate line (denoted simply by x^k), and observing that:

$$(\partial/\partial x^i)_p(x^k) = \delta_i^k, \tag{4.18}$$

it follows that necessarily $\alpha^k = 0$ for each k.

EXERCISE 4.5. **The traditional definition:** Show that when a new coordinate system $y^i = y^i(x^1, \ldots, x^m)$, $i = 1, \ldots, m$ is introduced, the components $\hat{v}_p^i$ of a vector in the new natural basis $(\partial/\partial y^i)_p$ are related to the old components v_p^i in the basis $(\partial/\partial x^i)_p$ by the formula:

$$\hat{v}_p^i = \left(\frac{\partial y^i}{\partial x^j}\right)_p v_p^j, \tag{4.19}$$

while the base vectors themselves are related by:

$$(\partial/\partial y^i)_p = \left(\frac{\partial x^j}{\partial y^i}\right)_p (\partial/\partial x^j)_p. \tag{4.20}$$

The components of vectors are, therefore, said to change *contravariantly*. It is possible to define tangent vectors (as was done in more traditional treatments) as indexed quantities that transform contravariantly under coordinate changes.

In conclusion, we have shown that at every point p of a manifold $\mathcal{M}$ there exists a well-defined vector space $T_p\mathcal{M}$, called the tangent space at p. The dimension of this vector space is equal to the dimension of the manifold itself. Given a coordinate system x^i, there exists a natural basis $\{(\partial/\partial x^i)_p\}$ associated with it. The components of tangent vectors in the natural basis change in a definite (contravariant) way with changes of coordinates.

4.7. The Tangent Bundle

In this section, we will introduce a very useful geometrical construct technically known as the tangent bundle $T\mathcal{M}$ of a differentiable manifold $\mathcal{M}$. This object is a particular case of a more general kind of differentiable manifolds, called *fibre bundles*, endowed with a map, called a *projection*, which allows us to identify points as belonging to smaller entities called *fibres*. Since a whole chapter will be devoted to the treatment of fibre bundles, we will limit ourselves here to characterize only the tangent bundle to the extent needed at this point. This partial characterization, on the other hand, may help to shed light on the general theory of fibre bundles to be presented later.

We already know that each point p of an m-dimensional differentiable manifold $\mathcal{M}$ is equipped with its own tangent space $T_p\mathcal{M}$, which is an m-dimensional vector space. All these tangent spaces are, therefore, isomorphic to each other, but in general there is no natural (or canonical) way to establish an isomorphism between tangent spaces at different points. A good way to think of this situation is to go back to our seemingly inexhaustible source of examples: the surface of the Earth, idealized as a sphere. Given a tangent vector in, say, Buenos Aires (namely, a small piece of a curve, to be more graphic, yet imprecise), can a corresponding tangent vector be chosen in, say, Rome? Obviously, there is no natural way to choose it (except for the zero vector). Only if we arbitrarily assign to a pair of linearly independent vectors in Buenos Aires a pair of linearly independent vectors in Rome, will we have established a correspondence between all the tangent vectors at both cities. Nevertheless, even without such an arbitrary assignment, it makes sense to think of the surface of the Earth as a sphere each of whose points carries its own tangent plane. This new entity will be of a higher dimension, since in order to specify an object therein we will have to give a point on the sphere (the Umbilicus Urbis, official centre of Rome, say) and a vector attached to it (say, the approximately 600-meter long segment joining the Umbilicus Urbis to the Arch of Constantine, at the other end of the Forum). We will need in this case four pieces of information: two coordinates (in a local chart, like the map of Italy) for Rome, and two additional data, namely, the components of the vector in the natural basis of this chart. Moreover, if these four pieces of information were given, we could isolate the first two (Rome), but not the second two, since the vector "Umbilicus Urbis–Arch of Constantine" does not have an independent existence outside of Rome. The vector presupposes its point of attachment. The operation of extracting the first half of the information (Rome, in this case) is called a *projection*.

Consider the collection $T\mathcal{M}$ of pairs $(p, \mathbf{v}_p)$, where $p \in \mathcal{M}$ and $\mathbf{v}_p \in T_p\mathcal{M}$. We remark that this collection is *not* a Cartesian product. Define the *projection map*:

$$\tau : T\mathcal{M} \longrightarrow \mathcal{M}, \tag{4.21}$$

by:

$$\tau(p, \mathbf{v}_p) = p. \tag{4.22}$$

We will endow the set $T\mathcal{M}$ with the structure of a differentiable manifold. Let $(\mathcal{U}, \phi)$ be a chart of some atlas of $\mathcal{M}$, and let the corresponding coordinates be denoted by x^i. We proceed to construct the natural basis $\{\partial/\partial x^i\}$ at each point of this chart. We are now in a position to assign to each point $(p, \mathbf{v}_p) \in \tau^{-1}(\mathcal{U}) \subset T\mathcal{M}$ the $2m$ numbers $(x^1, \ldots, x^m, v_p^1, \ldots, v_p^m)$, where v_p^i are the components of $\mathbf{v}_p$ in the natural basis at p. Wherever two charts intersect, the last m numbers change (point by point of the intersection) according to the linear transformation (4.19). We infer that the covering of $T\mathcal{M}$ by means of the open sets of the form $\tau^{-1}(\mathcal{U})$ and with the coordinates just introduced endows $T\mathcal{M}$ with an atlas and, therefore, with a differentiable structure. We conclude that $T\mathcal{M}$ is a $2m$-dimensional differentiable manifold. In the terminology of fibre bundles, the set $\tau^{-1}(p)$ is called the *fibre at p*. In the case of the tangent bundle, the fibre at p is precisely the tangent space $T_p\mathcal{M}$. Fibre bundles, like $T\mathcal{M}$, whose fibres are vector spaces are called *vector bundles*. The manifold $\mathcal{M}$ of departure (namely, the carrier of the fibres) is called the *base manifold* of the fibre bundle. Since all the fibres are m-dimensional vector spaces, they are all (noncanonically) isomorphic to $\mathbb{R}^m$, which may thus be called a *typical fibre of T$\mathcal{M}$*.

EXERCISE 4.6. **The projection map:** Show that the projection is a C^∞ map. [Hint: express it in coordinates.]

EXERCISE 4.7. **A vector space as base manifold:** Show that if the base manifold $\mathcal{M}$ is a finite-dimensional vector space, the tangent bundle can be seen as a Cartesian

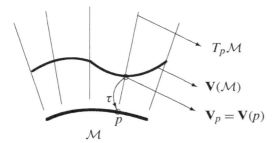

Figure 4.3. A cross section

product. Of what? [Hint: establish a canonical isomorphism between the tangent spaces and the base manifold.]

A *vector field* **V** on a manifold $\mathcal{M}$ is a smooth assignment to each point p of $\mathcal{M}$ of a tangent vector $\mathbf{V}_p$. The smoothness manifests itself (in an atlas) in that each of the components of the vector field is a differentiable function. A vector field is, therefore, a smooth map:

$$\mathbf{V} : \mathcal{M} \longrightarrow T\mathcal{M}, \tag{4.23}$$

satisfying the condition: $\tau \circ \mathbf{V} = id_{\mathcal{M}}$, where $id_{\mathcal{M}}$ is the identity map of $\mathcal{M}$. This condition guarantees that to each point of the base manifold a vector is assigned *in its own tangent space*. Such a map is also called a *cross section* of the bundle. Every tangent bundle has a canonical section: the *zero section*, assigning to each point the zero vector of its tangent space. The existence of an everywhere nonzero section of a vector bundle is not always guaranteed. The classical example is the sphere (no matter how much you try, you cannot comb a hairy sphere).

The terminology "cross section" is due to the following pictorial representation (Figure 4.3) : If we draw a shallow arc for the base manifold, with the fibres (in this case, the tangent spaces) as straight lines hovering over it, then a cross section (in this case, a vector field) looks indeed like a curve cutting smoothly through the fibres. The projection τ literally projects a point of a fibre to that point of $\mathcal{M}$ upon which the fibre hovers.

4.8. The Lie Bracket

If **V** is a (smooth) vector field on a manifold $\mathcal{M}$ and $f : \mathcal{M} \to \mathbb{R}$ is a smooth function, then the map:

$$\mathbf{V}f : \mathcal{M} \longrightarrow \mathbb{R}, \tag{4.24}$$

defined as:

$$p \mapsto \mathbf{V}_p(f) \tag{4.25}$$

is again a smooth map. It assigns to each point $p \in \mathcal{M}$ the directional derivative of the function f in the direction of the vector field at p. In other words, a vector field assigns to each smooth function another smooth function. Given, then, two vector fields **V** and **W** over $\mathcal{M}$, the iterated evaluation:

$$h = \mathbf{W}(\mathbf{V}f) : \mathcal{M} \longrightarrow \mathbb{R}, \tag{4.26}$$

gives rise to a legitimate smooth function h on $\mathcal{M}$.

On the basis of the above considerations, one may be tempted to define a composition of vector fields by declaring that the composition $\mathbf{W} \circ \mathbf{U}$ is the vector field which assigns to each function f the function h defined by Equation (4.26). This wishful thinking, however, does not work. To see why, it is convenient to work in components in some chart with coordinates x^i. Let:

$$\mathbf{V} = V^i \frac{\partial}{\partial x^i} \qquad \mathbf{W} = W^i \frac{\partial}{\partial x^i}, \tag{4.27}$$

where the components V^i and W^i $(i = 1, \ldots, n)$ are smooth real-valued functions defined over the n-dimensional domain of the chart. Given a smooth function $f : \mathcal{M} \to \mathbb{R}$, the function $g = \mathbf{V}f$ is evaluated at a point $p \in \mathcal{M}$ with coordinates x^i $(i = 1, \ldots, n)$ as:

$$g(p) = V^i \frac{\partial f}{\partial x^i}, \tag{4.28}$$

as prescribed by Equation (4.15). Notice the slight abuse of notation we incur into by identifying the function f with its representation in the coordinate system.

We now apply the same prescription to calculate the function $h = \mathbf{W}g$ and obtain:

$$h(p) = W^i \frac{\partial g}{\partial x^i} = W^i \frac{\partial \left(V^j \frac{\partial f}{\partial x^j} \right)}{\partial x^i} = \left(W^i \frac{\partial V^j}{\partial x^i} \right) \frac{\partial f}{\partial x^j} + W^i V^j \frac{\partial^2 f}{\partial x^i \partial x^j}. \tag{4.29}$$

The last term of this expression, by involving second derivatives, will certainly not transform as the components of a vector should under a change of coordinates. Neither will the first. This negative result, on the other hand, suggests that the offending terms could perhaps be eliminated by subtracting from the composition **WV** the opposite composition **VW**, namely:

$$(\mathbf{WV} - \mathbf{VW})(f) = \left(W^i \frac{\partial V^j}{\partial x^i} - V^i \frac{\partial W^j}{\partial x^i} \right) \frac{\partial f}{\partial x^j}. \tag{4.30}$$

The vector field thus obtained, is called the *Lie bracket* of **W** and **V** (in that order) and is denoted by [**W**, **V**]. More explicitly, its components in the coordinate system x^i are given by:

$$[\mathbf{W}, \mathbf{V}]^j = W^i \frac{\partial V^j}{\partial x^i} - V^i \frac{\partial W^j}{\partial x^i}. \tag{4.31}$$

Upon a coordinate transformation, these components transform according to the rules of transformation of a vector.

EXERCISE 4.8. Prove the above assertion by means of Equation (4.19).

The following properties of the Lie bracket are worthy of notice:

(1) Skew-symmetry:
$$[\mathbf{W}, \mathbf{V}] = -[\mathbf{V}, \mathbf{W}] \tag{4.32}$$

(2) Jacobi identity:
$$[[\mathbf{W}, \mathbf{V}], \mathbf{U}] + [[\mathbf{V}, \mathbf{U}], \mathbf{W}] + [[\mathbf{U}, \mathbf{W}], \mathbf{V}] = 0 \tag{4.33}$$

EXERCISE 4.9. Prove the above two properties of the Lie bracket.

The collection of all vector fields over a manifold has the natural structure of an infinite dimensional vector space, where addition and multiplication by a scalar are defined in the obvious way. In this vector space, the Lie bracket operation is bilinear. A vector space endowed with a bilinear operation satisfying conditions (1) and (2) is called a *Lie algebra*.

Vector fields can be multiplied by functions to produce new vector fields. Indeed, for a given function f and a given vector field $\mathbf{V}$, we can define the vector field $f\mathbf{V}$ by:
$$(f\mathbf{V})_p := f(p)\mathbf{V}_p. \tag{4.34}$$

EXERCISE 4.10. Prove that:
$$[g\mathbf{W}, f\mathbf{V}] = gf[\mathbf{W}, \mathbf{V}] + g(\mathbf{W}f)\,\mathbf{V} - f(\mathbf{V}g)\,\mathbf{W}, \tag{4.35}$$

where g, f are smooth functions and $\mathbf{W}, \mathbf{V}$ are vector fields over a manifold $\mathcal{M}$.

EXAMPLE 4.6. **Continuous distributions of dislocations in the limit of a crystal lattice:** For simplicity of the exposition, we will work in $\mathbb{R}^2$, although most of the reasoning can be carried through (at least locally) to any finite-dimensional differentiable manifold. Let an atomic lattice be given by, say, all points with integer coordinates in $\mathbb{R}^2$. To each atom we can associate two vectors (in this instance unit and orthogonal) determined by joining it to its immediate neighbours to the right and above, respectively. If the lattice is deformed regularly, these vectors will deform accordingly, changing in length and angle, but always remaining linearly independent at each atom. In the (not precisely defined) continuous limit, we can imagine that each point of $\mathbb{R}^2$ has been endowed with a basis or frame, the collection of which is called a *moving frame* (or *repère mobile*).[6]

Returning to the discrete picture, if there is a dislocation (e.g., a half-line of atoms is missing, as shown on the right-hand side of Figure 4.4), the local bases will be altered differently from the case of a mere deformation. The engineering way to recognize this is the so-called *Burgers' circuit*, which consists of a four-sided path made of the same number of atomic spacings in each direction. The failure of such

---

[6] This idea was introduced mathematically by E. Cartan and, in a physical context, by the brothers Cosserat.

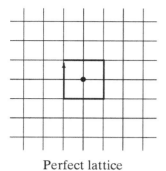
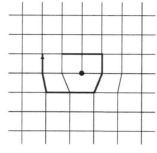

Perfect lattice   Dislocated lattice

Figure 4.4. Dislocation in a crystal lattice

a path to close is interpreted as the presence of a local dislocation in the lattice. We want to show that in the putative continuous limit this failure is represented by the nonvanishing of a Lie bracket. What we have in the continuous case as the only remnant of the discrete picture is a smoothly distributed collection of bases, which we have called a moving frame, and which can be seen as two vector fields $\mathbf{E}_\alpha$ ($\alpha = 1,2$) over $\mathbb{R}^2$.

From the theory of ordinary differential equations, we know that each vector field gives rise, at least locally, to a well-defined family of parametrized integral curves, where the parameter is determined up to an additive constant. More specifically, these curves are obtained as the solutions $\mathbf{r} = \mathbf{r}(s^\alpha)$ of the systems of equations:

$$\frac{d\mathbf{r}(s^\alpha)}{ds^\alpha} = \mathbf{E}_\alpha[\mathbf{r}(s^\alpha)], \quad (\alpha = 1,2; \text{ no sum on } \alpha), \quad (4.36)$$

where $\mathbf{r}$ represents the natural position vector in $\mathbb{R}^2$. The parameter s^α (one for each of the two families of curves) can be pinned down in the following way. Select a point p_0 as origin and draw the (unique) integral curve γ_1 of the first family passing through this origin. Adopting the value $s^1 = 0$ for the parameter at the origin, the value of s^1 becomes uniquely defined for all the remaining points of the curve. Each of the curves of the second family must intersect this curve of the first family. We adopt, therefore, for each of the curves of the second family the value $s^2 = 0$ at the corresponding point of intersection with that reference curve (of the first family). In this way we obtain (at least locally) a new coordinate system s^1, s^2 in $\mathbb{R}^2$. By construction, the second natural base vector of this coordinate system is $\mathbf{E}_2$. But there is no guarantee that the first natural base vector will coincide with $\mathbf{E}_1$, except at the curve γ_1 through the adopted origin. In fact, if we repeat the previous construction in reverse, that is, with the same origin but adopting the curve γ_2 of the *second* family as a reference, we obtain in general a different system of coordinates, which is well adapted to the basis vectors $\mathbf{E}_1$, but not necessarily to $\mathbf{E}_2$.

Assume now that, starting at the adopted origin, we move an amount of Δs^1 along γ_1 to arrive at a point p' and thereafter we climb an amount of Δs^2 along the encountered curve of the second family through p'. We arrive at some point p_1,

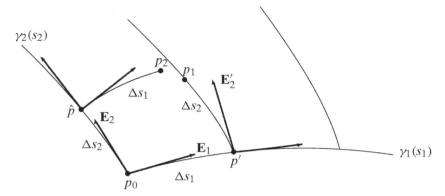

Figure 4.5. The continuous case

as shown in Figure 4.5. Incidentally, this is the point with coordinates $(\Delta s^1, \Delta s^2)$ in the coordinate system obtained by the first construction. If, however, starting at the same origin we move by Δs^2 along the curve γ_2 to a point $\hat{p}$ and then move by Δs^1 along the encountered curve of the first family, we will arrive at a point p_2 (whose coordinates are $(\Delta s^1, \Delta s^2)$ in the *second* construction) which is, in general, different from p_1. Thus, we have detected the failure of a four-sided circuit to close! The discrete picture has, therefore, its continuous counterpart in the noncommutativity of the flows along the two families of curves.

Let us calculate a first-order approximation to the difference between p_2 and p_1. For this purpose, let us evaluate, to the first order, the base vector $\mathbf{E}_2$ at the auxiliary point p'. The result is:

$$\mathbf{E}_2' = \mathbf{E}_2(p_0) + \frac{\partial \mathbf{E}_2}{\partial x^i} \frac{dx^i}{ds^1} \Delta s^1, \tag{4.37}$$

where derivatives are calculated at p_0. The position vector of p_1, always to first-order approximation, is obtained, therefore, as:

$$\mathbf{r}_1 = \Delta s^1 \mathbf{E}_1(p_0) + \Delta s^2 \left(\mathbf{E}_2(p_0) + \frac{\partial \mathbf{E}_2}{\partial x^i} \frac{dx^i}{ds^1} \Delta s^1 \right). \tag{4.38}$$

In a completely analogous fashion, we calculate the position vector of p_2 as:

$$\mathbf{r}_2 = \Delta s^2 \mathbf{E}_2(p_0) + \Delta s^1 \left(\mathbf{E}_1(p_0) + \frac{\partial \mathbf{E}_1}{\partial x^i} \frac{dx^i}{ds^2} \Delta s^2 \right). \tag{4.39}$$

By virtue of (4.36), however, we have:

$$\frac{dx^i}{ds^\alpha} = E_\alpha^i, \tag{4.40}$$

where E_α^i is the i-th component in the natural basis of $\mathbb{R}^2$ of the base vector $\mathbf{E}_\alpha$. From the previous three equations we obtain:

$$\mathbf{r}_2 - \mathbf{r}_1 = \left(\frac{\partial \mathbf{E}_1}{\partial x^i} E_2^i - \frac{\partial \mathbf{E}_2}{\partial x^i} E_1^i \right) \Delta s^1 \Delta s^2 = [\mathbf{E}_1, \mathbf{E}_2] \Delta s^1 \Delta s^2. \tag{4.41}$$

We thus confirm that the closure of the infinitesimal circuits generated by two vectors fields is tantamount to the vanishing of their Lie bracket. This vanishing, in turn, is equivalent to the commutativity of the flows generated by these vector fields. For this reason, the Lie bracket is also called the *commutator* of the two vector fields. In physical terms, we may say that the vanishing of the Lie brackets between the vector fields representing the limit of a lattice is an indication of the absence of dislocations.

In the above example, we have introduced the notion of a moving frame, that is, a smooth field of bases $\mathbf{E}_i$ ($i = 1, \ldots, n$) over an n-dimensional manifold. Since a Lie bracket of two vector fields is itself a vector field, there must exist unique scalar fields c_{ij}^k such that:

$$[\mathbf{E}_i, \mathbf{E}_j] = c_{ij}^k \mathbf{E}_k \quad (i, j, k = 1, \ldots, n). \tag{4.42}$$

These scalars are known as the *structure constants* of the moving frame. The structure constants vanish identically if, and only if, the frames can be seen locally as the natural base vectors of a coordinate system.

4.9. The Differential of a Map

We have defined the notion of *differentiability* of a map between manifolds by checking that the representation of the map in some charts of the manifolds is differentiable. This same idea could be pursued to actually define the differential of a differentiable map working first in charts and then bringing the result back to the manifolds. This technique, although correct, should not be abused. In fact, the main guiding principle of Differential Geometry is the definition of objects that are *intrinsic*, that is, independent of their coordinate representations. Accordingly, whenever possible, one should attempt to provide a definition of a geometric object in a manner that is independent of its coordinate representations. If such a manner cannot be found or turns out to be too complicated, one should then carefully check that the coordinate definition provides an object which is independent of the particular coordinate system used. We have already used both techniques. The definition of the manifold itself and of the differentiability of maps followed the coordinate route, while the definition of tangent vector (as an equivalence class of curves or as a linear operator on functions) followed the intrinsic method. Whenever possible, this last method should be preferred, not just because of its mathematical appeal, but also because an intrinsic definition usually reveals the deep meaning of an object with total clarity. We will now provide an intrinsic definition of the differential of a map between manifolds, although some of the properties of the object will be checked using coordinates.

Given a differentiable map:

$$g : \mathcal{M} \longrightarrow \mathcal{N} \tag{4.43}$$

between two manifolds, $\mathcal{M}$ and $\mathcal{N}$, of dimensions m and n, respectively, we focus attention on a particular point $p \in \mathcal{M}$ and its image $q = g(p) \in \mathcal{N}$. Let $\mathbf{v}_p \in T_p\mathcal{M}$

be a tangent vector at p and let $\gamma : I \to \mathcal{M}$ be one of its representative curves. The composite map:

$$g \circ \gamma : I \longrightarrow \mathcal{N} \tag{4.44}$$

is then a smooth curve in $\mathcal{N}$ passing through q. This curve (the image of γ by g) is, therefore, the representative of a tangent vector at q which we will denote $(g_*)_p(\mathbf{v}_p)$.

EXERCISE 4.11. **Independence of curve and linearity:** Check that the vector $(g_*)_p(\mathbf{v}_p)$ is independent of the representative curve γ chosen for $\mathbf{v}_p$. Moreover, show that $(g_*)_p$ is a linear map on vectors at p. You may use coordinate charts.

The map $(g_*)_p$ just defined is called the *differential of g at p*. It is a linear map between the tangent spaces $T_p\mathcal{M}$ and $T_{g(p)}\mathcal{N}$. Since this construction can be carried out at each and every point of $\mathcal{M}$, we obtain a map g_* between the tangent bundles, namely:

$$g_* : T\mathcal{M} \longrightarrow T\mathcal{N}, \tag{4.45}$$

called the *differential of g*. Alternative notations for this map are Dg and Tg, and it is also known as the *tangent map*. One should note that the map g_* includes the map g between the base manifolds, since it maps vectors at a point p linearly into vectors at the image point $q = g(p)$, and not just to any vector in $T\mathcal{N}$. It is, therefore, a *fibre-preserving map*. This fact is best illustrated in the following commutative diagram:

$$
\begin{array}{ccc}
T\mathcal{M} & \xrightarrow{\;g_*\;} & T\mathcal{N} \\
{\scriptstyle \tau_{\mathcal{M}}}\downarrow & & \downarrow{\scriptstyle \tau_{\mathcal{N}}} \\
\mathcal{M} & \xrightarrow{\;g\;} & \mathcal{N}
\end{array}
\tag{4.46}
$$

where $\tau_{\mathcal{M}}$ and $\tau_{\mathcal{N}}$ are the projection maps of $T\mathcal{M}$ and $T\mathcal{N}$, respectively.

In the particular case of a function $f : \mathcal{M} \to \mathbb{R}$, the differential f_* can be interpreted somewhat differently. Indeed, the tangent space $T_r\mathbb{R}$ can be canonically identified with $\mathbb{R}$ itself (see Exercise 4.7), so that f_* can be seen as a real-valued function on $T\mathcal{M}$. This function is denoted by $df : T\mathcal{M} \to \mathbb{R}$.

EXERCISE 4.12. **The differential of a function:** Show that:

$$df(\mathbf{v}) = \mathbf{v}(f). \tag{4.47}$$

[Hint: just follow all the definitions.]

The following three exercises contain important results:

EXERCISE 4.13. **Coordinate representation:** Show that in local systems of coordinates x^i ($i = 1,\ldots,m$) and y^α ($\alpha = 1,\ldots,n$) around p and $g(p)$, respectively, the differential of g at p maps the vector with components v^i into the vector with components:

$$[(g_*)_p(\mathbf{v}_p)]^\alpha = \left(\frac{\partial g^\alpha}{\partial x^i}\right)_p v^i, \tag{4.48}$$

where $g^\alpha = g^\alpha(x^1,\ldots,x^n)$ is the coordinate representation of g in the given charts. The $(m \times n)$-matrix with entries $\left\{\left(\frac{\partial g^\alpha}{\partial x^i}\right)_p\right\}$ is the *Jacobian matrix* at p of the map g in the chosen coordinate systems. Show that the *rank* of the Jacobian matrix is independent of the coordinates used. It is called the *rank of g at p*.

EXERCISE 4.14. **Operating on functions:** Let $f : \mathcal{N} \to \mathbb{R}$ be a differentiable function and let $g : \mathcal{M} \to \mathcal{N}$ be a differentiable map between manifolds. Show that:

$$((g_*)_p\mathbf{v}_p)(f) = \mathbf{v}_p(f \circ g), \quad p \in \mathcal{M}. \tag{4.49}$$

Draw a graph involving $\mathcal{M}, \mathcal{N}$, and $\mathbb{R}$ and interpret your result.

EXERCISE 4.15. **Differential of a composition:** Show that the differential of a composition of maps is equal to the composition of the differentials. More precisely, if $g : \mathcal{M} \to \mathcal{N}$ and $h : \mathcal{N} \to \mathcal{P}$ are differentiable maps, then:

$$((h \circ g)_*)_p(\mathbf{v}_p) = (h_*)_{g(p)}((g_*)_p(\mathbf{v}_p)). \tag{4.50}$$

How does this formula look in coordinates?

EXAMPLE 4.7. **The deformation gradient:** In Appendix A we define a material body $\mathcal{B}$ as a 3-dimensional trivial differentiable manifold, and a configuration as a differentiable map:

$$\kappa : \mathcal{B} \longrightarrow \mathbb{E}^3. \tag{4.51}$$

The deformation gradient $\mathbf{F}$ is, therefore, nothing but the differential of κ. Choosing global charts of $\mathcal{B}$ and of $\mathbb{E}^3$ with coordinates X^I and x^i, respectively, we obtain the point-wise matrix representation:

$$F_I^i = \frac{\partial x^i}{\partial X^I}. \tag{4.52}$$

Alternatively, we may choose to regard the chart of $\mathcal{B}$ as induced by a chart of $\mathbb{E}^3$ via a reference configuration $\kappa_0 : \mathcal{B} \to \mathbb{E}^3$. In this interpretation, the above expression for the deformation gradient is understood as the coordinate representation of the deformation $\chi = \kappa \circ \kappa_0^{-1} : (\kappa_0(\mathcal{B}) \subset \mathbb{E}^3) \to \mathbb{E}^3$. From the formal point of view both interpretations are equivalent. Nevertheless, the use of the reference configuration permits to use (and abuse) the underlying affine structure of $\mathbb{E}^3$, particularly its distant parallelism, so as to compare (by shifting) objects in the body (namely, in the reference configuration) with objects in space. A clear example of this use can be seen in the interpretation of the orthogonal component $\mathbf{R}$, in the polar decomposition of $\mathbf{F}$, as a rotation. In fact, what we have is a linear map $\mathbf{R} : T_b\mathcal{B} \to T_{\kappa(b)}\mathbb{E}^3$ satisfying the conditions: $\mathbf{RR}^T = id_{\mathbb{E}^3}$ and $\mathbf{R}^T\mathbf{R} = id_{T_b\mathcal{B}}$. Naturally, for these expressions to make sense, $T_b\mathcal{B}$ has to be an inner-product space (otherwise, the transposed tensor, operating between the dual spaces, could not be composed with the tensor itself). It is this inner product that the reference configuration (or the chart) induces.

4.9.1. Push-forwards

We have seen that the differential of a map between manifolds carries tangent vectors to tangent vectors. This operation is sometimes called a *push-forward*. Does a map also push forward vector fields to vector fields? Let $\mathbf{V} : \mathcal{M} \to T\mathcal{M}$ be a vector field on $\mathcal{M}$ and let $g : \mathcal{M} \to \mathcal{N}$ be a smooth map. Since the differential of g is a map of the form $g_* : T\mathcal{M} \to T\mathcal{N}$, the composition $g_* \circ \mathbf{V}$ makes perfect sense, but it delivers a (well-defined) map $g_*\mathbf{V}$ from $\mathcal{M}$ (and *not* from $\mathcal{N}$) into $T\mathcal{N}$. This is not a vector field, nor can it in general be turned into one. If the dimension of $\mathcal{M}$ is larger than that of $\mathcal{N}$, points in $\mathcal{N}$ will end up being assigned more than one vector. If the dimension of the source manifold is less than that of the target, on the other hand, even if the function is one-to-one, there will necessarily exist points in $\mathcal{N}$ to which no vector is assigned.

Notwithstanding the above remark, let $g : \mathcal{M} \to \mathcal{N}$ be a smooth map. We say that the vector fields $\mathbf{V} : \mathcal{M} \to T\mathcal{M}$ and $\mathbf{W} : \mathcal{N} \to T\mathcal{N}$ are *g-related* if:

$$g_*\mathbf{V}(p) = \mathbf{W}(g(p)) \quad \forall p \in \mathcal{M}. \tag{4.53}$$

According to this definition, if g happens to be a diffeomorphism, then $\mathbf{V}$ and $g_*\mathbf{V}$ are automatically g-related. The pushed-forward vector field is then given by $g_* \circ \mathbf{V} \circ g^{-1}$.

PROPOSITION 4.1. *Let $\mathbf{V}_1$ be g-related to $\mathbf{W}_1$ and let $\mathbf{V}_2$ be g-related to $\mathbf{W}_2$. Then the Lie bracket $[\mathbf{V}_1, \mathbf{V}_2]$ is g-related to the Lie bracket $[\mathbf{W}_1, \mathbf{W}_2]$, that is:*

$$[g_*\mathbf{V}_1, g_*\mathbf{V}_2] = g_*[\mathbf{V}_1, \mathbf{V}_2]. \tag{4.54}$$

PROOF. Although the proof can be carried out in an elegant intrinsic manner, it may turn out to be a useful exercise to exhibit it in terms of local coordinate systems x^i ($i = 1, \ldots, m$) and y^α ($\alpha = 1, \ldots, n$) in $\mathcal{M}$ and $\mathcal{N}$, respectively. Let the function g be represented in these coordinate systems as the n smooth functions $y^\alpha = y^\alpha(x^1, \ldots, x^m)$. The fact that the vector fields are g-related means that, in components, we have:

$$w_1^\alpha = \frac{\partial y^\alpha}{\partial x^i} v_1^i, \quad w_2^\alpha = \frac{\partial y^\alpha}{\partial x^i} v_2^i. \tag{4.55}$$

According to Equation (4.31), we can write:

$$[\mathbf{W}_1, \mathbf{W}_2]^\alpha = w_1^\beta \frac{\partial w_2^\alpha}{\partial y^\beta} - w_2^\beta \frac{\partial w_1^\alpha}{\partial y^\beta}. \tag{4.56}$$

Combining (4.55) with (4.56), we obtain:

$$[\mathbf{W}_1, \mathbf{W}_2]^\alpha = \frac{\partial y^\beta}{\partial x^i} \, v_1^i \, \frac{\partial}{\partial y^\beta} \left(\frac{\partial y^\alpha}{\partial x^j} \, v_2^j \right) - \frac{\partial y^\beta}{\partial x^i} \, v_2^i \, \frac{\partial}{\partial y^\beta} \left(\frac{\partial y^\alpha}{\partial x^j} \, v_1^j \right), \qquad (4.57)$$

which, by the chain rule, can be written as:

$$[\mathbf{W}_1, \mathbf{W}_2]^\alpha = v_1^i \, \frac{\partial}{\partial x^i} \left(\frac{\partial y^\alpha}{\partial x^j} \, v_2^j \right) - v_2^i \, \frac{\partial}{\partial x^i} \left(\frac{\partial y^\alpha}{\partial x^j} \, v_1^j \right), \qquad (4.58)$$

or, cancelling out the symmetric terms:

$$[\mathbf{W}_1, \mathbf{W}_2]^\alpha = \frac{\partial y^\alpha}{\partial x^j} \left(v_1^i \, \frac{\partial v_2^j}{\partial x^i} - v_2^i \, \frac{\partial v_1^j}{\partial x^i} \right) = \frac{\partial y^\alpha}{\partial x^j} \, [\mathbf{V}_1, \mathbf{V}_2]^j. \qquad (4.59)$$

$\square$

4.10. Immersions, Embeddings, Submanifolds

4.10.1. Linear Maps of Vector Spaces

A review of the basic terminology and ideas pertaining to linear maps (or linear *operators*) between vector spaces can serve as a good introduction to this section. Let X and Y be two vector spaces of dimensions m and n, respectively. Denoting by $\{\mathbf{E}_J\}$ ($J = 1, \ldots, m$) and $\{\mathbf{e}_i\}$ ($i = 1, \ldots, n$) some bases in X and Y, respectively, a linear map $\mathbf{L} : X \to Y$ is represented uniquely as:

$$\mathbf{L} = L^{iJ} \, \mathbf{e}_i \otimes \mathbf{E}_J. \qquad (4.60)$$

Although the entries of the matrix $\{L^{iJ}\}$ depend on the bases chosen, its *rank* does not. We recall that the rank of a matrix is the order of the largest square submatrix (obtained by eliminating entire rows and columns) whose determinant does not vanish. We, therefore, define the *rank of the map* $\mathbf{L}$, denoted rank($\mathbf{L}$), as the rank of any of its representative matrices. The largest possible rank of a linear map is, obviously, the smaller of the dimensions m and n. A map with this rank is said to be of *maximal rank*. The image $\mathbf{L}(X) \subset Y$ is called the *range* of the operator. The inverse image of the zero vector of Y, namely, ker($\mathbf{L}$) = $\mathbf{L}^{-1}(\mathbf{0}) \subset X$ is called the *kernel* of the operator. Both the range and the kernel are *vector subspaces* (of Y and X, respectively). The dimension of the range is equal to the rank of the linear map. The dimension of the kernel is:

$$\dim(\ker(\mathbf{L})) = m - \mathrm{rank}(\mathbf{L}). \qquad (4.61)$$

This result is known as the *rank-nullity theorem*. A linear map is one-to-one (injective) if, and only if, the dimension of its kernel is zero (the kernel in this case consists of just the zero vector of X). It follows that if $m > n$ the map cannot be one-to-one, since (in this case) the rank cannot exceed n. When $m = n$, the map is invertible (or one-to-one and onto, or bijective) if, and only if, it is of maximal rank (m). The inverse of a linear map, if it exists, is again a linear map.

If $m < n$ we have that the map is one-to-one if, and only if, it is of maximal rank (m). The range will, therefore, be a vector subspace of dimension m. We can now change the basis in Y in such a way that the first m base vectors belong to this subspace, while the remaining ones don't. In such a basis, it is clear that the matrix representing the operator **L** will attain the special form:

$$\begin{bmatrix} \mathbf{A}_{m \times m} \\ \mathbf{0}_{(n-m) \times m} \end{bmatrix}, \tag{4.62}$$

where $\mathbf{A}_{m \times m}$ is a nonsingular matrix and $\mathbf{0}_{(n-m) \times m}$ denotes the (in general rectangular) zero matrix of the indicated order. A further change of base within the subspace can render **A** the unit matrix.

4.10.2. The Inverse Function Theorem of Calculus

As we have already learnt, Differential Geometry blends elements of algebra (e.g., the tangent space), topology (e.g., open coverings) and analysis (e.g., differentiability) in practically everything it does. This is not surprising, since a differentiable manifold is something that looks locally like an open piece of $\mathbb{R}^m$. Some of its strongest results, in fact, are generalizations to manifolds of analytic theorems valid in Euclidean spaces. The *inverse function theorem* and the *implicit function theorem* belong to this category.

The gist of the inverse function theorem is to extend to a general map (not necessarily linear) from $\mathbb{R}^m$ (or a piece thereof) to $\mathbb{R}^m$ the maximal rank theorem of linear maps (namely, that a linear map between spaces of equal dimension is invertible if, and only if, it is of maximal rank). More precisely, the inverse function theorem states that, given a smooth map f from some open subset of $\mathbb{R}^m$ to $\mathbb{R}^m$ such that at some point p of its domain the Jacobian matrix is nonsingular (namely, of maximal rank m), then there exists a neighbourhood U of p mapped by f onto some neighbourhood V of $f(p)$ on which f has a smooth inverse $f^{-1} : V \to U$. (The case $m = 1$ helps to visualize why this is true.) The theorem fits nicely within the intuitive baggage of differential calculus: since the derivative of a function (represented here by the Jacobian matrix) at a point is supposed to be a linear approximation of the function itself in a small neighbourhood of the point, it makes sense that the invertibility of the one will have a direct repercussion on the invertibility of the other, at least in some small neighbourhood.

4.10.3. Implications for Differentiable Manifolds

What is important to us here is that the inverse function theorem, being essentially local, can be transported directly into the realm of differentiable manifolds by using charts.

We have already defined (Exercise 4.13) the rank of a map between two differentiable manifolds $g : \mathcal{M} \to \mathcal{N}$, of dimensions m and n, at a point $p \in \mathcal{M}$ as the rank of the coordinate representation of the differential $g_* : T_p\mathcal{M} \to T_{g(p)}\mathcal{N}$ at that point. A direct application of the inverse function theorem (using charts) yields the following result: if $\mathcal{M}$ and $\mathcal{N}$ have the same dimension m, and if the rank of the smooth function g at p is equal to m, then there exists a neighbourhood U of p mapped by g into a neighbourhood V of $g(p)$ such that the restriction of g to U is a diffeomorphism between U and V.

If $m < n$, and if g is of maximal rank (i.e., m) at the point $p \in \mathcal{M}$, one can prove that there exist local coordinate systems (U, X^J) and (V, y^i) at p and $g(p)$, respectively, such that the map g in these coordinates reads:

$$y^J = X^J, \quad J = 1, \ldots, m$$
$$y^j = 0, \quad j = m+1, \ldots, n. \tag{4.63}$$

Injectivity, therefore, is preserved in a neighbourhood of p.

The map $g : \mathcal{M} \to \mathcal{N}$ (with $m \leq n$) is called an *immersion* if it has maximal rank (i.e., m) at every point of $\mathcal{M}$. Thus, an immersion is a locally one-to-one map between manifolds.

An *embedding* is an immersion satisfying the following two conditions: (i) It is (globally) one-to-one; (ii) assigning to the image (as a set) the topology induced by the ambient space, the map onto the image is a homeomorphism.

EXAMPLE 4.8. **The epicycloid:** Consider the curve (called the *epicycloid*) given by the map $\gamma : \mathbb{R}^1 \to \mathbb{R}^2$ expressed by the equations:

$$x^1(t) = rt - R\sin t$$
$$x^2(t) = -R\cos t, \tag{4.64}$$

where r and R are constants, with $r < R$. The tangent map γ_* is given by:

$$x'^1(t) = r - R\cos t$$
$$x'^2(t) = R\sin t, \tag{4.65}$$

which is of rank 1 for all t, since these expressions never vanish simultaneously. So γ is an immersion. Nevertheless, it fails to be an embedding, since the epicycloid crosses itself for t satisfying $\sin t = (r/R)t$ (and every 2π thereafter).

The image $g(\mathcal{M})$ of an immersion $g : \mathcal{M} \to \mathcal{N}$ is a subset of $\mathcal{N}$ called an *immersed submanifold*. Similarly, if g is an embedding the image is called an *(embedded) submanifold*. There are, as we have already pointed out, two generally different topologies at play here. These submanifolds have a topology induced by the charts (for example, in the case of the epicycloid we have a global chart, since each point of the submanifold can be assigned a distinct value of t, and the self-crossings are actually seen as two different points), but they also have a topology induced by the ambient space, by intersection (and we see in the example of the epicycloid that this is going to cause problems at the self-crossings). What all this means is that, in the coarser topology induced by the ambient space $\mathcal{N}$, the image of an immersion may not be a manifold at all. In the case of an embedding, this problem does not occur only because we have explicitly prevented it from happening by definition. If we had just defined an embedding as a globally one-to-one immersion, it would have been easy to show examples of failure of the image to be a manifold in the induced topology.[7]

REMARK 4.2. **Characterization of submanifolds**: Embedded submanifolds can be completely characterized by the following property: A subset $\mathcal{S}$ of a manifold $\mathcal{N}$ of dimension n is an embedded submanifold of dimension k if, and only if, for each point $s \in \mathcal{S}$ one can find a chart on an open set U of $\mathcal{N}$ with coordinates x^i $(i = 1, \ldots, n)$ such that: (i) $s \in U$; (ii) $\mathcal{S} \cap U = \{(x^1, \ldots, x^k, x^{k+1}, \ldots, x^n) \in U \mid x^{k+1} = \ldots = x^n = 0\}$. Thus, locally, an embedded submanifold looks just like a k-dimensional hyperplane of $\mathbb{R}^n$.

Let $\mathcal{U} \subset \mathcal{M}$ be an open subset of the m-dimensional manifold $\mathcal{M}$. Restricting the domain of the charts of $\mathcal{M}$ to $\mathcal{U}$, it is not difficult to prove that $\mathcal{U}$ acquires an induced m-dimensional differentiable structure and the inclusion is a smooth embedding. With this structure, $\mathcal{U}$ is called an *open submanifold* of $\mathcal{M}$.

EXAMPLE 4.9. **Elastic surfaces and lines:** In the modelling of certain thin-walled structures with small bending stiffness (membranes), one can define the body as a 2-dimensional manifold. Triviality (and even orientability) are now back on the table, since we can have such structures as a balloon or a Moebius band to deal with. A configuration of such a structure, at any rate, will have to be defined at least as an embedding, since self-crossings are physically inadmissible. Injectivity can be interpreted as impenetrability of matter (i.e., two material points should not occupy the same place at the same instant of time). If we go down to one dimension, thereby dealing with a thin wire, we see that even when it appears to have the shape of an epicycloid, the self-crossings will not be real since, in fact, the two pieces of the wire involved in the crossing will be lying on top of each other, rather than interpenetrating. These problems of soft contact (such as when deforming a rubber band by rolling a portion of it with a finger against a table) are very challenging. They appear to be of relevance in the packing of long molecules, such as DNA.[8]

EXAMPLE 4.10. **A one-to-one immersion which is not an embedding:** Consider the map from $\mathbb{R}$ to $\mathbb{R}^2$ given by:

$$\begin{aligned} x^1(t) &= 1/t \\ x^2(t) &= \sin(\pi t) \end{aligned} \tag{4.66}$$

The graph of this map oscillates more and more rapidly as we approach the x^2-axis. We now restrict the map to the interval $[1, +\infty)$, and replace the discarded part

---

[7] Some authors, in fact, distinguish between an embedding (which is just a one-to-one immersion) and a *regular* embedding (which, in addition, satisfies the homeomorphism condition).

[8] In this respect, see Coleman BD, Swigon D (2000), Theory of Supercoiled Elastic Rings with Self-Contact and Its Application to DNA Plasmids, *Journal of Elasticity* **60**, 173–221.

by two pieces: (i) the interval $(-\infty, -1]$ is mapped by $x^1(t) = 0, x^2(t) = t + 2$ into a portion of the x^2-axis; (ii) the interval $(-1, 1)$ is mapped so such as to produce a graph that smoothly joins the tail of the first curve (at $x^1 = 1, x^2 = 0$) from above with the chosen piece of the x^2-axis at the point with coordinates $x^1 = 0, x^2 = 1$. The resulting map is one-to-one, but fails to be an embedding because the image of, say, the open interval $(-3, -1)$ is not open in the induced topology. (Recall that in the induced topology, the open sets are obtained as intersections of open sets of $\mathbb{R}^2$ with the curve. Thus, the piece of the x^2-axis given by $x^1 = 0, -1 < x^2 < 1$ is not open in the induced topology, since an open set in $\mathbb{R}^2$ that contains it will also necessarily contain (an infinite number of) pieces of the rapidly oscillating part.)

EXERCISE 4.16. Produce a graph of the curve described in Example 4.10 and follow the reasoning therein to convince yourself that indeed the map is not an embedding.

A *submersion* is a maximal-rank smooth map $g : \mathcal{M} \to \mathcal{N}$, with $m > n$. It follows that the rank of a submersion equals the dimension (n) of the target manifold ($\mathcal{N}$), and it is never one-to-one. The projection map $\tau : T\mathcal{M} \to \mathcal{M}$ is a surjective (onto) submersion.

4.11. The Cotangent Bundle

Starting from an m-dimensional differentiable manifold $\mathcal{M}$, we have constructed at each point the tangent space $T_p\mathcal{M}$ in an intrinsic manner. Each tangent space is an m-dimensional vector space, its elements representing equivalence classes of differentiable curves. Denoting by $T_p^*\mathcal{M}$ the dual space of $T_p\mathcal{M}$ we obtain, without further ado, another intrinsic object associated with each point of the manifold. This space is also an m-dimensional vector space, and its elements $\omega_p \in T_p^*\mathcal{M}$, called *covectors*, are linear operators on the vectors of $T_p\mathcal{M}$. It seems natural now to follow the same line of thought that led us to the tangent bundle so as to reach the dual notion of *cotangent bundle*. To this effect, we form the collection $T^*\mathcal{M}$ of all pairs (p, ω_p), where $p \in \mathcal{M}$ and $\omega_p \in T_p^*\mathcal{M}$. We define the projection map: $\pi : T^*\mathcal{M} \to \mathcal{M}$ by $\pi(p, \omega_p) = p$. We want to show that $T^*\mathcal{M}$ is endowed with a natural structure of a differentiable manifold of dimension $2m$.

Let $f : \mathcal{U} \to \mathbb{R}$ be a smooth function defined in a neighbourhood $\mathcal{U} \subset \mathcal{M}$ of the point p, and let $df_p : T\mathcal{U} \to \mathbb{R}$ denote its differential at p. We can regard this differential as an element of $T_p^*\mathcal{M}$ by defining its value $\langle df_p, \mathbf{v}_p \rangle$ on any vector $\mathbf{v}_p \in T_p\mathcal{M}$ as:

$$\langle df_p, \mathbf{v}_p \rangle := df_p(\mathbf{v}_p) = \mathbf{v}_p(f). \tag{4.67}$$

In other words, the action of the evaluation of the differential of the function on a tangent vector is equal to the directional derivative of the function in the direction of the vector (see also Exercise 4.12).

EXERCISE 4.17. **Linearity:** Check that the operation defined in (4.67) is linear.

Consider now a coordinate system $x^1, \ldots, x^m$ defined in $\mathcal{U}$, and denote (abusing the notation) by x^i the real-valued function $x^i : \mathcal{U} \to \mathbb{R}$ which assigns to each point $q \in \mathcal{U}$ the value $x^i(q)$ of its i-th coordinate. The differential dx^i_p of this function at p is, as we have just seen, a covector. Using Equation (4.67), we obtain:

$$\langle dx^i_p, (\partial/\partial x^j)_p \rangle = (\partial/\partial x^j)_p(x^i) = \delta^i_j, \tag{4.68}$$

which shows that the covectors $\{dx^i\}$ are precisely the dual basis of $\{\partial/\partial x^i\}$ at each point of the chart! The covector ω_p can, accordingly, be expressed uniquely as $\omega_p = \omega_i dx^i$. We can, therefore, assign to each point $(p, \omega_p) \in \pi^{-1}(\mathcal{U}) \subset T^*\mathcal{M}$ the $2m$ numbers $(x^1, \ldots, x^m, \omega_1, \ldots, \omega_m)$. Upon a change of chart, $y^i = y^i(x^1, \ldots, x^n)$, whereby $\omega_p = \hat{\omega}_i \, dy^i$, we obtain the covariant rule:

$$\hat{\omega}_i = \frac{\partial x^j}{\partial y^i} \, \omega_j. \tag{4.69}$$

We have explicitly constructed charts of $T^*\mathcal{M}$, thus showing that $T^*\mathcal{M}$ is indeed a differentiable manifold of dimension $2m$.

A smooth assignment of a covector ω_p to each point $p \in \mathcal{M}$ is called a *differential 1-form on the manifold*. It can be regarded as a cross section of the cotangent bundle, namely, a map:

$$\Omega : \mathcal{M} \longrightarrow T^*\mathcal{M}, \tag{4.70}$$

such that $\pi \circ \Omega = id_{\mathcal{M}}$.

As we have seen, the differential of a function at a point defines a covector. It follows that a smooth scalar function $f : \mathcal{M} \to \mathbb{R}$ determines, by point-wise differentiation, a differential 1-form $\Omega = df$. It is important to remark that *not all differential 1-forms can be obtained as differentials of functions*. The ones that can are called *exact*.

A differential 1-form Ω (that is, a cross section of $T^*\mathcal{M}$) can be regarded as acting on vector fields $\mathbf{V}$ (cross sections of $T\mathcal{M}$) to deliver functions $\langle \Omega, \mathbf{V} \rangle : \mathcal{M} \to \mathbb{R}$, by point-wise evaluation of a covector on a vector.

4.12. Tensor Bundles

In Chapter 3 we introduced the algebra $\mathcal{T}(V)$ of tensors of all types over the vector space V. Moreover, we defined some important subalgebras of $\mathcal{T}(V)$. In particular, we denoted by $\mathcal{C}(V)$ and $\mathcal{C}(V^*)$ the contravariant and covariant subalgebras, respectively. Recall the notation $\mathcal{C}^k(V)$ for the vector space of all tensors of type $(k, 0)$, that is, the contravariant tensors of degree k. In a similar spirit, one can denote by $\mathcal{C}_k(V)$ the space of all tensors of type $(0, k)$, namely, the covariant tensors of degree k. Note that $\mathcal{C}_k(V) = \mathcal{C}^k(V^*)$. Of particular interest were also the spaces $\Lambda^k(V)$ and $\Lambda_k(V)$ of k-vectors and k-forms on V, respectively, which we regarded as the spaces of completely skew-symmetric tensors of the corresponding type. Given a point p of a manifold $\mathcal{M}$, we may identify the vector space V with the tangent space $T_p\mathcal{M}$ and construct the corresponding spaces of tensors of any fixed type. Following the same procedure as for the tangent and cotangent bundles, which will thus become particular cases, it is clear that one can define *tensor bundles* of any type by adjoining to each point of a manifold the tensor space of the corresponding type. A convenient notational scheme is: $C^k(\mathcal{M}), C_k(\mathcal{M})$, respectively, for the bundles of contravariant and covariant tensors of order k. Similarly, the bundles of k-vectors and of k-forms can be denoted, respectively, by $\Lambda^k(\mathcal{M}), \Lambda_k(\mathcal{M})$. Each of these bundles can be shown (by a procedure identical to that used in the case of the tangent and cotangent bundles) to have a natural structure of a differentiable manifold of the appropriate dimension. A (smooth) section of a tensor bundle is called a *tensor field over* $\mathcal{M}$, of the corresponding type. A (smooth) section of the bundle $\Lambda_k(\mathcal{M})$ of p-forms is also called a *differential p-form*. A scalar function on a manifold is also called a *differential 0-form*.

In a chart of the m-dimensional manifold $\mathcal{M}$ with coordinates x^i, a contravariant tensor field **T** of order r is given as:

$$\mathbf{T} = T^{i_1,\dots,i_r} \frac{\partial}{\partial x^{i_1}} \otimes \dots \otimes \frac{\partial}{\partial x^{i_r}}, \tag{4.71}$$

where $T^{i_1,\dots,i_r} = T^{i_1,\dots,i_r}(x^1,\dots,x^m)$ are r^m smooth functions of the coordinates. Similarly, a covariant tensor field **U** of order r is given by:

$$\mathbf{U} = U_{i_1,\dots,i_r} \, dx^{i_1} \otimes \dots \otimes dx^{i_r}, \tag{4.72}$$

and a differential r-form ω by:

$$\omega = \omega_{i_1,\dots,i_r} \, dx^{i_1} \wedge \dots \wedge dx^{i_r}. \tag{4.73}$$

Notice that, in principle, the indexed quantity $\omega_{i_1,\dots,i_r}$ need not be specified as skew-symmetric with respect to the exchange of any pair of indices, since the exterior product of the base-forms will do the appropriate skew-symmetrization job. Recall that, as an alternative, we may suspend the standard summation convention in (4.73) and consider only indices in ascending order. As a result, if $\omega_{i_1,\dots,i_r}$ *is* skew-symmetric *ab initio*, the corresponding components are to be multiplied by $r!$.

Of particular interest for the theory of integration on manifolds are differential m-forms, where m is the dimension of the manifold. From our treatment of the algebra of r-forms in Chapter 3, we know that the dimension of the space of m-covectors is exactly 1. In a coordinate chart, a basis for differential m-forms is, therefore, given by $dx^1 \wedge \dots \wedge dx^m$. In other words, the representation of a differential m-form ω in a chart is :

$$\omega = f(x^1,\dots,x^m) \, dx^1 \wedge \dots \wedge dx^m, \tag{4.74}$$

where $f(x^1,\dots,x^m)$ is a smooth scalar function of the coordinates in the patch. Consider now another coordinate patch with coordinates $y^1,\dots,y^m$, whose domain has a nonempty intersection with the domain of the previous chart. In this chart we have:

$$\omega = \hat{f}(y^1,\dots,y^m) \, dy^1 \wedge \dots \wedge dy^m. \tag{4.75}$$

We want to find the relation between the functions f and $\hat{f}$. Since the transition functions $y^i = y^i(x^1, \ldots, x^m)$ are smooth, we can write:

$$\omega = \hat{f}(y^1, \ldots, y^m)\, dy^1 \wedge \cdots \wedge dy^m = \hat{f}\, \frac{\partial y^1}{\partial x_{j_1}} \cdots \frac{\partial y^m}{\partial x_{j_m}}\, dx^{j_1} \wedge \cdots \wedge dx^{j_m} \qquad (4.76)$$

or, by definition of determinant:

$$\omega = \det\left\{\frac{\partial y^1, \ldots, y^m}{\partial x^1, \ldots, x^m}\right\} \hat{f}\, dx^1 \wedge \cdots \wedge dx^m = J_{y,x}\, \hat{f}\, dx^1 \wedge \cdots \wedge dx^m, \qquad (4.77)$$

where the Jacobian determinant $J_{y,x}$ does not vanish at any point of the intersection of the two coordinate patches. Comparing with Equation (4.74), we conclude that:

$$f = J_{y,x}\hat{f}. \qquad (4.78)$$

A nowhere vanishing differentiable m-form on a manifold $\mathcal{M}$ of dimension m is called a *volume form* on $\mathcal{M}$. It can be shown that a manifold is orientable (see Section 4.4) if, and only if, it admits a volume form.

4.13. Pull-backs

Let $g : \mathcal{M} \to \mathcal{N}$ be a smooth map between the manifolds $\mathcal{M}$ and $\mathcal{N}$, of dimensions m and n, respectively. Certain geometrical objects have the following property: If they are defined over the target manifold $\mathcal{N}$, the map g allows to define a *pulled-back* version on the source manifold $\mathcal{M}$. To this category of objects belong functions, differential 1-forms and differential forms of all higher orders and, more generally, covariant tensors of any order. It will be sufficient to show how functions and differential 1-forms are pulled back from $\mathcal{N}$ to $\mathcal{M}$.

Let $f : \mathcal{N} \to \mathbb{R}$ be a smooth function. We define its pull-back by g as the map $g^*f : \mathcal{M} \to \mathbb{R}$ given by the composition:

$$g^*f = f \circ g. \qquad (4.79)$$

For a differential 1-form Ω on $\mathcal{N}$, we define the pull-back $g^*\Omega : \mathcal{M} \to T^*\mathcal{M}$ by showing how it acts, point by point, on tangent vectors:

$$\langle [g^*\Omega](p), \mathbf{v}_p \rangle = \langle \Omega(g(p)), (g_*)_p \mathbf{v}_p \rangle, \qquad (4.80)$$

which can be more neatly written in terms of vector fields as:

$$\langle g^*\Omega, \mathbf{V} \rangle = \langle \Omega \circ g, g_* \mathbf{V} \rangle. \qquad (4.81)$$

Expressed in words, this means that the pull-back by g of a 1-form in $\mathcal{N}$ is the 1-form in $\mathcal{M}$ that assigns to each vector the value that the original 1-form assigns to the image of that vector by g_*.

It is important to notice that the pull-backs of functions and differential forms are always well-defined, regardless of the dimensions of the spaces involved. This should be contrasted with the push-forward of vector fields, which fail in general to be vector fields on the target manifold.

For a contravariant tensor field $\mathbf{U}$ of order r on $\mathcal{N}$ (and, in particular, for differential r-forms on $\mathcal{N}$), the pull-back by a smooth function $g : \mathcal{M} \to \mathcal{N}$ is a corresponding field on $\mathcal{M}$ obtained by an extension of the case $r = 1$, as follows:

$$g^*\mathbf{U}\,(\mathbf{V}_1, \ldots, \mathbf{V}_r) = (\mathbf{U} \circ g)\,(g_*\mathbf{V}_1, \ldots, g_*\mathbf{V}_r), \tag{4.82}$$

where $\mathbf{U}$ is regarded as a multilinear function of r vector fields $\mathbf{V}_i$.

We note that, regarded as an operator on the space of contravariant tensor fields of all orders, the pull-back is a linear operator which commutes with contractions and with the tensor product.

EXERCISE 4.18. Prove the above properties.

EXERCISE 4.19. Prove that $(h \circ g)^* = g^* \circ h^*$, where $g : \mathcal{M} \to \mathcal{N}$ and $h : \mathcal{N} \to \mathcal{P}$ are smooth maps between manifolds.

Let x^α ($\alpha = 1, \ldots, m$) and y^i ($i = 1, \ldots, n$) be coordinate systems in $\mathcal{M}$ and $\mathcal{N}$, respectively. The map $g : \mathcal{M} \to \mathcal{N}$ is then represented by n smooth functions $y^i = y^i(x^1, \ldots, x^m)$, which we abbreviate as $y = y(x)$. If a covariant tensor $\mathbf{T} \in \mathcal{C}_r(\mathcal{N})$ is given in components by:

$$\mathbf{T} = T_{i_1 \ldots i_r}(y)\ dy^{i_1} \otimes \cdots \otimes dy^{i_r}, \tag{4.83}$$

the pulled-back tensor $g^*(\mathbf{T})$ is given in components by the expression:

$$g^*\mathbf{T} = T_{i_1 \ldots i_r}(y(x))\ \frac{\partial y^{i_1}}{\partial x^{\alpha_1}} \cdots \frac{\partial y^{i_r}}{\partial x^{\alpha_r}}\ dx^{\alpha_1} \otimes \cdots \otimes dx^{\alpha_r}. \tag{4.84}$$

Notice that the summation convention with respect to both types of indices is, in effect, each index ranging over the appropriate domain (i.e., $1, \ldots, m$ for Greek indices and $1, \ldots, n$ for Latin indices).

EXAMPLE 4.11. **The right Cauchy-Green tensor:** In classical Continuum Mechanics, physical space is equated to $\mathbb{E}^3$ with the usual Cartesian inner product. In any given (curvilinear, say) coordinate system x^i ($i = 1, 2, 3$), we denote the components of the spatial metric tensor $\mathbf{g}$ by g_{ij}. Given a deformation $\kappa : \mathcal{B} \to \mathbb{E}^3$ from the body manifold to space, and given a coordinate system (or reference configuration) X^I ($I = 1, 2, 3$) in the body, the configuration is expressed in coordinates as a deformation, namely, as three smooth functions $x^i = x^i(X^1, X^2, X^3)$. The pull-back of the spatial metric tensor to the body is, according to Equation (4.84), the tensor:

$$\mathbf{C} = \kappa^*\mathbf{g} = g_{ij}\,\frac{\partial x^i}{\partial X^I}\,\frac{\partial x^j}{\partial X^J}\ dX^I \otimes dX^J. \tag{4.85}$$

In other words (see Equation (A.9)), the right Cauchy-Green tensor is precisely the pull-back of the spatial metric to the body manifold. Notice that Equation (4.85) remains valid whether or not the physical space is Euclidean, as long as it has a metric structure (i.e., as long as it is a Riemannian manifold, as defined below in Section 4.16).

EXAMPLE 4.12. **Densities and fluxes in Continuum Mechanics:** Many important objects in Continuum Mechanics are made to be integrated, since only then do they acquire a clear physical meaning. Mass density, internal energy density, body forces, surface tractions, and heat influx are just a few important examples. Objects that can be integrated are, by definition, differential forms. The relations between Eulerian and Lagrangian densities and fluxes are dictated by the laws of pull-back of differential 3- and 2-forms, respectively, between 3-dimensional manifolds.

EXERCISE 4.20. **Pull-back of 2-forms in** $\mathbb{R}^3$: Recall, from Exercise 3.19, that in $\mathbb{R}^3$ the 2-form $\omega = A\mathbf{e}^2 \wedge \mathbf{e}^3 - B\mathbf{e}^1 \wedge \mathbf{e}^3 + C\mathbf{e}^1 \wedge \mathbf{e}^2$ is expressible as the vector $\mathbf{w} = A\mathbf{e}_1 + B\mathbf{e}_2 + C\mathbf{e}_3$. Let ω be a differential 2-form defined over $\mathbb{R}^3$ and let $\kappa : U \to \mathbb{R}^3$ be an embedding of the open set $U \subset \mathbb{R}^3$. Show that the pull-back of ω by κ is expressible as the vector field:

$$\mathbf{W} = J\,\mathbf{F}^{-1}\,\mathbf{w}, \tag{4.86}$$

over U, where at each point of U we denote $\mathbf{F} = \kappa_*$ and $J = \det(\mathbf{F})$.

4.14. Exterior Differentiation of Differential Forms

The exterior derivative of differential forms is an operation that generalizes the gradient, curl, and divergence operators of classical vector calculus. It is not surprising, therefore, that the true meaning of this operation manifests itself in the context of integral calculus on manifolds, a topic that we will cover in a later chapter.

The exterior derivative of a differential r-form is a differential $(r+1)$-form defined over the same manifold. Instead of introducing, as one certainly could, the definition of exterior differentiation in an intrinsic axiomatic manner, we will proceed to define it in a coordinate system and show that the definition is, in fact, coordinate independent. Let, x^i $(i = 1, \ldots, m)$ be a coordinate chart and let ω be an r-form given as:

$$\omega = \omega_{i_1,\ldots,i_r}\,dx^{i_1} \wedge \cdots \wedge dx^{i_r}, \tag{4.87}$$

where $\omega_{i_1,\ldots,i_r} = \omega_{i_1,\ldots,i_r}(x^1,\ldots,x^m)$ are smooth functions of the coordinates. We define the exterior derivative of ω, denoted by $d\omega$, as the differential $(r+1)$-form obtained as:

$$d\omega = d\omega_{i_1,\ldots,i_r} \wedge dx^{i_1} \wedge \cdots \wedge dx^{i_r}, \tag{4.88}$$

where the d on the right-hand side denotes the ordinary differential of functions. More explicitly:

$$d\omega = \frac{\partial \omega_{i_1,\ldots,i_r}}{\partial x^k}\,dx^k \wedge dx^{i_1} \wedge \cdots \wedge dx^{i_r}. \tag{4.89}$$

Note that for each specific combination of (distinct) indices $i_1,\ldots,i_r$, the index k ranges only on the remaining possibilities, since the exterior product is skew-symmetric. Thus, in particular, if ω is a differential m-form defined over an m-dimensional manifold, its exterior derivative vanishes identically (as it should, being an $(m+1)$-form).

Let y^i $(i = 1, \ldots, m)$ be another coordinate chart with a nonempty intersection with the previous chart. We have:

$$\omega = \hat{\omega}_{i_1, \ldots, i_r} \, dy^{i_1} \wedge \cdots \wedge dy^{i_r}, \tag{4.90}$$

for some smooth functions $\hat{\omega}_{i_1, \ldots, i_r}$ of the y^i-coordinates. The two sets of components are related by:

$$\omega_{i_1, \ldots, i_r} = \hat{\omega}_{j_1, \ldots, j_r} \frac{\partial y^{j_1}}{\partial x^{i_1}} \cdots \frac{\partial y^{j_r}}{\partial x^{i_r}}. \tag{4.91}$$

Notice that we have not troubled to collect terms by, for example, prescribing a strictly increasing order. The summation convention is in effect. We now apply the prescription (4.88) and obtain:

$$d\omega = d \left(\hat{\omega}_{j_1, \ldots, j_r} \frac{\partial y^{j_1}}{\partial x^{i_1}} \cdots \frac{\partial y^{j_r}}{\partial x^{i_r}} \right) \wedge dx^{i_1} \wedge \cdots \wedge dx^{i_r}. \tag{4.92}$$

The crucial point now is that the terms containing the second derivatives of the coordinate transformation will evaporate due to their intrinsic symmetry, since they are contracted with an intrinsically skew-symmetric wedge product of two 1-forms. We have, therefore:

$$d\omega = \frac{\partial \hat{\omega}_{j_1, \ldots, j_r}}{\partial y^m} \frac{\partial y^m}{\partial x^k} \frac{\partial y^{j_1}}{\partial x^{i_1}} \cdots \frac{\partial y^{j_r}}{\partial x^{i_r}} \, dx^k \wedge dx^{i_1} \wedge \cdots \wedge dx^{i_r}, \tag{4.93}$$

or, finally:

$$d\omega = \frac{\partial \hat{\omega}_{j_1, \ldots, j_r}}{\partial y^m} \, dy^m \wedge dy^{j_1} \wedge \cdots \wedge dy^{j_r}, \tag{4.94}$$

which is exactly the same prescription in the coordinate system y^i as Equation (4.88) is in the coordinate system x^i. This completes the proof of independence from the coordinate system.

From this definition, we can deduce a number of important properties of the exterior derivative:

(1) Linearity: d is a linear operator, viz.:

$$d(a\,\alpha + b\,\beta) = a\,d\alpha + b\,d\beta \quad \forall a, b \in \mathbb{R} \quad \alpha, \beta \in \Lambda_r(\mathcal{M}). \tag{4.95}$$

(2) Quasi-Leibniz rule:

$$d(\alpha \wedge \beta) = d\alpha \wedge \beta + (-1)^r \alpha \wedge d\beta \quad \forall \alpha \in \Lambda_r(\mathcal{M}), \ \beta \in \Lambda_s(\mathcal{M}). \tag{4.96}$$

(3) Nilpotence:

$$d^2(.) := d(d(.)) = 0. \tag{4.97}$$

It is possible (and very elegant) to *define* the exterior derivative as a map $d : \Lambda_r \to \Lambda_{r+1}$ satisfying properties ((1)), ((2)) and ((3)) and reducing to the ordinary differential of a function for $r = 0$. It can be shown that our definition in terms of components is the only nontrivial operation satisfying all these four conditions.

EXERCISE 4.21. Prove the above three properties.

EXAMPLE 4.13. **Classical vector calculus:** Consider the case in which the manifold of interest is $\mathbb{R}^3$, with all its attendant structure. We want to calculate the exterior derivatives of forms of all possible orders (i.e., $r = 0, 1, 2, 3$) and, using the star operator defined in Section 3.8, interpret the results in terms of vectors and tensors in $\mathbb{R}^3$. A 0-form is a smooth function $f = f(x^1, x^2, x^3)$ of the coordinates. Its exterior derivative is the 1-form:

$$df = \frac{df}{dx^i} dx^i. \tag{4.98}$$

This *gradient* of a scalar function, which is not a vector but a 1-form, can be calculated in this way on any manifold and in any coordinate system therein. It is not difficult to verify that the annihilator of the gradient at a point p, that is, the collection of vectors on which the action of the gradient vanishes, is precisely the tangent space to the submanifold $f = $ constant passing through that point. In the particular case of $\mathbb{R}^3$, however, the presence of an inner product permits us to interpret it as a vector, namely, the *gradient vector*, which is perpendicular to the surface $f = $ constant through p.

Consider next a 1-form $\alpha = \alpha_1 dx^1 + \alpha_2 dx^2 + \alpha_3 dx^3$, with components α_i that are smooth functions of the coordinates. Its exterior derivative is the 2-form:

$$d\alpha = \frac{d\alpha_1}{dx^2} dx^2 \wedge dx^1 + \frac{d\alpha_1}{dx^3} dx^3 \wedge dx^1 + \frac{d\alpha_2}{dx^1} dx^1 \wedge dx^2 + \frac{d\alpha_2}{dx^3} dx^3 \wedge dx^2$$
$$+ \frac{d\alpha_3}{dx^1} dx^1 \wedge dx^3 + \frac{d\alpha_3}{dx^2} dx^2 \wedge dx^3, \tag{4.99}$$

or, collecting terms:

$$d\alpha = \left(\frac{\partial \alpha_2}{\partial x^1} - \frac{\partial \alpha_1}{\partial x^2} \right) dx^1 \wedge dx^2 + \left(\frac{\partial \alpha_3}{\partial x^1} - \frac{\partial \alpha_1}{\partial x^3} \right) dx^1 \wedge dx^3 + \left(\frac{\partial \alpha_3}{\partial x^2} - \frac{\partial \alpha_2}{\partial x^3} \right) dx^2 \wedge dx^3. \tag{4.100}$$

Using the star operator, we should be able to interpret this 2-form as a vector (or pseudovector) field in $\mathbb{R}^3$. To find this vector, we start by noticing that, due to the inner product of $\mathbb{R}^3$, we can interpret the 2-form (4.100) directly as a 2-vector with the same components. Thereafter, we can use the standard volume form to convert the 2-vector into a 1-form which, again, can be directly interpreted as a vector. The result is:

$$\nabla \times \alpha = \left(\frac{\partial \alpha_3}{\partial x^2} - \frac{\partial \alpha_2}{\partial x^3} \right) \frac{\partial}{\partial x^1} - \left(\frac{\partial \alpha_3}{\partial x^1} - \frac{\partial \alpha_1}{\partial x^3} \right) \frac{\partial}{\partial x^2} + \left(\frac{\partial \alpha_2}{\partial x^1} - \frac{\partial \alpha_1}{\partial x^2} \right) \frac{\partial}{\partial x^3}, \quad (4.101)$$

which can be recognized as the *curl* of the "vector" field α. Notice, incidentally, that the nilpotence of the exterior derivative implies that the curl of the gradient of a scalar field vanishes identically.

Finally, consider a 2-form $\beta = A\,dx^2 \wedge dx^3 - B\,dx^1 \wedge dx^3 + C\,dx^1 \wedge dx^2$, where A, B, C are smooth functions of the coordinates. Its exterior derivative is the 3-form:

$$d\beta = \left(\frac{\partial A}{\partial x^1} + \frac{\partial B}{\partial x^2} + \frac{\partial C}{\partial x^3} \right) dx^1 \wedge dx^2 \wedge dx^3, \quad (4.102)$$

a general result. But, in the case of $\mathbb{R}^3$, the 2-form β corresponds to the vector with components $\{A, B, C\}$. We thus interpret the result (4.102) as the *divergence* of this vector field. Again, the nilpotence of the operator d implies that the divergence of a curl vanishes identically.

4.15. Some Properties of the Exterior Derivative

PROPOSITION 4.2. *The exterior derivative commutes with pull-backs.*

PROOF. Since any form can be expressed as a sum of exterior products of functions and differentials of functions, and since the pull-back commutes with the exterior product (see Exercise 4.18), we need to prove the proposition only for 0-forms and for exact 1-forms, namely, for 1-forms which are differentials of functions. Let $g : \mathcal{M} \to \mathcal{N}$ be a smooth function between manifolds, and let $f : \mathcal{N} \to \mathbb{R}$ be a function (i.e., a 0-form) defined on $\mathcal{N}$. For any vector field $\mathbf{v}$ on $\mathcal{M}$, we have:

$$[g^*(df)](\mathbf{v}) = df(g_*\mathbf{v}) = (g_*\mathbf{v})f = \mathbf{v}(g^*f) = [d(g^*f)]\mathbf{v}, \quad (4.103)$$

which proves the proposition for the case of 0-forms. For the case of an exact 1-form $\alpha = df$, since $d\alpha = 0$, we have $g^*(d\alpha) = 0$. On the other hand, by Equation (4.103), we know that $g^*\alpha = g^*df = d(g^*f)$ and, therefore, $d(g^*\alpha) = d(d(g^*f)) = 0$, which completes the proof. □

The exterior derivative of a 1-form has an interesting interaction with the Lie bracket:

PROPOSITION 4.3. *If α is a differential 1-form and $\mathbf{u}$ and $\mathbf{v}$ are smooth vector fields on a manifold $\mathcal{M}$, then:*

$$\langle d\alpha \mid \mathbf{u} \wedge \mathbf{v} \rangle = \mathbf{u}(\langle \alpha \mid \mathbf{v} \rangle) - \mathbf{v}(\langle \alpha \mid \mathbf{u} \rangle) - \langle \alpha \mid [\mathbf{u}, \mathbf{v}] \rangle. \quad (4.104)$$

PROOF. Just as in the previous proposition, it is sufficient to consider the case in which $\alpha = h\,df$, where h and f are smooth functions defined on $\mathcal{M}$, since any 1-form can be expressed as a sum of terms of that nature. The rest is left as an exercise.[9]

$\square$

EXERCISE 4.22. Let $\mathbf{E}_i$ $(i = 1,\dots,n)$ be a moving frame on a manifold. The dual vectors $\mathbf{E}^i$ $(i = 1,\dots,n)$ are then uniquely defined and can be said to constitute a moving *coframe*. Notice that each field $\mathbf{E}^i$ is a differential 1-form. Show that:

$$d\mathbf{E}^i = -c_{ij}^k \mathbf{E}^i \wedge \mathbf{E}^j, \qquad (4.105)$$

where c_{ij}^k is the tensor of structure constants of the moving frame as defined in Equation (4.42).

4.16. Riemannian Manifolds

If each tangent space $T_x\mathcal{M}$ is endowed with an inner product, and if this inner product depends smoothly on $x \in \mathcal{M}$, we say that $\mathcal{M}$ is a *Riemannian manifold*. To clarify the concept of smoothness, let $\{\mathcal{U},\phi\}$ be a chart in $\mathcal{M}$ with coordinates $x^1,\dots,x^n$. This chart induces the (smooth) basis field $\frac{\partial}{\partial x^1},\dots,\frac{\partial}{\partial x^n}$. We define the *contravariant components of the metric tensor* $\mathbf{g}$ associated with the given inner product (indicated by $\cdot$) as:

$$g_{ij} := \left(\frac{\partial}{\partial x^i}\right) \cdot \left(\frac{\partial}{\partial x^j}\right). \qquad (4.106)$$

Smoothness means that these components are smooth functions of the coordinates within the patch. The *metric tensor* itself is given by:

$$\mathbf{g} = g_{ij}\,dx^i \otimes dx^j. \qquad (4.107)$$

We have amply discussed the properties and manipulations pertaining to inner products in Section 2.7. In particular, we saw how an inner product defines an isomorphism between a vector space and its dual. When translated to Riemannian manifolds, this result means that the tangent and cotangent bundles are naturally isomorphic (via the point-wise isomorphisms of the tangent and cotangent spaces induced by the inner product).

EXAMPLE 4.14. **Lagrangian Mechanics:** In Lagrangian Mechanics, the kinetic energy (assumed to be a positive-definite quadratic form in the generalized velocities) is used to view the configuration space $\mathcal{Q}$ as a Riemannian manifold.

The theory of Riemannian manifolds is very rich in results. Classical differential geometry was almost exclusively devoted to their study and, more particularly, to the study of 2-dimensional submanifolds (i.e., surfaces) embedded in $\mathbb{R}^3$, where the

---

[9] See Lee JM (2003), *Introduction to smooth manifolds*, Springer, p. 311. Aptly, Sternberg S (1983), *Lectures on Differential Geometry*, Chelsea, p. 103, describes this formula as "an infinitesimal analogue of Stokes' theorem."

Riemannian structure is derived from the Euclidean structure of the surrounding space. The development of modern shell theory is based on these ideas.

4.17. Manifolds with Boundary

We recall that, roughly speaking, an n-dimensional differentiable manifold $\mathcal{M}$ is an entity that looks locally like $\mathbb{R}^n$ (or an open piece thereof). We may say that $\mathcal{M}$ is *modelled after* $\mathbb{R}^n$. There exist, however, more general definitions of manifolds whereby the modelling space is not necessarily $\mathbb{R}^n$. An important example is the case in which the modelling space is the *closed upper half-space* of $\mathbb{R}^n$, defined as:

$$\mathbb{H}^n = \{(x^1,\ldots,x^n) \in \mathbb{R}^n \mid x^n \geq 0\}, \tag{4.108}$$

with the induced (or subset) topology. In other words, a subset of $\mathbb{H}^n$ is open if, and only if, it is the intersection of an open subset of $\mathbb{R}^n$ with $\mathbb{H}^n$. In particular, the topology of $\mathbb{H}^n$ thus defined can be generated by (i) all the open balls of $\mathbb{R}^n$ which happen to fall entirely within $\mathbb{H}^n$, and (ii) the intersections with $\mathbb{H}^n$ of all the open balls in $\mathbb{R}^n$ with centre in the hyperplane $x^n = 0$. Notice that each of these half-balls includes an (open) $(n-1)$-dimensional ball as its base. Thus, for example, for $n = 3$, we have a hemisphere including the circular base (but not its boundary).

We proceed now to define a *topological manifold with boundary* following the same idea as in Definition 4.3, but replacing $\mathbb{R}^n$ with $\mathbb{H}^n$:

DEFINITION 4.5. An *n-dimensional topological manifold with boundary* is a Hausdorff second-countable topological space $\mathcal{T}$ such that each of its points has a neighbourhood homeomorphic to an open set of $\mathbb{H}^n$.

A point $p \in \mathcal{T}$ is called a *boundary point* if its image under one (and, therefore, every) chart lies in the hyperplane $x^n = 0$. The collection of boundary points determines the *manifold boundary* of $\mathcal{T}$. Notice that every topological manifold is automatically a topological manifold with boundary (happening to have an empty manifold boundary).

The definition of a *differentiable manifold with boundary* also follows the same lines as in the case of ordinary differentiable manifolds (without boundary).[10]

EXAMPLE 4.15. **Bodies with boundary:** In the standard treatment of Continuum Mechanics, bodies are considered as (ordinary) 3-dimensional manifolds, such as the interior of a sphere. Implicitly, therefore, the boundary of the body is, disappointingly, excluded ab initio from the physical picture. The consideration of a material body as a manifold with boundary, such as a closed ball in $\mathbb{R}^3$, in addition to its physical appeal, also has some technical advantages. In particular, the collection of C^k-maps of a closed ball into $\mathbb{R}^3$ can be shown to have the structure of an infinite-dimensional (Banach) manifold[11] $\mathcal{Q}^k$. In particular, the collections of C^k injections

[10] For a detailed treatment of this point, see Lee JM (2003), op. cit.

[11] See Binz E, Śniatycki J, Fischer H (1988), *Geometry of Classical Fields*, North Holland. A classical treatise of Banach manifolds is Lang S (1972), *Differential Manifolds*, Addison-Wesley.

and embeddings of a closed ball are open submanifolds of $\mathcal{Q}^k$. On the other hand, the case of the corresponding maps of an *open* ball does not lead to such Banach manifolds. What this means, from the point of view of Continuum Mechanics, is that the geometrization of the discipline following the lines of Classical Analytical Mechanics, as described in Chapter 1, necessitates the inclusion of the boundary of the body for the configuration space $\mathcal{Q}$ to be a manifold. Thus, the mathematics is automatically consistent with the physics of the situation.

4.18. Differential Spaces and Generalized Bodies

In the most general physical context, a context that should encompass the ability of the theory to model the mechanical response of an arbitrary collection of particles, there should not be any particular reason to suppose that the only geometrical entity that one may use as a representation of the underlying material body is necessarily a differentiable manifold, with or without boundary. Indeed, there are many applications that suggest that the use of fractals for certain structures arising naturally in various fields (Geology, Crystal Growth, Biology, and so on) may be advantageous to capture the essential features of the phenomena of interest. One avenue of approach towards the generalization of the notion of material body consists of the consideration of the most general geometrical entity amenable to support the concept of continuous fluxes, bounded by both the area and the volume of the domain of interest. This avenue, leading naturally to the use of Whitney's Geometric Integration Theory, will be explored later in Chapter 6. A different avenue, whereby every subset of $\mathbb{R}^n$ can in principle be considered as a collection of material particles amenable to mechanical treatment, is provided by Sikorski's theory of differential spaces,[12] to be briefly reviewed in this section.[13] For the particular case of fractal structures, there exist various other methods of treatment[14] that take advantage of the peculiar nature of fractals (such as self-similarity).

Fractals are geometrical, not physical, objects. But so are differentiable manifolds. When using such geometric concepts to model physical entities, care should be exercised in terms of the physical interpretation. Thus, for instance, a "real" piece of matter does not strictly have a volume or an area. We talk, nevertheless, of forces per unit volume, pressure, mass density, and so on, and we have become so used to these abuses that we no longer question them. Analogously, if we wish to model a body

---

[12] See Sikorski R (1967), Abstract Covariant Derivative, *Colloquium Mathematicum*, **18**, 251–272. See also, Sikorski R (1971), Differential Modules, *Colloquium Mathematicum*, **24**, 45–79.

[13] The presentation is based on Epstein M, Śniatycki J (2006), Fractal Mechanics, *Physica D* **220**, 54–68.

[14] See, for example, Capitanelli R, Lancia MR (2002), Nonlinear Energy Forms and Lipschitz Spaces on the Koch Curve, *Journal of Convex Analysis*, **9**, 245–257; Carpinteri A, Chiaia B, Cornetti P (2001), Static-Kinematic Duality and the Principle of Virtual Work in the Mechanics of Fractal Media, *Computer Methods in Applied Mechanics and Engineering*, **191**, 3-19; Carpinteri A, Cornetti P (2002), A Fractional Calculus Approach to the Description of Stress and Strain Localization in Fractal Media, *Chaos, Solitons and Fractals*, **13**, 85–94; Epstein M, Adeeb S (2008) The stiffness of self-similar fractals, *International Journal of Solids and Structures* **45**, 3238–3254; Mosco U (2002), Energy Functionals on Certain Fractal Structures, *Journal of Convex Analysis*, **9**, 581–600.

by means of a fractal, we should be willing to accept a new conceptual framework. Since the dimension of a fractal is not an integer, its (Hausdorff) measure[15] will be expressed in terms of units of length raised to a fractional power. We should be prepared, therefore, to accept that the material properties of such an entity be expressed and understood in similar terms. Thus, for instance, if the fractal is made of a linearly elastic material, we should not be surprised if its modulus of elasticity is expressed in units of force divided by a fractional power of units of length. The physical context of each application will be the guide to interpret this (only apparently) curious fact accordingly.

4.18.1. Differential Spaces

Fractals are not the only possible, or even interesting, generalization of the concept of material body. Manifolds with corners and unions of manifolds of different dimensions (including zero) are two examples that immediately come to mind. In fact, there is no reason to exclude a priori any subset of $\mathbb{R}^3$ as a potential model for a material entity on which to formulate a mechanical theory.[16] In order to formulate a mechanical theory that is valid for bodies that are arbitrary subsets of $\mathbb{R}^3$, we will only require that the notion of *smooth function* over such sets be available and well defined. This is the domain of the theory of differential spaces.

A *differential space* consists of a topological space $\mathcal{B}$ and a collection, denoted as $C^\infty(\mathcal{B})$, of real-valued continuous functions. That is, out of the class $C^0(\mathcal{B})$ of all the continuous functions $f : \mathcal{B} \to \mathbb{R}$, a particular subclass $C^\infty(\mathcal{B})$ is singled out to be considered as deserving the name of "smooth." This subclass, however, cannot be arbitrarily chosen, but must satisfy the following three conditions: (i) it must be closed under composition with smooth functions in $\mathbb{R}^n$; (ii) it must be consistent with localized restrictions; (iii) the family of sets $\{f^{-1}((a,b)) | f \in C^\infty(\mathcal{B}), a, b \in \mathbb{R}\}$ must generate the topology of $\mathcal{B}$. More specifically, let $f : \mathcal{B} \to \mathbb{R}$ be a smooth function on $\mathcal{B}$ and let $\phi : \mathbb{R} \to \mathbb{R}$ be a smooth function (in the ordinary sense) on $\mathbb{R}$. Then the composition $\phi \circ f$ must be a smooth function on $\mathcal{B}$. In other words:

$$f \in C^\infty(\mathcal{B}), \phi \in C^\infty(\mathcal{R}) \Rightarrow \phi \circ f \in C^\infty(\mathcal{B}). \tag{4.109}$$

Similarly, if for any n and any smooth functions $f_1, \dots, f_n$ we form the function $F : \mathcal{B} \to \mathbb{R}^n$ given by $x \mapsto (f_1(x), \dots, f_n(x))$, the composition of F with any smooth function $\Phi : \mathbb{R}^n \to \mathbb{R}$ belongs to $C^\infty(\mathcal{B})$:

$$f_i \in C^\infty(\mathcal{B}) \ (i = 1, \dots, n), \ \Phi \in C^\infty(\mathbb{R}^n) \Rightarrow \Phi \circ (f_1, \dots, f_n) \in C^\infty(\mathcal{B}). \tag{4.110}$$

As far as condition (ii) is concerned, it expresses the fact that if a continuous function $g : \mathcal{B} \to \mathbb{R}$ has the property that for each $x \in \mathcal{B}$ there exists a neighbourhood

---

[15] A general introduction to fractals can be found in Falconer K (2003), *Fractal Geometry*, John Wiley.

[16] Even this restriction to subsets of $\mathbb{R}^3$ may be too stringent, as the various theories of media with internal structure, whereby the material body is identified with some fibre bundle over a standard body manifold, clearly demonstrates.

$\mathcal{U}$ and a smooth function $f \in C^\infty(\mathcal{B})$ such that $g|_\mathcal{U} = f|_\mathcal{U}$, then g must be in $C^\infty(\mathcal{B})$. In fact, given any initial choice of functions generating the topology of $\mathcal{B}$, these two properties (being closed under composition and being consistent with restrictions) allow us to complete a class of smooth functions on $\mathcal{B}$ by adding all other functions that fulfill these conditions. In this way, the set $\mathcal{B}$ is said to have acquired a *differential structure* and becomes a differential space. This structure enables us to use calculus to study the geometry of $\mathcal{B}$ even though $\mathcal{B}$ itself need not be smooth.

EXAMPLE 4.16. It is not difficult to verify that a differentiable manifold together with the class of smooth functions is a differential space. More striking, however, is the fact that any subset $\mathcal{B}$ of $\mathbb{R}^n$ can be endowed with a natural differential structure by adopting the topology induced by the metric topology of $\mathbb{R}^n$ and defining the smooth functions $C^\infty(\mathcal{B})$ to consist of all functions on $\mathcal{B}$ that locally extend to smooth functions on $\mathbb{R}^n$. In other words, $f \in C^\infty(\mathcal{B})$ if, for each $x \in \mathcal{B} \subset \mathbb{R}^n$, there exists a neighbourhood U of x in $\mathcal{B}$ and a smooth function F on $\mathbb{R}^n$ such that f restricted to U coincides with the restriction of F to U. Loosely speaking, we say that in this case the smooth functions in $C^\infty(\mathcal{B})$ are defined as restrictions to $\mathcal{B}$ of smooth functions on $\mathbb{R}^n$. We will call this the *induced differential structure on $\mathcal{B}$*.

Using functions $\Phi : \mathbb{R}^2 \to \mathbb{R}$ such as $(x,y) \mapsto x+y$ and $(x,y) \mapsto xy$, and invoking property (4.110), one can see that $C^\infty(\mathcal{B})$ is an (associative and commutative) algebra (with point-wise sum and product of functions). A *derivation* at $p \in \mathcal{B}$ is a linear map $v : C^\infty(\mathcal{B}) \to \mathbb{R}$ satisfying the product (or Leibniz) rule:

$$v(fg) = v(f)g(p) + f(p)v(g), \quad \forall f,g \in C^\infty(\mathcal{B}). \tag{4.111}$$

The collection of all derivations at p forms a vector space (under the obvious operations) which, by analogy with the case of differentiable manifolds, is called the *tangent space to $\mathcal{B}$ at p*, denoted by $T_p\mathcal{B}$. Note that, while in the case of differentiable manifolds the tangent spaces at all points have the same dimension, in the case of differential spaces the dimension of $T_p\mathcal{B}$ may depend on p. The set of derivations is never empty, since the zero operator is linear and satisfies the Leibniz rule. The collection of all tangent vectors at all points of $\mathcal{B}$ is denoted $T\mathcal{B}$ and called the *tangent bundle space* of $\mathcal{B}$ (in the literature the term *tangent pseudobundle* is also used).

EXAMPLE 4.17. Consider the case in which $\mathcal{B}$ is a bracket, namely, the union of two (closed) noncollinear line segments with one common end point (the corner). Seen as a subset of $\mathbb{R}^3$ and with the induced differential structure, this is a well-defined differential space. A smooth function f on $\mathcal{B}$ is the restriction to $\mathcal{B}$ of some smooth function on $\mathbb{R}^3$. On the other hand, given $f \in C^\infty(\mathcal{B})$ there exists an infinite number of extensions to smooth functions on $\mathbb{R}^3$, all of which have, point by point, the same directional derivative in the direction of the leg to which the point belongs. At the corner, all these extensions will automatically share two directional derivatives. These directional derivatives constitute a basis for all possible derivations at the point. The dimension of $T_p(\mathcal{B})$ is, therefore, 1 at all points p (including the ends of the bracket) except at the corner, where the dimension is 2.

A map κ from $\mathcal{B}$ to $\mathbb{R}^3$ is a *smooth embedding* if it is an embedding and, for every $\phi \in C^\infty(\mathbb{R}^3)$, the composition $\phi \circ \kappa$ is in $C^\infty(\mathcal{B})$. A smooth embedding κ of $\mathcal{B}$ into $\mathbb{R}^3$ gives rise to a map $T\kappa$ from $T\mathcal{B}$ to $T\mathbb{R}^3$. For every $x \in \mathcal{B}$ and $v \in T_x\mathcal{B}$, $T\kappa(v)$ is a vector tangent to $\mathbb{R}^3$ at $\kappa(x)$ such that, for every function $\phi \in C^\infty(\mathbb{R}^3)$,

$$T\kappa(v)(\phi) = v(\phi \circ \kappa). \tag{4.112}$$

REMARK 4.3. For the cases of interest, in which $\mathcal{B}$ is a compact subset of $\mathbb{R}^3$ endowed with the induced differential structure, it is sometimes intuitively useful to think of a smooth embedding $\kappa : \mathcal{B} \to \mathbb{R}^3$ as simply the restriction to $\mathcal{B}$ of a diffeomorphism Φ of $\mathbb{R}^3$. Similarly, the tangent map $T\kappa$ at $p \in \mathcal{B}$ can be thought of as the restriction of $T\Phi$ to the subspace $T_p\mathcal{B}$.

4.18.2. Mechanics of Differential Spaces

The most fundamental kinematic notion is that of *configuration space*, namely, the collection of all possible *configurations* of a mechanical system. In Classical Mechanics of a system of particles, a configuration is a 1-1 map from $\mathcal{B} = \{x_1, \ldots, x_n\}$ to $\mathbb{R}^3$. Under the assumption of smoothness (or absence) of constraints among particles, the configuration space of Classical Mechanics is a smooth m-dimensional manifold $\mathcal{Q}$, where m is the number of degrees of freedom. In Continuum Mechanics, on the other hand, a configuration is a smooth embedding of $\mathcal{B}$ into $\mathbb{R}^3$. If $\mathcal{B}$ is a compact manifold with boundary (say, a closed ball in $\mathbb{R}^3$), the configuration space $\mathcal{Q}$ is an infinite-dimensional manifold of maps. If $\mathcal{B}$ is not compact, then there are difficulties with defining a manifold structure on $\mathcal{Q}$. Nevertheless, even in this case it is convenient to use the notation and terminology of the theory of manifolds of maps. In what follows, we revisit some of the concepts informally introduced in Chapter 1.

As we have shown above, the notion of smooth embedding of $\mathcal{B}$ into $\mathbb{R}^3$ makes sense for arbitrary subsets of $\mathbb{R}^3$. Accordingly, we define a configuration of any given body $\mathcal{B} \subset \mathbb{R}^3$ as a smooth embedding $\kappa : \mathcal{B} \to \mathbb{R}^3$. The space $\mathcal{Q}$ of all smooth embeddings of $\mathcal{B}$ into $\mathbb{R}^3$ (satisfying, if necessary, any given kinematic constraints) is the configuration space of our system. We do not know whether $\mathcal{Q}$ is a manifold of maps. Nevertheless, following the tradition of Continuum Mechanics for noncompact bodies, we use the manifold notation and terminology. Given a configuration κ, the derived map $T\kappa : T\mathcal{B} \to T\mathbb{R}^3$ is called the *deformation gradient*.

Virtual displacements are tangent vectors to $\mathcal{Q}$. A virtual displacement of a configuration $\kappa \in \mathcal{Q}$ is a vector $\delta\kappa \in T_\kappa\mathcal{Q}$, which can be visualized as vector field in $\mathbb{R}^3$ defined on the range of κ.

REMARK 4.4. Intuitively, we may think of a tangent vector $\delta\kappa$ to a configuration $\kappa \in \mathcal{Q}$ as an equivalence class of curves in $\mathcal{Q}$ through κ. Somewhat more precisely, the construction of a tangent vector is the following: We start from a smooth one-parameter (β, say) family of configurations of $\mathcal{B}$, such that the given κ corresponds to $\beta = 0$. As β varies, each point of $\mathcal{B}$ describes an ordinary smooth curve in $\mathbb{R}^3$ with a well-defined tangent vector at $\beta = 0$. It is clear that the tangent vector $\delta\kappa$

corresponding to this one-parameter family can be identified with the collection of all these ordinary tangent vectors, that is, with a vector field on $\kappa(\mathcal{B})$. The collection of all tangent vectors $\delta\kappa$ at the configuration κ constitutes the (finite- or infinite-dimensional) tangent space $T_\kappa\mathcal{Q}$.

In Newtonian Mechanics, the concept of force is a primary notion. Using the underlying metric structure of Euclidean space, forces are defined as vectors at the same level as displacements, velocities, and accelerations. The point of view of Analytical Mechanics, on the other hand, considers forces (or generalized forces) to be linear operators on virtual displacements. In other words, forces are elements of the cotangent bundle of the configuration space. The evaluation of a covector (force) on a vector (virtual displacement) is called *virtual work*. No metric structure is, therefore, invoked to define forces. The Newtonian (vectorial) conditions of equilibrium of a system are replaced with a single scalar identity stating the vanishing of the virtual work for all virtual displacements. This approach, which assigns a primary status to the kinematics of a system, is not in standard use in Continuum Mechanics, which, with few exceptions,[17] tends to follow the Newtonian point of view. Nevertheless, its usefulness should be evident whenever we are confronted with a nonstandard mechanical system for which the kinematical setting is well defined. As we have shown, this is the case of the mechanics of differential spaces and, in particular, of fractals.

Following the lead of Analytical Mechanics, we define a *generalized force f* at a configuration $\kappa \in \mathcal{Q}$ as a bounded linear operator on $T_\kappa\mathcal{Q}$. The evaluation $\langle f, \delta\kappa \rangle$ of a generalized force f at a configuration κ on a virtual displacement $\delta\kappa$ at the same configuration is called the *virtual work of f on $\delta\kappa$*.

REMARK 4.5. This notion of generalized force is wide enough to encompass all the classical notions of external force, internal stress, nonlocal interactions, and so on. A distinction between internal and external forces can be postulated as part of a given mechanical theory (for instance, by demanding that the internal forces perform no virtual work on a class of virtual displacements). Such distinction, however, is not essential for the development of the general theory.

To completely define the response of a mechanical system, the generalized forces acting on it must be specified by means of a so-called *constitutive equation*. In principle, the generalized forces may depend on the whole past history of the configurations of the system. Excluding all such *memory effects*, we will consider the elastic case, in which the generalized forces depend only on the present configuration of the system, namely, $f = f(\kappa)$. From a differential geometric point of view, we observe that this constitutive equation is tantamount to choosing a particular cross section of the cotangent bundle $T^*\mathcal{Q}$.

[17] See Epstein M, Segev R (1980), Differentiable Manifolds and the Principle of Virtual Work in Continuum Mechanics, *Journal of Mathematical Physics* **21**, 1243–1245; Segev R (1986), Forces and the Existence of Stresses in Invariant Continuum Mechanics, *Journal of Mathematical Physics* **27**, 163–170.

REMARK 4.6. Under conditions of smoothness, an elastic constitutive equation is simply a differential 1-form on $\mathcal{Q}$. In other words:

$$f : \mathcal{Q} \to T^*\mathcal{Q}, \tag{4.113}$$

with the condition that $\pi \circ f = id_{\mathcal{Q}}$, where $\pi : T^*\mathcal{Q} \to \mathcal{Q}$ is the cotangent-bundle projection and $id_{\mathcal{Q}} : \mathcal{Q} \to \mathcal{Q}$ is the identity map. A system is called *conservative* (or hyperelastic) if this 1-form is exact, namely, if there exists a scalar function $W : \mathcal{Q} \to \mathbb{R}$ such that $f = dW$.

A configuration of a system is said to be an *equilibrium configuration* if the equation:

$$\langle f(\kappa), \delta\kappa \rangle = 0 \tag{4.114}$$

is satisfied identically for all virtual displacements.

REMARK 4.7. We are confining out attention to Statics alone. In this case, the identity (4.114) can be considered as an equation (or system of equations) to be solved for the unknown equilibrium configuration κ. The nature of this equation (algebraic, ODE, PDE, integro-differential, etc.) depends on the particular theory at hand.

EXAMPLE 4.18. In the framework of Continuum Mechanics, the standard theory of elasticity corresponds to the case in which the constitutive 1-form can be represented as an integral of the form:

$$\langle f, \delta\kappa \rangle = \int_{\mathcal{B}} [f_i \delta\kappa^i - f_i^J \delta\kappa^i{}_{,J}] dV + \int_{\partial\mathcal{B}} t_i \delta\kappa^i dA, \tag{4.115}$$

where the usual coordinate notation is used and where dV and dA denote, respectively, the ordinary Cartesian volume and area element. The quantities f_i, f_i^J, and t_i represent, respectively, the body force, the (Piola) stress, and the surface traction. A material is of the first grade (or simple) if the stress at a point p is a function only of the local value of the deformation gradient at that point, namely:

$$f_i^J = f_i^J(\kappa^j{}_{,K}(p), p). \tag{4.116}$$

Dependence on higher gradients of the deformation, particularly the second gradient, is also of practical interest.

Since the induced differential structure described in Example 4.16 is always available, it is clear that we can always formulate the mechanics of any subset B of $\mathbb{R}^3$ by adopting this canonical structure. From the physical point of view, we are attaching permanently the points of B to the background space and confining the possible configurations to those induced by a smooth deformation of the background. In the case of fractals, this naive policy may result in an undesirable overstiffening of the structure and may also lead to nonexistence and/or non-uniqueness of solutions of the field equations provided by the principle of virtual work. Clearly, the choice of admissible functions in the differential space is a crucial mathematical and physical part of the formulation.

Lie Derivatives, Lie Groups, Lie Algebras

The Norwegian mathematician Sophus Lie (1842–1899) is rightly credited with the creation of one of the most fertile paradigms in mathematical physics. Some of the material discussed in the previous chapter, in particular the relation between brackets of vector fields and commutativity of flows, is directly traceable to Lie's doctoral dissertation. Twentieth-century Physics owes a great deal to Lie's ideas, and so does Differential Geometry.

5.1. Introduction

We will revisit some of the ideas introduced in Example 4.6 from a more general point of view. Just as the velocity field of a fluid presupposes an underlying flow of matter, so can any vector field be regarded as the velocity field of the steady motion of a fluid, thereby leading to the mathematical notion of the *flow of a vector field*. Moreover, were one to attach a marker to each of two neighbouring fluid particles, representing the tail and the tip of a vector, as time goes on the flow would carry them along, thus yielding a rate of change of the vector they define. The rigorous mathematical counterpart of this idea is the *Lie derivative*.

Let $\mathbf{V} : \mathcal{M} \to T\mathcal{M}$ be a (smooth) vector field. A (parametrized) curve $\gamma : I \to \mathcal{M}$ is called an *integral curve* of the vector field if its tangent at each point coincides with the vector field at that point. In other words, denoting by s the curve parameter, the following condition holds:

$$\frac{d\gamma(s)}{ds} = \mathbf{V}(\gamma(s)) \qquad \forall s \in I \subset \mathbb{R}. \tag{5.1}$$

The hard facts of differential geometry tend to come from Analysis. In the case at hand, the main question to be answered is: Given the vector field $\mathbf{V}$ and a point $p \in \mathcal{M}$, is there a unique integral curve of $\mathbf{V}$ through p? Since, when written out in a coordinate chart, Equation (5.1) reveals itself as a system of ordinary differential equations (ODEs), it is not surprising that the answer to the question is provided by the fundamental theorem of existence and uniqueness of solutions of systems of

ODEs. It may not be a bad idea, therefore, to restate this theorem, albeit without proof.

5.2. The Fundamental Theorem of the Theory of ODEs

We consider a connected open set $U \subset \mathbb{R}^n$ on which n real-valued smooth[1] functions v^i $(i = 1, \ldots, n)$ have been defined. The system of equations:

$$\frac{dx^i(s)}{ds} = v^i(x^1(s), \ldots, x^n(s)), \tag{5.2}$$

with the initial conditions:

$$x^i(s_0) = x_0^i, \tag{5.3}$$

is known as the *initial-value problem* at "time" $s_0 \in \mathbb{R}$ through the initial point $x_0 = \{x_0^1, \ldots, x_0^n\} \in U$. A solution of this system consists of n functions $x^i = f^i(s)$ satisfying (5.2) and (5.3). For obvious reasons, a solution can be regarded as a parametrized curve in $\mathbb{R}^n$ passing through x_0. Notice that the image of the curve may consist of a single point in $\mathbb{R}^n$.

THEOREM 5.1. *For each $s_0 \in \mathbb{R}$ and each $x_0 \in U$, there exist: (i) an open interval $I_0 \in \mathbb{R}$ containing s_0; (ii) an open set $U_0 \subset U$ containing x_0; and (iii) n smooth functions $f^i : I_0 \to U_0$, such that $x^i = f^i(s)$ satisfy the initial-value problem (5.2) and (5.3). Moreover, two solutions $f^i(s)$ and $g^i(s)$ coincide in the intersection of their domains. The solution can, therefore, be said to be unique and extended to a largest possible interval I_0. Finally, the solution depends smoothly on the initial point x_0.*

The following corollary, known as the *translation property*, can be proved easily:

COROLLARY 5.2. *If the functions $f^i(s)$ constitute a solution of the initial-value problem (5.2) and (5.3), the functions $g^i(s) = f^i(s - a)$ constitute a solution of the initial-value problem (5.2) with the initial conditions:*

$$x^i(s_0 + a) = x_0^i, \tag{5.4}$$

for any fixed $a \in \mathbb{R}$.

PROOF. The proof is a direct consequence of the fact that the variable s does not appear in the right-hand side of (5.2). Such systems of ODEs are called *autonomous*.

□

The importance of this corollary is that it shows that the parametrization of the solution curves is determined uniquely up to an arbitrary additive constant. In view of this property, let us introduce the following notation. Let us denote by $f(s,x)$ the solution with the initial condition $f(0,x) = x$. In other words, apart from the fact that we represent the quantities f^i and x^i under the single symbols f and x, we adjust (by a simple translation) the initial value of the parameter so that it is zero at x.

---

[1] Much less than smoothness is required for the proof.

Moreover, let us denote by I_x the corresponding maximum interval for which the solution is defined, according to the fundamental theorem. An important corollary of the fundamental theorem is the so-called *pseudo-group* property:

COROLLARY 5.3. *With the above notation, we have:*

$$f(s, f(s', x)) = f(s + s', x), \tag{5.5}$$

whenever s, s', and $s + s'$ belong to I_x.

PROOF. The proof is also a straightforward (if somewhat confusing) consequence of the fact that s does not appear explicitly on the right-hand side of the ODE system.   □

The intuitive meaning of this corollary is that, no matter which point of a given solution curve is used as initial condition, one always obtains the same image curve and that, along this image curve, the parameter is additive. In other words, if we start at an initial point x (with $s = 0$) and we move s units along the corresponding solution curve we will arrive at a point $x' \in \mathbb{R}^n$. If we use this point x' as an initial condition, we obtain, in principle, a new solution curve. Moving along this curve s' units we arrive at a new point x''. This point turns out to be the same as if we had advanced $s + s'$ units along the first solution curve.

5.3. The Flow of a Vector Field

Rephrased in terms of the concepts of vector fields and integral curves, the fundamental theorem 5.1 implies that:

THEOREM 5.4. *If $\mathbf{V}$ is a vector field on a manifold $\mathcal{M}$, then for every $p \in \mathcal{M}$ there exists an integral curve $\gamma(s, p) : I_p \to \mathcal{M}$ such that (i) I_p is an open interval of $\mathbb{R}$ containing the origin $s = 0$; (ii) $\gamma(0, p) = p$; and (iii) I_p is maximal in the sense that there exists no integral curve starting at p and defined on an open interval of which I_p is a proper subset.*

Moreover,

COROLLARY 5.5.

$$\gamma(s, \gamma(s', x)) = \gamma(s + s', x) \quad \forall s, s', s + s' \in I_p. \tag{5.6}$$

DEFINITION 5.1. The map given by:

$$p, s \mapsto \gamma(s, p) \tag{5.7}$$

is called the *flow* of the vector field $\mathbf{V}$ whose integral curves are $\gamma(s, p)$.

In this definition, the map is expressed in terms of its action on pairs of points belonging to two different manifolds, $\mathcal{M}$ and $\mathbb{R}$, respectively. Not all pairs, however, are included in the domain, since I_p is not necessarily equal to $\mathbb{R}$. Moreover, since the intervals I_p are point dependent, the domain of the flow is not even a product manifold. One would be tempted to take the intersection of all such intervals so as to work with a product manifold given by $\mathcal{M}$ times the smallest interval I_p. Unfortunately, as we know from elementary calculus, this (infinite) intersection may consist of a single point. All that can be said about the domain of the flow is that it is an open submanifold of the Cartesian product $\mathcal{M} \times \mathbb{R}$. When the domain is equal to this product manifold, the vector field is said to be *complete* and the corresponding flow is called a *global flow*. It can be shown that if $\mathcal{M}$ is compact, or if the vector field is smooth and vanishes outside a compact subset of $\mathcal{M}$, the flow is necessarily global.

5.4. One-parameter Groups of Transformations Generated by Flows

Given a point $p_0 \in \mathcal{M}$, it is always possible to find a small enough neighbourhood $U(p_0) \subset \mathcal{M}$ such that the intersection of all the intervals I_p with $p \in U(p_0)$ is an open interval J containing the origin. For each value $s \in J$, the flow $\gamma(s,p)$ can be regarded as a map:

$$\gamma_s : U(p_0) \longrightarrow \mathcal{M}, \tag{5.8}$$

defined as:

$$\gamma_s(p) = \gamma(s,p), \quad p \in U(p_0). \tag{5.9}$$

This map is clearly one-to-one, since otherwise we would have two integral curves intersecting each other, against the statement of the fundamental theorem. Moreover, again according to the fundamental theorem, this is a smooth map with a smooth inverse over its image. The inverse is, in fact, given by:

$$\gamma_s^{-1} = \gamma_{-s}, \tag{5.10}$$

where γ_{-s} is defined over the image $\gamma_s(U(p_0))$. Notice that γ_0 is the identity map of $U(p_0)$. Finally, according to Corollary 5.5, for the appropriate range of values of s and r, we have the composition law:

$$\gamma_r \circ \gamma_s = \gamma_{r+s}. \tag{5.11}$$

Because of these properties, the set of maps γ_s is said to constitute the *one-parameter local pseudo-group* generated by the vector field (or by its flow). If the neighbourhood $U(p_0)$ can be extended to the whole manifold for some open interval J (no matter how small), each map γ_s is called a *transformation* of $\mathcal{M}$. In that case we speak of a *one-parameter pseudo-group of transformations* of $\mathcal{M}$. Finally, in the best of all possible worlds, if $J = \mathbb{R}$, the one-parameter subgroup of transformations becomes elevated to a *one-parameter group of transformations*. This is an Abelian (i.e., commutative) group, as is clearly shown by the composition law (5.11). In the terminology of the last section, we may say that every complete vector field generates a one-parameter group of transformations of the manifold.

The converse construction, namely, the generation of a vector field out of a given one-parameter pseudo-group of transformations, is also of interest. It is borne by the following theorem:

THEOREM 5.6. *Every one-parameter pseudo-group of transformations γ_s is generated by the vector field:*

$$\mathbf{V}(p) = \frac{d\gamma_s(p)}{ds}\big|_{s=0}. \tag{5.12}$$

EXERCISE 5.1. Prove the above theorem.[2]

5.5. Time-Dependent Vector Fields

In many physical situations, such as is the case in Continuum Mechanics, time (or, in principle, any other contextually meaningful scalar parameter) is part of the picture one intends to represent. Thus, the velocity field of a fluid may not be stationary but rather vary with time at any given position of an observer in physical space. Similarly, the heat flux vector field in a (rigid, say) heat-conducting medium may correspond to a transient regime rather than a steady state. At each instant of time, the instantaneous vector field $\mathbf{V}(p,t)$ generates a distinct flow. These flows, however, cannot be regarded individually as carriers of a time-dependent field, for the simple reason that they are frozen in time. If, ignoring this fact, we were to formulate a (nonautonomous) system of ODEs analogous to (5.1), we would obtain,

$$\frac{d\gamma(t)}{dt} = \mathbf{V}(\gamma(t),t), \tag{5.13}$$

where we have, naturally enough, required that the curve parameter be identified with the physically meaningful time-variable t. To be sure, a slightly more general version of the fundamental theorem of systems of ODEs guarantees the local existence and uniqueness of solutions of this problem. The solution curves, however, will not enjoy in general the nice properties that we have encountered for the integral curves of an autonomous vector field. In particular, their images may cross each other, since two solution curves starting at the same point in $\mathcal{M}$ but at different instants of time will, in general, be different.

One way to restore the orderly state of affairs peculiar to autonomous systems is to introduce a *space-time* picture (not necessarily relativistic, of course), whereby the underlying manifold is the $(n+1)$-dimensional product $\mathcal{M} \times \mathbb{R}$. In this picture, the system of equations (5.13) is replaced by the equivalent system of $n+1$ equations:

$$\begin{aligned}
\frac{d\gamma(s)}{ds} &= \mathbf{V}(\gamma(s),s) \\
\frac{dt}{ds} &= 1,
\end{aligned} \tag{5.14}$$

where the curve-parameter s is identified with t not directly but via an extra differential equation for the (now supposedly dependent) variable t. By means of this simple

---

[2] For this, as for much of the material in this and other chapters, you may wish to consult the book by Choquet-Bruhat Y, de Witt-Morette C, Dillard-Bleick M (1977), *Analysis, Manifolds and Physics*, North-Holland. Although generally rigorous, the treatment is not encumbered by excessive technical apparatus.

stratagem, the system of ODEs has become technically autonomous and its integral curves exist, are unique and induce a flow with the usual characteristics, except that it acts on a differentiable manifold one dimension higher. This concept of *extended flow* can often be applied with benefit even in those cases in which the vector field of departure is actually stationary. The solution curves of the original system become thus "unfolded" in the sense that, for example, a circular trajectory traversed time and again as the parameter increases (namely, a periodic solution) is now stretched (like a slinky toy) to become a helix (of constant unit slope) by taking advantage of the extra dimension.

5.6. The Lie Derivative

We have learned that a vector field determines at least a one-parameter pseudo-group in a neighbourhood of each point of the underlying manifold. For each value of the parameter s within a certain interval containing the origin, this neighbourhood is mapped diffeomorphically onto another neighbourhood. Having at our disposal a diffeomorphism, we can consider the pushed-forward or pulled-back versions of tensors of every type, including multivectors and differential forms. Physically, these actions represent how the various quantities are *convected* (or dragged) by the flow. To elicit a mental picture, we show in Figure 5.1 a vector $\mathbf{w}_p$ in a manifold as a small segment $\overrightarrow{pq}$ (a small piece of a curve, say), and we draw the integral curves of a vector field $\mathbf{V}$ emerging from each of its end points, p and q. These curves are everywhere tangent to the underlying vector field $\mathbf{V}$, which we do not show in the figure. If s denotes the (natural) parameter along these integral curves, an increment of Δs applied from each of these points along the corresponding integral curve, will result in two new points p' and q', respectively. The (small) segment $\overrightarrow{p'q'}$ can be seen as a vector $\mathbf{w}'$, which we regard as the convected counterpart of $\mathbf{w}_p$ as it is dragged by the flow of $\mathbf{V}$ by an amount Δs. If $\mathbf{w}_p$ happens to be part of a vector field $\mathbf{W}$ defined in a neighbourhood of p', so that $\mathbf{w}_p = \mathbf{W}(p)$, we have that at the point p' there is, in addition to the dragged vector $\mathbf{w}'$, a vector $\mathbf{W}(p')$. There is no reason why these two vectors should be equal. The difference $\mathbf{W}(p') - \mathbf{w}'$ (divided by Δs) gives us an idea of the meaning of the Lie derivative of $\mathbf{W}$ with respect to $\mathbf{V}$ at p'.

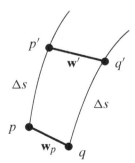

Figure 5.1. Dragging of a vector by a flow

The idea behind the definition of the Lie derivative of a tensor field with respect to a given vector field at a point p is the following. We consider a small value s of the parameter and convect the tensor field back to $s = 0$ by using the appropriate pull-back or push-forward. This operation will, in particular, provide a value of the convected tensor field at the point p. We then subtract from this value the original value of the field at that point (a legitimate operation, since both tensors operate on the same tangent and/or cotangent space), divide by s and compute the limit as $s \to 0$. To understand how to calculate a Lie derivative, it is sufficient to make the definition explicit for the case of functions, vector fields and 1-forms. The general case is then inferred from these three basic cases, as we shall demonstrate. We will also prove that the term "derivative" is justified. Notice that a Lie derivative is defined with respect to a given vector field. It is not an intrinsic property of the tensor field being differentiated. The Lie derivative of a tensor field at a point is a tensor of the same type.

5.6.1. The Lie Derivative of a Scalar

Let $g : \mathcal{M} \to \mathcal{N}$ be a map between two manifolds and let $f : \mathcal{N} \to \mathbb{R}$ be a function. Recall that, according to Equation (4.79), the pull-back of f by g is the map $g^*f : \mathcal{M} \to \mathbb{R}$ defined as the composition:

$$g^*f = f \circ g. \tag{5.15}$$

Let a (time-independent, for now) vector field $\mathbf{V}$ be defined on $\mathcal{M}$ and let $\gamma_s : U \to \mathcal{M}$ denote the action of its flow on a neighbourhood of a point $p \in \mathcal{M}$. If a function $f : \mathcal{M} \to \mathbb{R}$ is defined, we can calculate:

$$\gamma_s^*f := f \circ \gamma_s. \tag{5.16}$$

The Lie derivative at the point p is given by:

$$L_V f(p) := \lim_{s \to 0} \frac{(\gamma_s^*f)(p) - f(p)}{s} = \lim_{s \to 0} \frac{f(\gamma_s(p)) - f(p)}{s} \tag{5.17}$$

Comparing this with Equation (4.6), we obtain:

$$L_V f(p) = \mathbf{v}_p(f). \tag{5.18}$$

In simple words, the Lie derivative of a scalar field with respect to a given vector field coincides, at each point, with the directional derivative of the function in the direction of the field at that point.

5.6.2. The Lie Derivative of a Vector Field

Vectors are pulled forward by mappings. Thus, given the map $g : \mathcal{M} \to \mathcal{N}$, to bring a tangent vector from $\mathcal{N}$ back to $\mathcal{M}$, we must use the fact that g is invertible and that the inverse is differentiable, such as when g is a diffeomorphism. Let $\mathbf{W} : \mathcal{N} \to T\mathcal{N}$ be a vector field on $\mathcal{N}$. The corresponding vector field on $\mathcal{M}$ is then given by $g_*^{-1} \circ \mathbf{W} \circ g$: $\mathcal{M} \to T\mathcal{M}$. Accordingly, the Lie derivative of the vector field $\mathbf{W}$ with respect to the vector field $\mathbf{V}$, with flow γ_s, at a point $p \in \mathcal{M}$ is defined as:

$$L_V \mathbf{W}(p) = \lim_{s \to 0} \frac{\gamma_{s*}^{-1} \circ \mathbf{W} \circ \gamma_s(p) - \mathbf{W}(p)}{s}. \tag{5.19}$$

It can be shown[3] that the Lie derivative of a vector field coincides with the Lie bracket:

$$L_V \mathbf{W} = [\mathbf{V}, \mathbf{W}]. \tag{5.20}$$

EXERCISE 5.2. Convince yourself of the plausibility of the above result by means of a simple drawing of the 2-dimensional case.

5.6.3. The Lie Derivative of a One-Form

Since 1-forms are pulled back by a map, we define the Lie derivative of the 1-form $\omega : \mathcal{M} \to T^*\mathcal{M}$ at the point p as:

$$L_V \omega(p) = \lim_{s \to 0} \frac{\gamma_s^* \circ \omega \circ \gamma_s(p) - \omega(p)}{s}. \tag{5.21}$$

5.6.4. The Lie Derivative of Arbitrary Tensor Fields

It is clear that, by virtue of their definition by means of limits, the Lie derivatives defined so far are linear operators. To extend the definition of the Lie derivative to tensor fields of arbitrary order, we need to make sure that the Leibniz rule with respect to the tensor product is satisfied. Otherwise, we wouldn't have the right to use the term "derivative" to describe it. It is enough to consider the case of a monomial such as:

$$\mathbf{T} = \omega_1 \otimes \cdots \otimes \omega_m \otimes \mathbf{W}_1 \otimes \cdots \otimes \mathbf{W}_n, \tag{5.22}$$

where ω_i are m 1-forms and $\mathbf{W}_j$ are n vector fields. We define:

$$L_V \mathbf{T}(p) = \lim_{s \to 0} \frac{\gamma_s^* \circ \omega_1 \circ \gamma_s(p) \otimes \cdots \otimes \gamma_{s*}^{-1} \circ \mathbf{W}_1 \circ \gamma_s(p) \otimes \cdots - \mathbf{T}(p)}{s}. \tag{5.23}$$

Let us verify the satisfaction of the Leibniz rule for the case of the tensor product of a 1-form by a vector:

$$L_V(\omega \otimes \mathbf{W})(p) = \lim_{s \to 0} \frac{\gamma_s^* \circ \omega \circ \gamma_s(p) \otimes \gamma_{s*}^{-1} \circ \mathbf{W} \circ \gamma_s(p) \otimes - \omega(p) \otimes \mathbf{W}(p)}{s}. \tag{5.24}$$

Subtracting and adding to the denominator the expression $\omega(p) \otimes \gamma_{s*}^{-1} \circ \mathbf{W} \circ \gamma_s(p)$ the Leibniz rule follows suit. It is not difficult to check that the Lie derivative commutes with contractions.

---

[3] See: Kobayashi S, Nomizu K (1963): *Foundations of Differential Geometry*, John Wiley.

An important property of the Lie derivative is the following:

PROPOSITION 5.7. *The Lie derivative of a differential form (of any order) commutes with the exterior derivative, that is:*

$$L_V(d\omega) = d(L_V\omega), \tag{5.25}$$

for all vector fields **V** *and for all differential forms* ω.

PROOF. The proof is a direct consequence of the definition of the Lie derivative (by means of pull-backs) and of Proposition 4.2.   □

5.6.5. The Lie Derivative in Components

Taking advantage of the Leibniz rule, it is not difficult to calculate the components of the Lie derivative of a tensor in a given coordinate system x^i, provided the components of the Lie derivative of the base vectors $\frac{\partial}{\partial x^i}$ and the base 1-forms dx^i are known. A direct application of the formula (4.31) for the components of the Lie bracket, yields:

$$L_V\left(\frac{\partial}{\partial x^i}\right) = -\frac{\partial V^k}{\partial x^i}\frac{\partial}{\partial x^k}. \tag{5.26}$$

To obtain the Lie derivative of dx^i we recall that the action of dx^i (as a covector) on $\frac{\partial}{\partial x^j}$ is simply δ^i_j, whose Lie derivative vanishes. This action can be seen as the contraction of their tensor product. We therefore obtain:

$$0 = L_V\left\langle dx^i, \frac{\partial}{\partial x^j}\right\rangle = \left\langle dx^i, -\frac{\partial V^k}{\partial x^j}\frac{\partial}{\partial x^k}\right\rangle + \left\langle L_v dx^i, \frac{\partial}{\partial x^j}\right\rangle, \tag{5.27}$$

whence:

$$L_V dx^i = \frac{\partial V^i}{\partial x^k}dx^k. \tag{5.28}$$

EXAMPLE 5.1. **The rate-of-deformation tensor:** Let us calculate the Lie derivative of a symmetric twice covariant tensor field $\mathbf{g} = g_{ij}dx^i \otimes dx^j$ with respect to a vector field **v**. We have:

$$L_v\mathbf{g} = L_v(g_{ij}dx^i \otimes dx^j) = L_v(g_{ij})dx^i \otimes dx^j + g_{ij}L_v(dx^i) \otimes dx^j + g_{ij}dx^i \otimes L_v(dx^j). \tag{5.29}$$

Using the formulas derived above for the Lie derivative of the bases, we obtain:

$$L_v\mathbf{g} = \left(\frac{\partial g_{pq}}{\partial x^k}v^k + g_{iq}\frac{\partial v^i}{\partial x^p} + g_{pj}\frac{\partial v^j}{\partial x^q}\right)dx^p \otimes dx^q. \tag{5.30}$$

Imagine now, for simplicity, that the underlying manifold is $\mathbb{R}^3$ with its natural coordinate system in which the metric properties are governed by the unit matrix $\{g_{ij}\} = \{\delta_{ij}\}$, and that v^i are the components of the velocity vector of a medium. Under these assumptions, we obtain:

$$[L_v\mathbf{g}]_{ij} = \frac{\partial v_i}{\partial x^j} + \frac{\partial v_j}{\partial x^i}, \tag{5.31}$$

with the identification of upper and lower indices permitted by the structure of $\mathbb{R}^3$. In other words, the rate-of-deformation tensor of continuum mechanics (Equation (A.33)) measures (one-half) the instantaneous rate of change of the spatial metric as convected by an observer moving with the medium. The identical vanishing of this Lie derivative would imply that the moving observer carries this metric as an invariant of the motion. It follows that the identical vanishing of this Lie derivative implies the rigidity of the motion.

5.6.6. The Nonautonomous Lie Derivative

In the case of a time-dependent tensor field $\mathbf{T}(p,t)$, we can augment the base manifold $\mathcal{M}$ to the Cartesian product $\mathcal{M} \times \mathbb{R}$. In a pictorial representation (in which the manifold $\mathcal{M}$ can be imagined to be a 2-dimensional horizontal plane, for clarity, and the time axis aligned with the vertical), we see that each instant of time corresponds to a floor of a hypothetical skyscraper with an infinite number of infinitesimally tall floors. On each floor there appears the instantaneous picture of the current tensor field. If we now consider the vector field $\mathbf{V}$, with respect to which the Lie derivative is to be calculated, we realize that its flow should be able to encounter the present value of the tensor field. In order to do so, we must unfold it in time, just as we have described in Section 5.5. This should be done whether or not the vector field $\mathbf{V}$ is time independent. As we have already seen, this unfolded flow is tantamount to adding a unit component in the time direction. In other words, we now have a time-dependent tensor field $\mathbf{T}$, extended in space-time to a tensor field $\hat{\mathbf{T}}$ with space-time components $\{\mathbf{T}, 0\}$, and an extended space-time vector field $\hat{\mathbf{V}}$ with space-time components $\{\mathbf{V}, 1\}$. We define the *nonautonomous Lie derivative* of $\mathbf{T}$ with respect to $\mathbf{V}$ as:

$$\hat{L}_V \mathbf{T} := L_{\hat{V}} \hat{\mathbf{T}}. \tag{5.32}$$

A straightforward calculation reveals that the result is:

$$\hat{L}_V \mathbf{T} = \frac{\partial \mathbf{T}}{\partial t} + L_V \mathbf{T}, \tag{5.33}$$

where the last term is evaluated after freezing time at its current value.

EXERCISE 5.3. Show that for a scalar field in the Eulerian description of Continuum Mechanics, the nonautonomous Lie derivative with respect to the spatial velocity field coincides with the material time derivative. Note that this is not the case for nonscalar fields. What is the nonautonomous Lie derivative of the velocity field with respect to itself?

5.7. Invariant Tensor Fields

A diffeomorphism of $\phi : \mathcal{M} \to \mathcal{M}$ of a manifold $\mathcal{M}$ with itself is called a *transformation* of $\mathcal{M}$. We have already used this terminology before in Section 5.4. Because a transformation is smoothly invertible, tensor fields of all types can be pushed forward and pulled back in any direction at will. More precisely, we may say that the transformation ϕ induces an isomorphism $\tilde{\phi} : \mathcal{T}(T_p\mathcal{M}) \to \mathcal{T}(T_{\phi(p)}\mathcal{M})$ of the tensor algebra at each point $p \in \mathcal{M}$ with the tensor algebra at the image point $\phi(p) \in \mathcal{M}$. A given tensor field $\mathbf{T}$ is said to be *invariant* under the transformation ϕ of $\mathcal{M}$ if:

$$\tilde{\phi}\mathbf{T}(p) = \mathbf{T}(\phi(p)) \quad \forall p \in \mathcal{M}. \tag{5.34}$$

Note that a scalar function $f : \mathcal{M} \to \mathbb{R}$ is invariant under ϕ if, and only if:

$$\phi^*(f) = f, \tag{5.35}$$

or, equivalently:

$$f \circ \phi(p) = f(p) \quad \forall p \in \mathcal{M}. \tag{5.36}$$

EXERCISE 5.4. **Periodicity:** Let $\mathbf{n}$ be a unit vector in $\mathbb{R}^2$. Define the *translation* $\tau_{s,n}$ of extent s in the direction of $\mathbf{n}$ as the transformation of $\mathbb{R}^2$ given by:

$$\tau_{s,n}(\mathbf{r}) = \mathbf{r} + s\mathbf{n}, \tag{5.37}$$

where $\mathbf{r} = (x, y)$ is the generic point of $\mathbb{R}^2$. What does the graph of a function $f : \mathbb{R}^2 \to \mathbb{R}$ invariant under $\tau_{s,n}$ look like? How does this graph look if the function is invariant under $\tau_{s,n}$ for all values of s? How does it look if it is invariant only for all values of s in a small interval? Consider another unit vector $\mathbf{m}$ not collinear with $\mathbf{n}$ and assume that the function is simultaneously invariant under $\tau_{s,n}$ and $\tau_{q,m}$, for some fixed nonzero values of s and q. How does the graph of the function look? Speculate about what would happen if you were to require further invariance with respect to a translation of some extent in yet a third direction. It may help to first look at the 1-dimensional case.

For a vector field $\mathbf{V} : \mathcal{M} \to T\mathcal{M}$, invariance under a transformation $\phi : \mathcal{M} \to \mathcal{M}$ translates into:

$$\phi_* \circ \mathbf{V} \circ \phi^{-1} = \mathbf{V}. \tag{5.38}$$

In plain language, the vector field is invariant under ϕ if the pushed-forward vector finds itself already at the target point. A differential form ω is invariant whenever:

$$\phi^* \circ \omega \circ \phi = \omega. \tag{5.39}$$

REMARK 5.1. Notice that with the tilde notation, we could have defined the Lie derivative of a tensor $\mathbf{T}$ of any type with respect to a vector field $\mathbf{V}$, according to Equation (5.23), as follows:

$$L_V\mathbf{T}(p) = \lim_{s \to 0} \frac{\tilde{\gamma}_s^{-1} \circ \mathbf{T} \circ \gamma_s(p) - \mathbf{T}(p)}{s}, \tag{5.40}$$

or, what is the same:

$$L_V\mathbf{T}(p) = \lim_{s \to 0} \frac{\mathbf{T}(p) - \tilde{\gamma}_s \circ \mathbf{T} \circ \gamma_{-s}(p)}{s}, \tag{5.41}$$

We say that a tensor field is *invariant under a local one-parameter pseudo-group of transformations* γ_s if it is invariant under each of the transformations involved in the definition of the local pseudo-group. If **V** is the vector field that generates the transformations, it is clear that an invariant tensor field **T** will automatically have a vanishing Lie derivative, that is: $L_V\mathbf{T} = 0$. This result follows directly from Equation (5.40). The converse is also true, although more difficult to prove, namely: If the Lie derivative $L_V\mathbf{T}$ of a tensor field with respect to a given vector field **V** vanishes, then **T** is invariant under the local pseudo-group of transformations generated by **V**.

EXERCISE 5.5. **The Lie derivative in a Riemannian manifold:** Let **g** be the metric tensor of a Riemannian manifold, as defined in components by Equation (4.106). Because of the isomorphism between each tangent space and its dual induced by the existence of this metric, any tensor field **T** of type $(1,1)$ can also be regarded as a tensor field of type, say, $(0,2)$, as discussed in Exercise 2.34, with components T^i_j and T_{ij}, respectively. Using the formulas derived in Section 5.6.5 to calculate the components of the Lie derivative with respect to a given vector field **V**, obtain the expression for the Lie derivative of **T** independently as a tensor of type $(1,1)$ and of type $(0,2)$. In this way, you will have generated two tensors, each of one of the types of departure, and you might expect that their components are related by the same relation as the original components. Check, however, that this is not the case. Explain the reason for this unexpected discrepancy. Conclude that, even in the presence of a metric structure, it is not correct to talk about covariant and contravariant components of *the same* vector or tensor. [Hint: notice that the metric tensor is not necessarily invariant under the action of the flow of **V**.]

EXAMPLE 5.2. **Isochoric motions:** A motion is *isochoric* or *volume preserving* if the spatial volume form is invariant under the flow of its velocity field. Adopting a Cartesian coordinate system (x^1, x^2, x^3), the standard volume form ω is given by:

$$\omega = dx^1 \wedge dx^2 \wedge dx^3. \tag{5.42}$$

Denoting by v^i the components of the velocity field **v** and following the prescription of Equation (5.28), we obtain:

$$L_v\omega = \left(\frac{\partial v^1}{\partial x^1} + \frac{\partial v^2}{\partial x^2} + \frac{\partial v^3}{\partial x^3} \right) dx^1 \wedge dx^2 \wedge dx^3. \tag{5.43}$$

It follows that a motion is isochoric if, and only if, the divergence of the velocity field vanishes identically.

EXAMPLE 5.3. **Rigid-body motions:** A motion is *rigid* if the standard spatial inner product is invariant under the flow of its velocity field. More generally, a transformation that preserves (i.e., leaves invariant) a metric tensor field is called an *isometry*. In a previous example, we already calculated the corresponding Lie derivative, Equation (5.31), and concluded that a motion is rigid if, and only if, the rate-of-deformation tensor vanishes identically.

5.8. Lie Groups

Recall that a *group* is a set $\mathcal{G}$ endowed with an operation:

$$\mathcal{G} \times \mathcal{G} \longrightarrow \mathcal{G}, \tag{5.44}$$

usually called the *group multiplication* or *product* and indicated simply by:

$$(g,h) \mapsto gh \quad g,h \in \mathcal{G}, \tag{5.45}$$

with the following properties:

(1) Associativity:
$$(gh)k = g(hk) \quad \forall g,h,k \in \mathcal{G}; \tag{5.46}$$

(2) Existence of identity: $\exists e \in \mathcal{G}$ such that:
$$eg = ge \quad \forall g \in \mathcal{G}; \tag{5.47}$$

(3) Existence of inverse: For each $g \in \mathcal{G}$ there exists an *inverse*, denoted by g^{-1}, such that:
$$gg^{-1} = g^{-1}g = e. \tag{5.48}$$

EXERCISE 5.6. Prove that the identity element e (also called the *unit element*, and sometimes denoted as 1) is unique. Prove that inverses are unique.

If the group operation is also *commutative*, namely, if $gh = hg$ for all $g,h \in \mathcal{G}$, the group is said to be *commutative* or *Abelian*. In this case, it is customary to call the operation *group addition* and to indicate it as $(g,h) \mapsto g+h$. The identity is then called the *zero element* and is often denoted as 0. Finally, the inverse of g is denoted as $-g$. This notation is easy to manipulate as it is reminiscent of the addition of numbers, which is indeed a particular case.

EXERCISE 5.7. Provide a few examples of multiplicative (i.e., noncommutative) and additive (i.e., commutative) groups. Be inventive in the sense of providing one or two examples of groups with elements and operations that are not necessarily numerical. Can you provide at least one example of a set with a binary internal operation which is not associative?

DEFINITION 5.2. A *subgroup* of a group $\mathcal{G}$ is a subset $\mathcal{H} \subset \mathcal{G}$ closed under the group operations of multiplication and inverse. Thus, a subgroup is itself a group.

DEFINITION 5.3. Given two groups, $\mathcal{G}_1$ and $\mathcal{G}_2$, a *group homomorphism* is a map $\phi : \mathcal{G}_1 \to \mathcal{G}_2$ that preserves the group multiplication, namely:

$$\phi(gh) = \phi(g)\,\phi(h) \quad \forall\, g,h \in \mathcal{G}_1, \tag{5.49}$$

where the multiplications on the left- and right-hand sides are, respectively, the group multiplications of $\mathcal{G}_1$ and $\mathcal{G}_2$.

The group structure is a purely algebraic concept, whereby nothing is assumed as far as the nature of the underlying set is concerned. The concept of *Lie group*[4] arises from making such an assumption. More specifically:

DEFINITION 5.4. A *Lie group* is a (smooth) manifold $\mathcal{G}$ with a group structure that is compatible with the differential structure, namely, such that the multiplication $\mathcal{G} \times \mathcal{G} \to \mathcal{G}$ and the inversion $\mathcal{G} \to \mathcal{G}$ are smooth maps.

A homomorphism ϕ between two Lie groups is called a *Lie-group homomorphism* if ϕ is C^∞. If ϕ happens to be a diffeomorphism, we speak of a *Lie-group isomorphism*. An isomorphism of a Lie group with itself is called a *Lie-group automorphism*.

EXAMPLE 5.4. **The general linear group of a vector space:** Let V be an n-dimensional vector space. The collection $L(V,V)$ of all linear operators from V to V can be considered as a differentiable manifold. Indeed, fixing a basis in V, we obtain a global chart in $\mathbb{R}^{n^2}$. The collection $GL(V)$ of all *invertible* linear operators (i.e., the automorphisms of V) is an open subset of $L(V,V)$.[5] Thus, $GL(V)$ is an open submanifold. Defining the operation of multiplication as the composition of linear operators, the manifold $GL(V)$ becomes a Lie group, known as the *general linear group of V*.

EXAMPLE 5.5. **The matrix groups:** The general linear group of $\mathbb{R}^n$ is usually denoted by $GL(n;\mathbb{R})$. Its elements are the nonsingular square matrices of order n, with the unit matrix I acting as the group unit. Its various Lie subgroups are known as the *matrix groups*. The group operation is the usual matrix multiplication. The general linear group $GL(n;\mathbb{R})$ is not connected (in the topological sense). It consists of two disjoint connected components (those with positive and those with negative determinant, respectively). Only the first of these connected components, denoted as $GL^+(n;\mathbb{R})$, contains the unit matrix. It is itself a Lie subgroup of dimension n^2. The collection $SL(3;\mathbb{R})$ of all matrices whose determinant has a unit absolute value[6] is also a subgroup with two disjoint connected components. The connected component $SL^+(3;\mathbb{R})$ containing the unit (i.e., the collection of all matrices with determinant 1) is known as the *special linear group*. Its dimension is $n^2 - 1$. The *orthogonal group* $\mathcal{O}(n;\mathbb{R})$ consists of all orthogonal matrices that is, all matrices Q such that $QQ^T = I$. Its dimension is $n(n-1)/2$. Its connected component of the unit, denoted by $\mathcal{O}^+(n;\mathbb{R})$, is called the *special orthogonal group*. Interpreted as geometrical transformations of $\mathbb{R}^n$, the special linear group consists of all volume-preserving transformations, while the special orthogonal group consists of all rotations. The (full) orthogonal group

---

[4] The classical treatise on Lie groups is Chevalley C (1946), *Theory of Lie Groups*, Princeton University Press.

[5] This conclusion follows from the fact that, choosing a basis in V, the invertible linear maps are those with a nonsingular matrix A, namely, those satisfying the analytic condition $\det A \neq 0$. Since the determinant function is clearly continuous, the result follows suit.

[6] In Continuum Mechanics it has become customary to call this group the *unimodular group*, a terminology that is in conflict with the usual meaning of this term in group theory.

also includes reflections. Notice the inclusions:

$$\mathcal{O}^+(n;\mathbb{R}) \subset SL^+(n;\mathbb{R}) \subset GL^+(n;\mathbb{R}). \tag{5.50}$$

5.9. Group Actions

Let $\mathcal{G}$ be a group (not necessarily a Lie group) and let X be a set (not necessarily a differentiable manifold). We say that the group $\mathcal{G}$ *acts on the right* on the set X if for each $g \in \mathcal{G}$ there is a map $R_g : X \to X$ such that (i) $R_e(x) = x$ for all $x \in X$, where e is the group identity; (ii) $R_g \circ R_h = R_{hg}$ for all $g, h \in \mathcal{G}$. When there is no room for confusion, we also use the notation xg for $R_g(x)$.

EXERCISE 5.8. Prove that each of the maps R_g is a bijection of X (i.e., one-to-one and onto) and that $R_{g^{-1}} = (R_g)^{-1}$. Check that we could have replaced condition (i) with the condition that each R_g be a bijection of X.

DEFINITION 5.5. The action of $\mathcal{G}$ on X is said to be *effective* if the condition $R_g(x) = x$ for every $x \in X$ implies $g = e$. The action is *free* if $R_g(x) = x$ for *some* $x \in X$ implies $g = e$. Finally, the action is *transitive* if for every $x, y \in X$ there exists $g \in \mathcal{G}$ such that $R_g(x) = y$.

DEFINITION 5.6. Given the right action R_g of $\mathcal{G}$ on X, the *orbit* through $x \in X$ is the following subset of X:

$$x\mathcal{G} = \{R_g(x) \mid g \in \mathcal{G}\}. \tag{5.51}$$

In a completely analogous manner, we can say that $\mathcal{G}$ *acts on the left* on X if for each $g \in \mathcal{G}$ there is a map $L_g : X \to X$ such that (i) $L_e(x) = x$ for all $x \in X$, where e is the group identity; (ii) $L_g \circ L_h = L_{gh}$ for all $g, h \in \mathcal{G}$. The order of the composition is the essential difference between a right and a left action. We may also use the notation gx for $L_g(x)$. The orbit through x is denoted as $\mathcal{G}x$.

EXAMPLE 5.6. **Constitutive laws:** Let Φ be the set of all possible hyperelastic constitutive laws of a material point. Thus, each element of Φ is a function:

$$\phi : GL(3;\mathbb{R}) \longrightarrow \mathbb{R}$$
$$\mathbf{F} \mapsto \phi(\mathbf{F}), \tag{5.52}$$

where $\mathbf{F}$ is the deformation gradient. We define the right action of $GL(3;\mathbb{R})$ on Φ by:

$$R_\mathbf{G}\, \phi(\mathbf{F}) := \phi(\mathbf{F}\mathbf{G}) \quad \forall \mathbf{F} \in GL(3;\mathbb{R}), \tag{5.53}$$

where $\mathbf{G} \in GL(3;\mathbb{R})$. This action is certainly not transitive. If it were, all hyperelastic constitutive laws would be essentially the same, via a change of reference configuration. The action is not free either, since there exist constitutive laws with nontrivial symmetries. The action is effective only, since there exist functions with just the trivial symmetry. Note, however, that if the set Φ were restricted to those functions that satisfy the principle of frame-indifference (Section A.10.1, Appendix A), the action would not even be effective, since all such functions have the nontrivial symmetry $-\mathbf{I}$. What is the physical meaning of the orbit of an element ϕ of Φ under the action of $GL(3;\mathbb{R})$?

The notion of group action can naturally be applied when $\mathcal{G}$ is a Lie group. In this instance, a case of particular interest is that for which the set on which $\mathcal{G}$ acts is a differentiable manifold and the induced bijections are transformations of this manifold. Recall that a *transformation* of a manifold $\mathcal{M}$ is a diffeomorphism $\phi : \mathcal{M} \to \mathcal{M}$. The definition of the action is then supplemented with a smoothness condition. More explicitly, we have:

DEFINITION 5.7. A Lie group $\mathcal{G}$ is said to *act on the right* on a manifold $\mathcal{M}$ if:

 (1) Every element $g \in \mathcal{G}$ induces a transformation $R_g : \mathcal{M} \to \mathcal{M}$.
 (2) $R_g \circ R_h = R_{hg}$, namely, $(ph)g = p(hg)$ for all $g,h \in \mathcal{G}$ and $p \in \mathcal{M}$.
 (3) The *right action* $R : \mathcal{G} \times \mathcal{M} \to \mathcal{M}$ is a smooth map. In other words, $R_g(p)$ is differentiable in *both* variables (g and p).

With these conditions, the Lie group $\mathcal{G}$ is also called a *Lie group of transformations of* $\mathcal{M}$. Just as in the general case, we have used the alternative notation pg for $R_g(p)$, with $p \in \mathcal{M}$, wherever convenient.

A similar definition can be given for the *left action* of a Lie group on a manifold, namely:

DEFINITION 5.8. A Lie group $\mathcal{G}$ is said to *act on the left* on a manifold $\mathcal{M}$ if:

 (1) Every element $g \in \mathcal{G}$ induces a transformation $L_g : \mathcal{M} \to \mathcal{M}$. When there is no room for confusion, we use the notation gp for $L_g(p)$, with $p \in \mathcal{M}$.
 (2) $L_g \circ L_h = R_{gh}$, namely, $g(hp) = (gh)p$ for all $g,h \in \mathcal{G}$ and $p \in \mathcal{M}$.
 (3) The *left action* $L : \mathcal{G} \times \mathcal{M} \to \mathcal{M}$ is a smooth map. In other words, $L_g(p)$ is differentiable in *both* variables (g and p).

If e is the group identity, then R_e and L_e are the identity transformation of $\mathcal{M}$. Indeed, since a transformation is an invertible map, every point $p \in \mathcal{M}$ can be expressed as qg for some $q \in \mathcal{M}$ and some $g \in \mathcal{G}$. Using Property ((2)) of the right action we have: $R_e(p) = pe = (qg)e = q(ge) = qg = p$, with a similar proof for the left action.

It is convenient to introduce the following (useful, though potentially confusing) notation. We denote the right action *as a map from* $\mathcal{G} \times \mathcal{M}$ to $\mathcal{M}$ by the symbol R. Thus, $R = R(g,p)$ has two arguments, one in the group and the other in the manifold. Therefore, fixing, a particular element g in the group, we obtain a function of the single variable x which we have already denoted by $R_g : \mathcal{M} \to \mathcal{M}$. But we can also fix a particular element p in the manifold and thus obtain another function of the single variable g. We will denote this function by $R_p : \mathcal{G} \to \mathcal{M}$. A similar scheme of notation can be adopted for a left action L. Notice that the image of R_p (respectively, L_p) is nothing but the orbit $p\mathcal{G}$ (respectively, $\mathcal{G}p$). The potential for confusion arises when the manifold $\mathcal{M}$ happens to coincide with the group $\mathcal{G}$, as described below.

Whenever an ambiguous situation arises, we will resort to the full action function of two variables.

Recall that a Lie group is both a group and a manifold. Thus, it is not surprising that every Lie group $\mathcal{G}$ induces two canonical groups of transformations on itself, one by right action and one by left action, called, respectively, *right translations* and *left translations* of the group. They are defined, respectively, by $R_g(h) = hg$ and $L_g(h) = gh$, with $g, h \in \mathcal{G}$, where the right-hand sides are given by the group multiplication itself. For this reason, it should be clear that these actions are both free (and, hence, effective) and transitive.

5.10. One-Parameter Subgroups

So far, the fact that a Lie group is also a manifold has played only a small role in the definitions, thus perhaps creating the false impression that Lie groups involve only some technical subtleties as compared with the general counterpart. On the contrary, we will discover now that the differentiable structure of the underlying set of a Lie group is crucial to the development of a new conceptual structure, not available in the case of ordinary groups. In particular, since $\mathcal{G}$ is a differentiable manifold, we have available the notion of differentiable curves in $\mathcal{G}$. This gives rise to the following definition:

DEFINITION 5.9. A one-parameter subgroup of a Lie group $\mathcal{G}$ is a differentiable curve:

$$\gamma : \mathbb{R} \longrightarrow \mathcal{G}$$
$$t \mapsto g(t), \tag{5.54}$$

satisfying:

$$g(0) = e, \tag{5.55}$$

and

$$g(t_1)\, g(t_2) = g(t_1 + t_2) \quad \forall t_1, t_2 \in \mathbb{R}. \tag{5.56}$$

If the group $\mathcal{G}$ acts (on the left, say) on a manifold $\mathcal{M}$, the composition of this action with a one-parameter subgroup determines a one-parameter group of transformations of $\mathcal{M}$, namely:

$$\gamma_t(p) = L_{g(t)}(p) \quad p \in \mathcal{M}. \tag{5.57}$$

From Theorem 5.6, we know that associated with this flow there exists a unique vector field $\mathbf{v}^\gamma$. More precisely, we have:

$$\mathbf{v}^\gamma(p) = \left. \frac{d\gamma_t(p)}{dt} \right|_{t=0} \tag{5.58}$$

Fixing the point p, we obtain the map L_p from the group to the manifold. The image of the curve γ under this map is obtained by composition as:

$$t \mapsto L_p(g(t)) = L(g(t), p) = L_{g(t)}(p) = \gamma_t(p), \tag{5.59}$$

where we have used Equation (5.57). In other words, the image of the curve γ (defining the one-parameter subgroup) by the map L_p is nothing but the integral curve of the flow passing through p. By definition of derivative of a map between manifolds, we conclude that the tangent $\mathbf{g}$ to the the one-parameter subgroup γ at the group identity e is mapped by L_{p*} to the vector $\mathbf{v}^{\gamma}(p)$:

$$\mathbf{v}^{\gamma}(p) = \left(L_{p*}\right)_e \mathbf{g}. \tag{5.60}$$

This means that a one-parameter subgroup $g(t)$ appears to be completely characterized by its tangent vector $\mathbf{g}$ at the group identity. We will shortly confirm this fact more fully. The vector field induced on $\mathcal{M}$ by a one-parameter subgroup is called the *fundamental vector field*[7] associated with the corresponding vector $\mathbf{g}$ at the group identity.

Let us now identify the manifold $\mathcal{M}$ with the group $\mathcal{G}$ itself. In this case, we have, as already discussed, two canonical actions giving rise to the left and right translations of the group. We want to reinterpret Equation (5.60) in this particular case. For this purpose, and to avoid the notational ambiguity alluded to above, we restore the fully fledged notation for the action as a function of two variables. We thus obtain:

$$\mathbf{v}^{\gamma}(h) = \left(\frac{\partial L(g,h)}{\partial g}\right)_{g=e} \mathbf{g}. \tag{5.61}$$

Notice that, somewhat puzzlingly, but consistently, this can also be written as:

$$\mathbf{v}^{\gamma}(h) = (R_{h*})_e \mathbf{g}. \tag{5.62}$$

Thus, when defining the action of a one-parameter subgroup from the left, it is the right action whose derivative delivers the corresponding vector field, and vice versa.

5.11. Left- and Right-Invariant Vector Fields on a Lie Group

DEFINITION 5.10. A vector field $\mathbf{v} : \mathcal{G} \to T\mathcal{G}$ is said to be *left invariant* if:

$$\mathbf{v}(L_g h) = L_{g*}\mathbf{v}(h) \qquad \forall g, h \in \mathcal{G}. \tag{5.63}$$

In other words, vectors at one point are dragged to vectors at any other point by the derivative of the appropriate left translation. A similar definition, but replacing L with R, applies to *right-invariant vector fields*.

PROPOSITION 5.8. *A vector field is left invariant if, and only if:*

$$\mathbf{v}(g) = \left(L_{g*}\right)_e \mathbf{v}(e) \quad \forall g \in \mathcal{G}. \tag{5.64}$$

---

[7] We are using the largely accepted terminology (see Chern SS, Chen WH, Lam KS (2000), *Lectures on Differential Geometry*, World Scientific). Choquet-Bruhat *et al.* (op. cit.) prefer the name *Killing vector field*, which is usually reserved for groups of isometries.

PROOF. If the vector field $\mathbf{v}$ is left invariant, substituting e for h in Equation (5.63), Equation (5.64) follows suit. Conversely, let Equation (5.64) be valid. Then we have:

$$\mathbf{v}(L_g h) = \mathbf{v}(gh) = \left(L_{(gh)*}\right)_e \mathbf{v}(e) = \left((L_g \circ L_h)_*\right)_e \mathbf{v}(e) = \left(L_{g*}\right)_h \mathbf{v}(h). \tag{5.65}$$

$\square$

Another way of expressing this result is by saying that there exists a one-to-one correspondence between the set of left- (or right-) invariant vector fields on $\mathcal{G}$ and the tangent space $T_e\mathcal{G}$ at the group identity.

COROLLARY 5.9. *The vector field induced by a one-parameter subgroup acting on the left (right) is right (left) invariant.*

The correspondence between left- (or right-) invariant vector fields established by Proposition 5.8 is clearly linear, since a linear combination of left- (right-) invariant vector fields is again a left- (right-) invariant vector field whose associated vector at the unit is the corresponding linear combination of the vectors associated with the individual fields of departure. This conclusion is easily verified by the very definition of invariant vector field. Thus, the left- (or right-) invariant vector fields form a vector space isomorphic to $T_e\mathcal{G}$. We will show now that this vector space is endowed canonically with a *bracket operation*.

PROPOSITION 5.10. *The Lie bracket $[\mathbf{u},\mathbf{v}]$ of two left- (right-) invariant vector fields is left- (right-) invariant.*

PROOF. By Proposition 4.1 of Chapter 4, we obtain:

$$L_{g*}[\mathbf{u},\mathbf{v}] = [L_{g*}\mathbf{u}, L_{g*}\mathbf{v}] = [\mathbf{u},\mathbf{v}], \tag{5.66}$$

where Equation (5.63) was used. $\square$

We have already established that the vector fields on $\mathcal{G}$ associated with one-parameter subgroups of $\mathcal{G}$ are right or left invariant according to whether the action considered is, respectively, from the left or from the right. We now want to strengthen this result by proving the following proposition.

PROPOSITION 5.11. *The one-parameter subgroups of $\mathcal{G}$ are precisely the integral curves through the unit e of the right- (or left-) invariant vector fields.*

PROOF. Let a one-parameter subgroup γ be given in terms of the function $g = g(t)$. We want to show that this is the integral curve through the unit e of the (right- invariant) vector field it generates by left translation. But this result is already available in the very definition of the flow, Equation (5.57), namely:

$$\gamma_t(e) = L_{g(t)}(e) = g(t), \tag{5.67}$$

where we have taken into consideration that the manifold $\mathcal{M}$ is now identified with $\mathcal{G}$ and, thus, we can choose the point p as the group identity e. Conversely, if a right-invariant vector field is given, its integral curve through e is unique and satisfies the one-parameter subgroup conditions (5.55) and (5.56), as we have learned in earlier sections of this chapter.   □

Notice that the integral curve through the group unit e is one and the same, whether the vector field is right or left invariant. This is not the case for other integral curves. Left and right invariant vector fields corresponding to the same tangent vector at e coincide over $\mathcal{G}$ if, and only if, the Lie group $\mathcal{G}$ is commutative. It is also worthy of notice that the flow generated on $\mathcal{G}$ by a one-parameter subgroup is indeed a global one-parameter group (not pseudo-group) of transformations, namely, defined for all real values of the parameter t.

5.12. The Lie Algebra of a Lie Group

The results of the previous section can be summarized as follows: (i) There exists an isomorphism between the set of right- (or left-) invariant vector fields of a group and the tangent space at the group identity; (ii) there is a one-to-one correspondence between one-parameter subgroups and integral curves of invariant vector fields through the group identity; (iii) the Lie bracket (or *commutator*) of two left- (right-) invariant vector fields is itself a left- (right-) invariant vector field. It follows from this summary that, not only can one completely characterize one-parameter subgroups and their associated right- and left-invariant vector fields by just one piece of information (namely, the tangent vector to the subgroup at the group identity e), but also that the tangent space thereat, $T_e\mathcal{G}$, is endowed with a bilinear operation satisfying conditions (4.32) (i.e., skew-symmetry) and (4.33) (i.e., the Jacobi identity). Thus, the set of left-invariant vector fields (or, equivalently, the tangent space $T_e\mathcal{G}$) becomes a Lie algebra $\mathfrak{g}$ called the *Lie algebra of the group* $\mathcal{G}$. From an intuitive point of view, the elements of the Lie algebra of a Lie group represent infinitesimal approximations, which Sophus Lie himself called *infinitesimal generators* of the elements of the group. Although the infinitesimal generators are in principle commutative (sum of vectors), the degree of noncommutativity of the actual group elements is captured, to first order, by the Lie bracket.

An important technical point needs to be made. We have defined the Lie algebra $\mathfrak{g}$ by means of the Lie bracket operation applied to *left*-invariant vector fields. More specifically, to find the Lie bracket of two vectors $\mathbf{U}, \mathbf{V} \in T_e\mathcal{G}$, one constructs the corresponding left-invariant fields, $\mathbf{u}^L(g)$ and $\mathbf{v}^L(g)$ by:

$$\mathbf{u}^L(g) = \left(L_{g*}\right)_e \mathbf{U}, \tag{5.68}$$

and

$$\mathbf{v}^L(g) = \left(L_{g*}\right)_e \mathbf{V}. \tag{5.69}$$

The Lie bracket of the two *vectors*, $[\mathbf{U}, \mathbf{V}]$, is then defined by:

$$[\mathbf{U}, \mathbf{V}] = [\mathbf{u}^L(g), \mathbf{v}^L(g)]_e, \tag{5.70}$$

where the right-hand side is the Lie bracket of two *fields*, evaluated at the group identity. Had we used instead the *right*-invariant vector fields induced by **U** and **V**, namely:

$$\mathbf{u}^R(g) = (R_{g*})_e \, \mathbf{U}, \tag{5.71}$$

and

$$\mathbf{v}^R(g) = (R_{g*})_e \, \mathbf{V}, \tag{5.72}$$

the Lie bracket would have resulted in:

$$[\mathbf{U}, \mathbf{V}] = [\mathbf{u}^R(g), \mathbf{v}^R(g)]_e. \tag{5.73}$$

There is no reason to expect that the two result be the same. Therefore, strictly speaking, there are two different Lie algebras,[8] the *left Lie algebra* $\mathfrak{g}^L = \mathfrak{g}$ and the *right Lie algebra* $\mathfrak{g}^R$, defined by the bracket operation (5.73). The relation between them is the subject of the following proposition.

PROPOSITION 5.12. *The right and left Lie algebras are isomorphic, their Lie brackets differing only in sign.*

PROOF. We start by defining the *inversion* map $\mathcal{I} : \mathcal{G} \to \mathcal{G}$ defined by $\mathcal{I}(g) = g^{-1}$. This map is a diffeomorphism of manifolds (by definition of Lie group). It maps the left multiplication into the right multiplication, viz.:

$$\mathcal{I}(g)\,\mathcal{I}(h) = \mathcal{I}(hg). \tag{5.74}$$

Let **v** be a left-invariant vector field. Then, for any given $h \in \mathcal{G}$:

$$R_{h*}(\mathcal{I}_* \mathbf{v}) = (R_h \mathcal{I})_* \, \mathbf{v} = (\mathcal{I} L_{h^{-1}})_* \, \mathbf{v} = \mathcal{I}_* \left(L_{h^{-1}*} \mathbf{v} \right) = \mathcal{I}_* \mathbf{v}. \tag{5.75}$$

In other words, $\mathcal{I}$ maps left-invariant vector fields into right-invariant vector fields, and vice versa. On the other hand, as a diffeomorphism, $\mathcal{I}$ preserves the Lie bracket of vector fields. In particular, if **u** and **v** are left-invariant vector fields, we have:

$$[\mathcal{I}_* \mathbf{u}, \mathcal{I}_* \mathbf{v}] = \mathcal{I}_*[\mathbf{u}, \mathbf{v}]. \tag{5.76}$$

This equation is valid at any point, so we can evaluate it at the group identity e, where $\mathcal{I}_* = -\mathbf{I}$ (where **I** is the identity of $T_e \mathcal{G}$), since for any curve (one-parameter subgroup) through e we must have $g^{-1}(t) = g(-t)$. Therefore, denoting $\mathbf{U} = \mathbf{u}(e)$ and $\mathbf{V} = \mathbf{v}(e)$, the previous equation can be written as:

$$[-\mathbf{U}, -\mathbf{V}]_R = -[\mathbf{U}, \mathbf{V}]_L, \tag{5.77}$$

where the brackets are now understood as operations in $T_e \mathcal{G}$, and where we indicate by the subscripts R, L the brackets for the right and left Lie algebras, respectively. Since the bracket is a bilinear operation, we obtain the desired result:

$$[\mathbf{U}, \mathbf{V}]_R = -[\mathbf{U}, \mathbf{V}]_L, \tag{5.78}$$

□

---

[8] See: Olver PJ (1995) *Equivalence, Invariants and Symmetry*, Cambridge University Press.

In other words, the two Lie algebras are isomorphic, and the isomorphism is given by changing the sign of the Lie bracket. Unless indicated otherwise, when we speak of *the* Lie algebra of a Lie group, we mean the one induced by the *left*-invariant vector fields.

EXAMPLE 5.7. **The Lie algebra of** $GL(n;\mathbb{R})$**:** We have already defined the general linear group of $\mathbb{R}^n$ as the set of all nonsingular square matrices of order n with real entries, under the operation of matrix multiplication. That this is a differentiable manifold is mods clear by using as an atlas the natural global chart of $\mathbb{R}^{n^2}$. Thus, an element $Z \in GL(n;\mathbb{R})$ is identified with the entries Z^i_j of its matrix. We place the row index as a superscript and apply the summation convention, unless otherwise indicated. The group identity is given by the identity matrix I whose components are given by the Kronecker symbol δ^i_j. The left action is explicitly given by:

$$(L_G Z)^i_k = G^i_j Z^j_k. \tag{5.79}$$

The tangent space at any point is given by the vector space of *all* square matrices, not necessarily nonsingular, of order n. The natural basis consists of the n^2 matrices obtained by successively setting one entry to 1 and all the rest to 0. The derivative of the left action is expressed by the same formula as the left action. If A is a tangent vector at the group identity, the corresponding left-invariant vector field is given by:

$$(A(Z))^i_k = Z^i_j A^j_k. \tag{5.80}$$

The Lie bracket of two left-invariant vector fields $A(Z)$ and $B(Z)$ is the left-invariant vector field with components:

$$[A(Z), B(Z)]^p_q = (A(Z))^i_j \frac{\partial (B(Z))^p_q}{\partial Z^i_j} - (B(Z))^i_j \frac{\partial (A(Z))^p_q}{\partial Z^i_j}, \tag{5.81}$$

where we have used Equation (4.31). In view of (5.80), we can write:

$$[A(Z), B(Z)]^p_q = Z^p_m A^m_j B^j_q - Z^p_m B^m_j A^j_q. \tag{5.82}$$

Evaluating this expression at $Z = I$, we obtain:

$$[A(Z), B(Z)]^p_q = A^p_j B^j_q - B^p_j A^j_q. \tag{5.83}$$

or, using matrix notation:

$$[A, B] = AB - BA. \tag{5.84}$$

This is, therefore, the Lie bracket of the (left) Lie algebra $\mathfrak{g}$. Repeating, however, the above calculations for right-invariant vector fields, we obtain a different result. Indeed, the right-invariant vector field $\hat{A}(Z)$ corresponding to the tangent vector A at the group identity is given by:

$$(\hat{A}(Z))^i_k = A^i_j Z^j_k. \tag{5.85}$$

The Lie bracket of two right-invariant vector fields $\hat{A}$ and $\hat{B}$ is:

$$[\hat{A}(Z),\hat{B}(Z)]^p_q = (\hat{A}(Z))^i_j \frac{\partial(\hat{B}(Z))^p_q}{\partial Z^i_j} - (\hat{B}(Z))^i_j \frac{\partial(\hat{A}(Z))^p_q}{\partial Z^i_j}. \qquad (5.86)$$

Following the same steps as before, we obtain the Lie bracket of the right Lie algebra as:

$$[A,B]_R = BA - AB = -[A,B]. \qquad (5.87)$$

5.12.1. The Structure Constants of a Lie Group

We have already encountered the concept of structure constants when dealing with moving frames. Equation (4.42) defines the structure constants of a moving frame as the components (in the local frame) of all the possible Lie brackets of the frame elements. If, in the case of the Lie algebra of a Lie group, we adopt a basis in $T_e\mathcal{G}$, we clearly induce a moving frame on $\mathcal{G}$ by means of the corresponding right- (or left-) invariant vector fields. The corresponding structure constants at e are known as the *structure constants of the Lie group* in the chosen basis.

EXERCISE 5.9. Evaluate the structure constants of $GL(n;\mathbb{R})$. Comment on the possible disadvantage arising from having used mixed upper and lower indices to represent the matrix components, instead of just subscripts. [Hint: the components of the base vector $E^b_a := \frac{\partial}{\partial Z^a_b}$ are $\delta^i_a \delta^b_j$.]

EXERCISE 5.10. Let g^i $(i=1,\ldots,n)$ be a coordinate system on $\mathcal{G}$. Then, the group multiplication gh will be given in terms of n smooth functions of $2n$ variables:

$$(gh)^i = L^i(g^1,\ldots,g^n,h^1,\ldots,h^n). \qquad (5.88)$$

A vector field:

$$\mathbf{v} = v^i(g^1,\ldots,g^n)\frac{\partial}{\partial g^i} \qquad (5.89)$$

is left invariant if:

$$v^i(g^1,\ldots,g^n) = \left[\frac{\partial L^i(g^1,\ldots,g^n,h^1,\ldots,h^n)}{\partial h^j}\right]_{g^1,\ldots,g^n,0\ldots,0} v^j(0,\ldots,0), \qquad (5.90)$$

where we have assumed that the coordinate patch contains the group identity and the coordinates vanish thereat. Show[9] that the structure constants of the group in this coordinate patch are given by:

$$c^i_{jk} = \left[\frac{\partial^2 L^i}{\partial g^i \partial h^k} - \frac{\partial^2 L^i}{\partial g^k \partial h^j}\right]_{0,\ldots,0,\ldots,0}. \qquad (5.91)$$

Verify the result of Exercise 5.9.

[9] Choquet-Bruhat et al., op. cit.

5.13. Down-to-Earth Considerations

We have attempted to introduce at least some of the basic notions of Lie groups and their Lie algebras in an intrinsic (coordinate free) manner. A good example is the notion of left-invariant vector fields and of the induced algebra on the tangent space to the group at the identity. Our intention in this section is to try to motivate similar notions while enslaved to a coordinate system, in the hope of securing a further insight into this subject matter.

Assume, therefore, that we are in possession of a "continuum" $\mathcal{G}$ and that a (global) coordinate system g^i ($i = 1, \ldots, n$) therein is at our disposal to describe some physical phenomenon. A point in this continuum will be denoted interchangeably by a single letter, g say, or by the ordered collection of its coordinates: $(g^1, \ldots, g^n)$. The coordinate system is assumed (without any loss of generality) to include the point $e = (0, \ldots, 0)$ with all coordinates equal to zero. The salient feature of this set is the existence of an internally valued function of two points (and, therefore, of $2n$ arguments) with very peculiar properties, justified in each case by the physical context. Let us denote this function by $L = L(g, h)$. Since it takes its values internally (that is, the result of the evaluation of the function is itself a point in $\mathcal{G}$), the symbol L stands, in fact, for n (component) functions of $2n$ arguments:

$$L^i = L^i(g^1, \ldots, g^n, h^1, \ldots, h^n) \quad (i = 1, \ldots, n). \tag{5.92}$$

For these functions to qualify as definers of a Lie group, they must satisfy the following properties:

(1) Each function L^i must be smooth in all its arguments.
(2) The (rectangular) Jacobian matrix $\left\{ \left\{ \frac{\partial L^i}{\partial g^j} \right\} \left\{ \frac{\partial L^i}{\partial h^j} \right\} \right\}$ is such that each of the two $n \times n$ blocks shown is invertible for all values of the variables.
(3) $L(g, e) = L(e, g) = g$ for each $g \in \mathcal{G}$.
(4) $L(L(g, h), k) = L(g, L(h, k))$ for all $g, h, k \in \mathcal{G}$.

Of these four properties, perhaps the most striking is the last one, which is an expression of associativity. Although associativity is very easy to believe in when written in the abstract fashion $(gh)k = g(hk)$, it becomes quite a surprising property when written explicitly in terms of the properties of n functions of $2n$ variables:

$$L^i \left(L^1(g^1, \ldots, h^1, \ldots), \ldots, L^n(g^1, \ldots, h^1, \ldots), k^1, \ldots, k^n \right)$$
$$= L^i \left(g^1, \ldots, g^n, L^1(h^1, \ldots, k^1, \ldots), \ldots, L^n(h^1, \ldots, k^1, \ldots) \right). \tag{5.93}$$

One may say that it is this innocent-looking property that is responsible,[10] more than any other, for the strong results delivered by the theory of Lie groups. We will not pursue this delicate point here. We only remark that, given a candidate group

---

[10] In this respect, see O'Raifeartaigh L (1985), *Group Structure of Gauge Theories*, Cambridge University Press.

operation, one can write a system of partial differential equations acting as some sort of integrability conditions that this operation must satisfy to be actually associative.

Let us now expand the structure functions L^i in a Taylor series around the origin:

$$L^i(g,h) = L^i(e,e) + \left(\frac{\partial L^i}{\partial g^j}\right)_{e,e} g^j + \left(\frac{\partial L^i}{\partial h^j}\right)_{e,e} h^j$$

$$+ \frac{1}{2}\left(\left(\frac{\partial^2 L^i}{\partial g^j \partial g^k}\right)_{e,e} g^j g^k + 2\left(\frac{\partial^2 L^i}{\partial g^j \partial h^k}\right)_{e,e} g^j h^k + \left(\frac{\partial^2 L^i}{\partial h^j \partial h^k}\right)_{e,e} h^j h^k\right) + \cdots$$

$$(5.94)$$

The time has come now to take full advantage of the properties of the functions L^i. By Property (3), we conclude immediately that:

$$L^i(e,e) = 0. \tag{5.95}$$

Moreover, again by Property (3), the first partial derivatives at the origin are evaluated as:

$$\left(\frac{\partial L^i}{\partial g^j}\right)_{e,e} = \left(\frac{\partial L^i(g,e)}{\partial g^j}\right)_{g=e} = \left(\frac{\partial g^i}{\partial g^j}\right)_{g=e} = \delta^i_j, \tag{5.96}$$

with a similar expression for the derivative with respect to the second variable. What we have learned so far is that the first-order approximation of the basic operation of any Lie group around the origin is given by:

$$L^i(g,h) \approx g^i + h^i. \tag{5.97}$$

This remarkable property is the main idea behind the concept of the Lie algebra as the tangent space at the group identity. Indeed, a vector thereat can be considered as a small perturbation around the identity. These infinitesimal perturbations, in contradistinction with their finite counterparts, are always commutative. Moreover, the group operation, usually a complicated combination of the variables, boils down to a simple addition of components.

EXAMPLE 5.8. **The infinitesimal version of the polar decomposition:** The linearized version of many statements and theories in Solid Mechanics is sometimes known informally as the *small-displacement* version of an otherwise *large-displacement* counterpart. Examples are provided by the polar decomposition theorem, which is a statement in $GL(n;\mathbb{R})$, and by theories of multiplicative decomposition of the deformation gradient into components with physical meaning (such as in the theory of finite plasticity or in theories of growth and remodelling or, more generally, in the theory of continuous distributions of defects). To obtain the corresponding infinitesimal expressions in the Lie algebra, we change coordinates by translation so as to assign to the unit matrix the origin of coordinates. In other words, we express every matrix A as $A = I + \Delta A$. With this notation, the polar decomposition theorem (Theorem A.1), for example, can be expressed as:

$$(I + \Delta F) = (I + \Delta R)(I + \Delta U). \tag{5.98}$$

In the Lie algebra, by neglecting the quadratic term, we obtain:

$$\Delta F = \Delta R + \Delta U. \tag{5.99}$$

Moreover, in this particular case, ΔR and ΔU can be shown to be, respectively, skew-symmetric and symmetric matrices.

Returning to Equation (5.94) we notice that, continuing the reasoning leading to Equation (5.96), we obtain:

$$\left(\frac{\partial^2 L^i}{\partial g^j \partial g^k}\right)_{e,e} = \left(\frac{\partial^2 L^i}{\partial h^j \partial h^k}\right)_{e,e} = 0. \tag{5.100}$$

Introducing the results (5.95), (5.96) and (5.100) into Equation (5.94), we can write:

$$L^i(g,h) = g^i + h^i + a^i_{jk} g^j h^k + \cdots, \tag{5.101}$$

where we have notated:

$$a^i_{jk} = \left(\frac{\partial^2 L^i}{\partial g^j \partial h^k}\right)_{e,e}. \tag{5.102}$$

We want to use this quadratic approximation formula to derive the quadratic approximation to the inverse map. By definition of inverse, we have:

$$L^i(g,g^{-1}) = 0 = g^i + g^{-i} + a^i_{jk} g^j g^{-k} + \dots, \tag{5.103}$$

where, for the sake of economy, we have denoted $\left(g^{-1}\right)^i = g^{-i}$. Since we are interested only in the quadratic approximation, we are justified in using repeated substitutions, as long as the quadratic part is preserved. Thus, starting from Equation (5.103) rewritten as:

$$g^{-i} = -g^i - a^i_{jk} g^j g^{-k} + \cdots, \tag{5.104}$$

we write:

$$g^{-i} = -g^i - a^i_{jk} g^j(-g^k + \dots) + \cdots = -g^i + a^i_{jk} g^j g^k + \cdots, \tag{5.105}$$

which is the desired result.

EXERCISE 5.11. Verify, by direct substitution into Equation (5.103), that the second-order approximation to the inverse given by the extreme right-hand side of (5.105), is indeed correct.

One of the salient features of a group operation is that it is, in general, noncommutative. In other words, for two arbitrary elements g and h in $\mathcal{G}$, we have in general that:

$$gh \neq hg. \tag{5.106}$$

Recalling that the only operations available are group multiplication and inversion, a measure of the discrepancy between the left- and right-hand sides of (5.106) can be defined as:

$$C(g,h) := (gh)(hg)^{-1}, \tag{5.107}$$

a quantity known as the *commutator* of g and h. Clearly, the commutator is equal to the identity e if, and only if, $gh = hg$, in which case we say that the two elements commute. Following the general philosophy of this section, we are interested in the infinitesimal version of the commutator in the vicinity of the identity. Using the above expressions, we obtain:

$$C(g,h) = (gh)^i + (hg)^{-i} + a^i_{jk}(gh)^i(hg)^{-i} + \cdots, \qquad (5.108)$$

or, after repeated substitutions:

$$C(g,h) = (a^i_{jk} - a^i_{kj})g^j h^k + \cdots = c^i_{jk}g^j h^k + \cdots \qquad (5.109)$$

In other words, the commutator is approximated by (twice) the skew-symmetric part of a^i_{jk}. The skew-symmetric quantities c^i_{jk} constitute the tensor of structure constants of the group. This tensor is the carrier of the lowest-order approximation to the noncommutativity of a Lie group.

EXERCISE 5.12. Show that the change of coordinates[11]:

$$\hat{g}^i = g^i - \frac{1}{2}a^i_{jk}g^j g^k, \qquad (5.110)$$

eliminates the symmetric part of a^i_{jk} altogether from (5.101).

EXERCISE 5.13. By direct computation of their transformation under coordinate changes, show that $\{c^i_{jk}\}$ indeed constitute a tensor. Is that the case for a^i_{jk} as well?

At the group identity e (namely, at the origin of coordinates) we have the natural base vectors:

$$\mathbf{e}_i = \frac{\partial}{\partial g^i}. \qquad (5.111)$$

Given two vectors, $\mathbf{u} = u^i\mathbf{e}_i$ and $\mathbf{v} = v^i\mathbf{e}_i$ in $T_e\mathcal{G}$, we *define* their Lie product $[\mathbf{u},\mathbf{v}]$ as the vector with components:

$$[\mathbf{u},\mathbf{v}]^i := c^i_{jk}u^j v^k. \qquad (5.112)$$

Since $c^i_{jk} = -c^i_{kj}$, we immediately obtain:

$$[\mathbf{u},\mathbf{v}] = -[\mathbf{v},\mathbf{u}]. \qquad (5.113)$$

Moreover,

$$[[\mathbf{u},\mathbf{v}],\mathbf{w}]^i + [[\mathbf{w},\mathbf{u}],\mathbf{v}]^i + [[\mathbf{v},\mathbf{w}],\mathbf{u}]^i = \left(c^i_{mn}c^m_{jk} + c^i_{mk}c^m_{nj} + c^i_{mj}c^m_{kn}\right)u^j v^k w^n. \qquad (5.114)$$

In order for the Jacobi identity (4.33) to be satisfied, we need the identical vanishing of the expression within parentheses on the right-hand side of Equation (5.114). The verification that this is the case, is left as a not very pleasant exercise for the reader. It

---

[11] For this exercise, as well as for a no-nonsense treatment of Lie groups, see Belinfante JGF, Kolman B (1972), *A Survey of Lie Groups and Lie Algebras with Applications and Computational Methods*, Society for Industrial and Applied Mathematics.

is based this time on the enforcement of the associative Property (4). In this section we have learned how all the important features of Lie groups and their Lie algebras can be constructed from the bottom up.

5.14. The Adjoint Representation

Fix an element $g \in \mathcal{G}$ and associate to it the map:

$$ad(g) = L_g R_g^{-1} : \mathcal{G} \longrightarrow \mathcal{G}, \tag{5.115}$$

given more explicitly[12] by:

$$h \mapsto ghg^{-1} \quad \forall h \in \mathcal{G}. \tag{5.116}$$

The map $ad(g)$ is easily seen to be a bijection (in fact, a diffeomorphism of the group onto itself). Since it preserves the group multiplication and the taking of inverses, it is also a *group automorphism*. In particular, it maps one-parameter subgroups into one-parameter subgroups.

EXERCISE 5.14. Verify that the above properties are satisfied.

The (nonsingular) tangent map:

$$Ad(g) := ad(g)_*|_{h=e} : T_e\mathcal{G} \longrightarrow T_e\mathcal{G}, \tag{5.117}$$

is called the *adjoint map* associated with $g \in \mathcal{G}$. Recalling that the tangent space $T_e\mathcal{G}$ is canonically isomorphic to the Lie algebra $\mathfrak{g}$, we can regard $Ad(g)$ as an automorphism of the Lie algebra. This is schematically represented in Figure 5.2, where the generic element h is close to the group unit e, so that one can, albeit imprecisely, regard it as a vector in the Lie algebra, that is, in $T_e\mathcal{G}$. For each $g \in \mathcal{G}$ we have thus created an element $Ad(g)$ of the general linear group $GL(\mathfrak{g})$. Moreover,

$$Ad(g_1g_2) = ad(g_1g_2)_*|_e = ad(g_1)_*|_e\, ad(g_2)_*|_e = Ad(g_1)Ad(g_2), \tag{5.118}$$

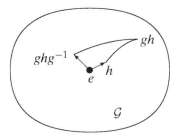

Figure 5.2. Schematic depiction of the map $Ad(g)$

[12] Notice that if the group is Abelian, this map reduces (for every g) to the identity $id_{\mathcal{G}}$ of $\mathcal{G}$. If the group is not commutative, then the nonempty collection of elements g for which the map $ad(g)$ equals $id_{\mathcal{G}}$ is a subgroup of $\mathcal{G}$ called the *centralizer* of $\mathcal{G}$.

where the right-hand side is the ordinary composition in $GL(\mathfrak{g})$. A similar calculation can be carried out for the inverse g^{-1}. If we let g vary, we may regard $Ad(g)$ as a map from the group $\mathcal{G}$ to the general linear group $GL(\mathfrak{g})$. Since this map preserves multiplications and inverses,[13] it is a group homomorphism.

DEFINITION 5.11. The group homomorphism:

$$Ad : \mathcal{G} \longrightarrow GL(\mathfrak{g}), \tag{5.119}$$

is called the *adjoint representation* of $\mathcal{G}$.

Notice that, choosing a basis in $\mathfrak{g}$, we have obtained a representation of the Lie group as a subgroup of $GL(n;\mathbb{R})$, that is, a matrix group.

LEMMA 5.13. *If we regard, as we may, the map $Ad(g)$ as a map between left-invariant vector fields $\mathbf{V}$ (rather than just elements of $T_e\mathcal{G}$), we have:*

$$Ad(g)\mathbf{V} = \left(R_{g^{-1}}\right)_* \mathbf{V} \quad \forall\, \mathbf{V} \in \mathfrak{g}. \tag{5.120}$$

More precisely,

$$Ad(g)\mathbf{V}(h) = \left(R_{g^{-1}}\right)_* \mathbf{V}(gh) \quad \forall\, \mathbf{V} \in \mathfrak{g},\ h \in \mathcal{G}. \tag{5.121}$$

PROOF. The result is a direct consequence of the fact that we have defined $\mathfrak{g}$ as the algebra of the *left* invariant vector fields on $\mathcal{G}$. □

We would like to have an analogous result for fundamental vector fields associated with the right action of $\mathcal{G}$ on a manifold $\mathcal{M}$. It is easier to work with the flows themselves. Let $h(t)$ be the one-parameter subgroup corresponding to a vector $\mathbf{v} \in T_e\mathcal{G}$. The value of the corresponding fundamental vector field $\mathbf{X_v}(p)$ at a point $p \in \mathcal{M}$ is nothing but the tangent vector to the curve $ph(t)$ evaluated at $t = 0$. Let $g \in \mathcal{G}$ be a fixed element in the group. Then, at the point $q = pg$ of $\mathcal{M}$, the value of the same fundamental vector field $\mathbf{V}(q)$ is the tangent vector at $t = 0$ of the curve in $\mathcal{M}$ given by:

$$qh(t) = pgh(t) = p(gh(t)g^{-1})g = R_g\left(p(gh(t)g^{-1})\right). \tag{5.122}$$

But the tangent at p to the curve $p(gh(t)g^{-1})$ (which clearly passes through p for $t = 0$) is the value of the fundamental vector field corresponding to the vector $Ad(g)\mathbf{v}$, by definition of the adjoint map. We then conclude that the image by R_{g*} of this vector is precisely the value of the fundamental vector field corresponding to $\mathbf{v}$ at q. We have thus proved the following:

LEMMA 5.14.

$$\mathbf{X_v}(pg) = \left(R_g\right)_* \mathbf{X}_{Ad(g)\mathbf{v}}(p). \tag{5.123}$$

---

[13] The preservation of inverses is, in fact, a direct consequence of the preservation of the product.

6 Integration and Fluxes

6.1. Integration of Forms in Affine Spaces

As a first step towards a theory of integration of differential forms on manifolds, we will present the particular case of integration on certain subsets of an affine space (without necessarily having an inner-product structure). In Section 2.8.3 we introduced the rigorous concept of an *affine simplex* and, later, in Section 3.4, we developed the idea of the multivector uniquely associated to an oriented affine simplex. Moreover, we have already advanced, on physical grounds, the notion that the evaluation of an r-form on an r-vector conveys the meaning of the calculation of the physical content of the quantity represented by the form within the volume represented by the multivector. For this idea to be of any practical use, we should be able to pursue it to the infinitesimal limit. Namely, given an r-dimensional domain $\mathcal{D}$ in an affine space $\mathcal{A}$ and, given at each point of this domain a (continuously varying, say) r-form ω, we would like to be able to subdivide the domain into small r-simplexes and define the total content as the limit of the sum of the evaluations of the r-forms on a point of each of the simplexes. In this way, we would have a generalization of the concept of Riemann integral.

6.1.1. Simplicial Complexes

A domain of integration within an n-dimensional affine space $\mathcal{A}$ may be a rather general set. It may, for instance, be an embedded submanifold. Nevertheless, we will content ourselves first with a class of domains of integration consisting of a collection of affine simplexes that *fit together* perfectly. Formally, we have:

DEFINITION 6.1. A simplicial complex K is a finite collection of affine simplexes such that: (i) each face of a simplex of K is itself in K; (ii) the intersection of any two simplexes of K is either empty or a common face (and, hence, a simplex in K).

Moreover,

DEFINITION 6.2. The set of all points contained in the simplexes of the complex K is called the *polyhedron* $|K|$ associated with K.[1] Conversely, the simplicial complex K is called a *(simplicial) subdivision* or a *triangulation* of the polyhedron $|K|$.

Definition 6.2 is general enough to accommodate as a polyhedron, for example, the union of a solid triangle and a segment sticking out of one of its vertices, as one can easily verify. Let us, therefore, restrict attention to complexes that are generated by a finite collection of distinct r-dimensional simplexes, with a fixed r. Such a complex will consist precisely of these generating simplexes and all their faces of all orders. Any two r-simplexes either have no common points or they intersect along a common face of any order. We will call such a complex an *r-dimensional complex*, and the associated polyhedron an *r-dimensional polyhedron*.[2] Moreover, to simplify further, we will assume that every simplex of dimension $r - 1$ (i.e., every face of order $r - 1$ of the generating r-simplexes) is a face of no more than two r-simplexes. Notice that this condition does not guarantee either that there are no "strangulation" points, or that the underlying polyhedron is connected. Nevertheless, with this condition (satisfied automatically if $r = n$) the polyhedron $|K|$ looks quite close to an r-dimensional manifold with boundary (which it is, after all, our intention to eventually approximate). Finally, we will assume that each of the generating r-simplexes has been oriented and that, with the consistent orientation of the faces of order $r - 1$, the $(r - 1)$-faces common to two r-simplexes have opposite orientations in the two intersecting simplexes. This last condition one may or may not be able to satisfy. If one can, the polyhedron is said to be *orientable* and, with one particular choice of consistent orientation, it becomes an *oriented* r-dimensional polyhedron. The same terminology is used for the corresponding complex. As a classical counterexample, a (triangulated) Moebius band can be seen as a *non-orientable* 2-dimensional polyhedron.

It is intuitively clear that, just as in the finite element method, any domain of physical interest may be approximated by a suitably chosen r-dimensional simplicial complex consisting of sufficiently small simplexes. Before entering such conceptual ground, however, we can consider a domain that is *exactly* covered by a simplicial complex or, more specifically, an oriented r-dimensional polyhedron. We now define the notion of *simplicial refinement*, also akin to the mesh refinement idea of the finite element method.

DEFINITION 6.3. A simplicial complex K' is a *refinement* of the simplicial complex K if: (i) $|K| = |K'|$, and (ii) every simplex of K' is contained in a simplex of K. It follows that every simplex of K is a union of simplexes of K'.

Since our point of departure is an r-dimensional-oriented simplicial complex K, we will consider refinements which are also of this kind and with the same orientation as K, a notion that should be intuitively clear.

---

[1] See Pontryagin LS (1999), *Foundations of Combinatorial Topology*, Dover.
[2] See Alexandroff P (1961), *Elementary Concepts of Topology*, Dover.

6.1.2. The Riemann Integral of an r-Form

Let $|K|$ be the oriented r-dimensional polyhedron associated with the oriented r-dimensional simplicial complex $K = K_0$ in the n-dimensional affine space $\mathcal{A}$ ($r \leq n$). Let $\omega = \omega(p)$ be a (continuous) assignment of an r-form to each point $p \in |K|$. Then the Riemann integral of ω over $|K|$ is defined as:

$$\int_{|K|} \omega = \lim_{n \to \infty} \sum_{K_n} \langle \omega(p_i), \{\sigma_i\} \rangle, \tag{6.1}$$

where K_n is a refinement of K_{n-1} for each $n > 0$, $\{\sigma_i\}$ is the multivector of the i-th oriented of r-simplex σ_i in K_n, and p_i is an arbitrary interior point of σ_i.

EXERCISE 6.1. Identifying $\mathcal{A}$ with $\mathbb{R}^2$, consider the rectangular domain $|K| = [a,b] \times [c,d]$. Let $\omega = f(x,y)dx \wedge dy$. Show that:

$$\int_{|K|} \omega = \int_{|K|} f \, dx \, dy, \tag{6.2}$$

where the right-hand side is the ordinary Riemann integral of calculus.

6.1.3. Simplicial Chains and the Boundary Operator

Let K be an r-dimensional-oriented simplicial complex and let σ_i^s denote the (oriented) s-simplexes in K ($s \leq r$). The number of such simplexes is clearly finite, since we have generated the complex out of a finite number of r-simplexes. Consider the (infinite) collection of formal linear combinations:

$$A^s = \sum_i a_i \sigma_i^s, \quad a_i \in \mathbb{R}. \tag{6.3}$$

Each A^s so defined is called a *(simplicial) s-chain of the complex K*. We would like to endow the collection $\mathcal{A}^s$ of all s-chains of K with the structure of a vector space. The *addition* of s-chains is defined trivially by:

$$A^s + B^s = \sum_i a_i \sigma_i^s + \sum_i b_i \sigma_i^s := \sum_i (a_i + b_i)\sigma_i^s. \tag{6.4}$$

We define the *zero s-chain* 0^s as that for which all the coefficients are zero. The *multiplication* by a scalar $\alpha \in \mathbb{R}$ is defined as:

$$\alpha A^s = \alpha \sum_i a_i \sigma_i^s := \sum_i (\alpha a_i)\sigma_i^s. \tag{6.5}$$

With these two operations, $\mathcal{A}^s$ becomes a vector space of dimension equal to the number of s-simplexes in K. For convenience of notation, if a term in a chain has a zero coefficient, we may omit it. Moreover, in the physical interpretation, we identify the chain $1\sigma_i^s$ with the oriented simplex σ_i^s, and the chain $-1\sigma_i^s$ with the simplex having the opposite orientation. For $s = 0$ (that is, for the vertices of the complex), this interpretation is irrelevant.

The Riemann integral of an s-form ω over the simplicial chain A^s (6.3) is defined by linearity as:

$$\int_{A^s} \omega := \sum_i a_i \int_{|\sigma_i^s|} \omega. \tag{6.6}$$

Recall (Section 3.4) that to each oriented s-simplex σ^s in an affine space we can associate a unique s-vector, which we denote by $\{\sigma^s\}$. Given a simplicial chain $A^s = \sum_i a_i \sigma_i^s$, we define the s-vector $\{A^s\}$ associated with A^s as the linear combination:

$$\{A^s\} := \sum_i a_i \{\sigma_i^s\}. \tag{6.7}$$

Of particular interest for an oriented r-dimensional simplicial complex is the r-chain with all coefficients equal to 1, for obvious reasons.

The *boundary* $\partial \sigma^s$ of the oriented s-simplex σ^s is the oriented $(s-1)$-simplicial chain defined as:

$$\partial \sigma^s := \sum_{i=0}^{s} \sigma_i^{s-1}, \tag{6.8}$$

where σ_i^{s-1} is the consistently oriented $(s-1)$-face opposite the i-the vertex of the s-simplex. By convention, the boundary of a 0-simplex vanishes. For the oriented 1-simplex (p_0, p_1) the boundary is the 0-chain $p_1 - p_0$. The boundary of a chain is defined by regarding ∂ as a linear operator:

$$\partial \sum_i a_i \sigma_i^s := \sum_i a_i \partial \sigma_i^s. \tag{6.9}$$

A direct computation shows that the boundary of the boundary of an oriented s-simplex vanishes. We conclude, by linearity, that this is also true for an arbitrary simplicial chain:

$$\partial^2 A^s := \partial \partial A^s = 0. \tag{6.10}$$

This result is consistent with the fact that, as discussed in Section 3.5, the sum of the multivectors of the consistently oriented faces of a simplex vanishes.

EXERCISE 6.2. Prove (6.10) for an oriented s-simplex.[3]

Notice how Equation (6.10) formally resembles Equation (4.97). This resemblance is more than coincidental, and we will show later how, in a precise sense connected to the theory of integration, the exterior differentiation and the boundary operators are dual of each other.

---

[3] See Pontryagin, op. cit.

6.1.4. Integration of n-Forms in $\mathbb{R}^n$

We have already suggested (Exercise 6.1) that our definition of Riemann integral of a form over a polyhedron in an affine space $\mathcal{A}$ reduces, when $\mathcal{A} = \mathbb{R}^n$, to the ordinary multiple Riemann integral of the coordinate representation of the form. Following this lead, we will now *define* the Riemann integral of a smooth n-form ω over an arbitray domain in $\mathbb{R}^n$ as follows. Let $x^1, \ldots, x^n$ be the standard global chart of $\mathbb{R}^n$, and let ω be defined over some open set $\mathcal{D} \subset \mathbb{R}^n$. There exists, then, a smooth function $f : \mathcal{D} \to \mathbb{R}$ such that:

$$\omega = f \, dx^1 \wedge \cdots \wedge dx^n. \tag{6.11}$$

For any regular domain of integration $\mathcal{A} \subset \mathcal{D}$ we define:

$$\int_{\mathcal{A}} \omega := \underbrace{\int \int \ldots \int}_{\mathcal{A}} f dx^1 dx^2 \ldots dx^n, \tag{6.12}$$

where the right-hand side is the ordinary n-fold Riemann integral in $\mathbb{R}^n$.

It is important to check that this definition is independent of the coordinate system adopted in $\mathcal{D}$. For this purpose, let:

$$\phi : \mathcal{D} \longrightarrow \mathbb{R}^n \tag{6.13}$$

be a coordinate transformation expressed in components as the n smooth functions:

$$x^1, \ldots, x^n \mapsto y^1(x^1, \ldots, x^n), \ldots, y^n(x^1, \ldots, x^n). \tag{6.14}$$

Recall that for (6.14) to qualify as a coordinate transformation, the Jacobian determinant:

$$J = \det \left[\frac{\partial(y^1, \ldots, y^n)}{\partial(x^1, \ldots, x^n)} \right], \tag{6.15}$$

must be nonzero throughout $\mathcal{D}$. For definiteness, we will assume that it is strictly positive (so that the change of coordinates is orientation preserving). According to the formula of transformation of variables under a multiple Riemann integral, we must have:

$$\underbrace{\int \int \ldots \int}_{\mathcal{A}} f(x^i) \, dx^1 \ldots dx^n = \underbrace{\int \int \ldots \int}_{\mathcal{A}} f(x^i(y^j)) \, J^{-1} dy^1 \ldots dy^n. \tag{6.16}$$

But, according to Equation (4.78), the representation of ω in the new coordinate system is precisely:

$$\omega = f(x^i(y^j)) \, J^{-1} dy^1 \wedge \ldots \wedge dy^n, \tag{6.17}$$

which shows that the definition (6.12) is indeed independent of the coordinate system adopted in $\mathcal{D}$.

A more fruitful way to exploit the coordinate independence property is to regard $\phi : \mathcal{D} \to \mathbb{R}^n$ not as a mere coordinate transformation but as an actual change of the domain of integration. In this case, the transformation formula is interpreted readily in terms of the pull-back of ω as:

$$\int_{\phi(\mathcal{A})} \omega = \int_{\mathcal{A}} \phi^*(\omega), \tag{6.18}$$

for every n-form ω defined over an open set containing $\phi(\mathcal{A})$.

6.2. Integration of Forms on Chains in Manifolds

6.2.1. Singular Chains in a Manifold

We have so far introduced the notion of a simplex and of simplicial chains in an affine space. In the very definition of an affine simplex (Section 2.8.3) the fact that we could form linear combinations of points was crucial. In an arbitrary n-dimensional differentiable manifold $\mathcal{M}$, on the other hand, we cannot afford that luxury and we have to content ourselves with a definition that makes use of the fact that a manifold is locally diffeomorphic to $\mathbb{R}^n$ and, thus, it is possible to map a simplex in $\mathbb{R}^n$ into $\mathcal{M}$ differentiably.

It is convenient to start by introducing the notion of *standard s-simplex* in $\mathbb{R}^s$. To this effect, we define a collection of $s+1$ points a_i^s ($i = 0, \ldots, n$) in $\mathbb{R}^s$ as follows:

$$a_0^s = \{0, \ldots, 0\} \quad a_i^s = \{0, ..0, \underbrace{1}_{i}, 0, \ldots 0\} \quad (i = 1, \ldots, s). \tag{6.19}$$

In other words, a_0^s is the origin of $\mathbb{R}^s$ while a_i^s is the point located at one unit along the positive i-th axis, as shown in Figure 6.1. The standard s-simplex is defined as the affine (positively) oriented simplex with these points as vertices. We will denote it as $\Delta^s = (a_0^s, a_1^s, \ldots, a_s^s)$. The standard 0-simplex is the origin singleton $\{0\}$.

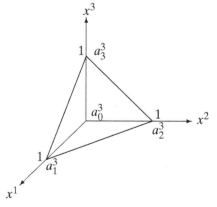

Figure 6.1. The standard 3-simplex (solid tetrahedron) in $\mathbb{R}^3$

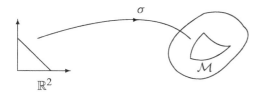

Figure 6.2. A 2-simplex in a manifold $\mathcal{M}$

DEFINITION 6.4. A (differentiable) *singular s-simplex* σ in the n-dimensional manifold $\mathcal{M}$ is the restriction to Δ^s of a smooth map of a neighbourhood of Δ^s into $\mathcal{M}$.

Figure 6.2 represents a singular 2-simplex. We shall often use the shorter terminology *s-simplex* instead of singular s-simplex. A (singular) *s-chain* in $\mathcal{M}$ is a finite linear combination of s-simplexes with real coefficients.

REMARK 6.1. The reason for the terminology "singular" is that we are only requiring smoothness (and not requiring, for example, that the map be an embedding), so that the image $\sigma(\Delta^s)$ may exhibit self-intersections or other such "singular" features.

The definition of the boundary of an s-chain in $\mathcal{M}$ requires some subtlety. Notice that we cannot simply say that the various faces of σ are the images by σ of the corresponding faces of Δ^s. The reason for this is that, if we want a face of σ to be an $(s-1)$-chain in $\mathcal{M}$, we need to exhibit a map from Δ^{s-1} (rather than from a face of Δ^s) to $\mathcal{M}$. To fix this subtle issue, therefore, we introduce the *face maps* F_i^s $(i = 0,\ldots,s)$ as the affine maps (see Definition 2.1) $F_i^s : \mathbb{R}^{s-1} \to \mathbb{R}^s$ defined as follows: each vertex a_j of Δ^{s-1} with $j < i$, if any, is mapped into the vertex a_j of Δ^s, while each vertex a_j of Δ^{s-1} with $j \geq i$, if any, is mapped into the vertex a_{j+1} of Δ^s. Under these conditions, the affine map is uniquely defined. In practical terms, the face map F_i^s maps the standard simplex Δ^{s-1} onto the face of Δ^s opposite to the i-th vertex. Figure 6.3 shows the meaning of the face maps for the case $s = 2$. With these maps in place, we can now define rigorously the i-th face of the s-simplex σ in $\mathcal{M}$ as the map $\sigma^i : \Delta^{s-1} \to \mathcal{M}$ given by:

$$\sigma^i = (-1)^i \sigma \circ F_i^{s-1}. \tag{6.20}$$

The factor $(-1)^i$ is there to guarantee that the faces are consistently oriented according to our definition in Section 3.5.

EXERCISE 6.3. Coordinate expressions of the face maps: Let $x^1,\ldots,x^s$ and $y^1,\ldots,y^{s-1}$ denote the standard coordinates in $\mathbb{R}^s$ and $\mathbb{R}^{s-1}$, respectively. Show that

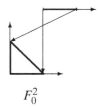
F_0^2

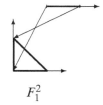
F_1^2

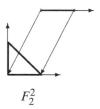
F_2^2

Figure 6.3. Face maps of Δ^2

the coordinate expressions of $F_i^s : \mathbb{R}^{s-1} \to \mathbb{R}^s$ are:

$$
F_0^s : \begin{cases}
x^1 = 1 - (y^1 + \cdots + y^{s-1}) \\
x^2 = y^1 \\
x^3 = y^2 \\
\quad . \\
\quad . \\
x^s = y^{s-1}
\end{cases}
\tag{6.21}
$$

$$
F_1^s : \begin{cases}
x^1 = 0 \\
x^2 = y^1 \\
x^3 = y^2 \\
\quad . \\
\quad . \\
x^s = y^{s-1}
\end{cases}
\quad
F_2^s : \begin{cases}
x^1 = y^1 \\
x^2 = 0 \\
x^3 = y^2 \\
\quad . \\
\quad . \\
x^s = y^{s-1}
\end{cases}
\quad \cdots \quad
F_s^s : \begin{cases}
x^1 = y^1 \\
x^2 = y^2 \\
x^3 = y^3 \\
\quad . \\
\quad . \\
x^s = 0
\end{cases}
\tag{6.22}
$$

[Hint: verify that each of these maps is an affine map and that it satisfies the conditions established in terms of mapping vertices to vertices.]

The boundary $\partial\sigma$ of a singular s-simplex σ in $\mathcal{M}$ is defined as the $(s-1)$-simplex:

$$
\partial\sigma = \sum_{i=0}^{s} \sigma^i.
\tag{6.23}
$$

The boundary of an s-chain in $\mathcal{M}$ is defined similarly by linearity.

6.2.2. Integration of Forms over Chains in a Manifold

The Riemann integral of an s-form ω over the s-simplex σ in the n-dimensional manifold $\mathcal{M}$ is defined as:

$$
\int_\sigma \omega := \int_{\Delta^s} \sigma^*(\omega).
\tag{6.24}
$$

Note that this definition is clearly inspired by and consistent with Equation (6.18). The integral over an s-chain $c = \sum_i \alpha_i \sigma_i$ in $\mathcal{M}$ is defined by linearity as:

$$
\int_c \omega := \sum_i \alpha_i \int_{\sigma_i} \omega.
\tag{6.25}
$$

6.2.3. Stokes' Theorem for Chains

A fundamental result of the theory of integration is the following:

THEOREM 6.1. **Stokes' theorem**: *For every s-chain c (with $s \geq 1$) in a manifold $\mathcal{M}$ and every smooth $(s-1)$-form ω defined on a neighbourhood of the image of c:*

$$\int_{\partial c} \omega = \int_c d\omega. \tag{6.26}$$

REMARK 6.2. This important theorem is a generalization of the fundamental theorem of calculus ("integration is the inverse of differentiation"). It establishes the precise relation between the operators d and ∂. With the proper interpretation in various contexts, it reproduces all the theorems of classical vector calculus (such as the divergence theorem, Green's formulas, and so on).

PROOF. Because of linearity, it is clear that it is sufficient to prove the theorem for the case of a single s-simplex σ in $\mathcal{M}$. Most presentations[4] simplify the proof even further by considering the image under σ of the unit cube (rather than the standard simplex). We will, however, follow schematically the lines of the proof[5] based on the s-simplex, which we have already toiled to define. Although somewhat tedious, the proof is quite elementary. Thus, we need to prove that for a singular s-simplex σ:

$$\int_{\partial \sigma} \omega = \int_\sigma d\omega. \tag{6.27}$$

Using (6.20) and (6.24), we may rewrite this as:

$$\sum_{i=0}^{s} (-1)^i \int_{\Delta^{s-1}} F_i^{s-1*} \circ \sigma^*(\omega) = \int_{\Delta^s} d\left(\sigma^*(\omega)\right), \tag{6.28}$$

where we have invoked Exercise 4.19 and Proposition 4.2. The case $s = 1$ is trivial, so we assume $s \geq 2$. The pull-back of ω by σ is an $(s-1)$-form in $\mathbb{R}^s$. It will therefore, have, a coordinate expression given by:

$$\sigma^*(\omega) = \sum_{k=1}^{s} f_k \, dx^1 \wedge \ldots \wedge \widehat{dx^k} \wedge \ldots \wedge dx^s, \tag{6.29}$$

where a hat over a term indicates its absence. The components $f_k = f_k(x^1, \ldots, x^s)$ are smooth functions of the coordinates. By linearity, it is clear that we need to prove the theorem for just one term in this sum, which we may choose to be the first. In other words, we replace (6.29) by:

$$\sigma^*(\omega) = f_1 \, dx^2 \wedge \ldots \wedge dx^s, \tag{6.30}$$

[4] See, e.g., Spivak M (1965), *Calculus on Manifolds*, Westview Press.
[5] See Warner FW (1983), *Foundations of Differentiable Manifolds and Lie Groups*, Springer.

and note that the other terms can be dealt with analogously by an appropriate renaming of the variables. Under these conditions, the right-hand side of (6.28) becomes:

$$\int_{\Delta^s} d\left(\sigma^*(\omega)\right) = \int_{\Delta^s} \frac{\partial f_1}{\partial x^1}\, dx^1 \wedge \ldots \wedge dx^s. \tag{6.31}$$

By virtue of Equation (6.12), we obtain:

$$\int_{\Delta^s} d\left(\sigma^*(\omega)\right) = \int_0^1 \cdots \int_0^{1-(x^3+\cdots+x^s)} \left(\int_0^{1-(x^2+\cdots+x^s)} \frac{\partial f_1}{\partial x^1}\, dx^1 \right) dx^2 \ldots dx^s, \tag{6.32}$$

or, finally:

$$\int_{\Delta^s} d\left(\sigma^*(\omega)\right) = \tag{6.33}$$

$$\int_0^1 \cdots \int_0^{1-(x^3+\cdots+x^s)} \left(f_1(1 - (x^2 + \cdots + x^s), x^2, \ldots, x^s) - f_1(0, x^2, \ldots, x^s) \right) dx^2 \ldots dx^s.$$

Our job is now to prove that the left-hand side of (6.28) gives the same result. For any function $\phi : \mathbb{R}^{s-1} \to \mathbb{R}^s$, expressed in coordinates as some functions $x^i = x^i(y^j)$ with $i = 1, \ldots, s$ and $j = 1, \ldots, s-1$, we have:

$$\phi^*(\sigma^*(\omega)) = f_1(x^i(y^j)) \frac{\partial x^2}{\partial y^{m_1}} \cdots \frac{\partial x^2}{\partial y^{m_{s-1}}}\, dy^{m_1} \wedge \cdots \wedge dy^{m_{s-1}}, \tag{6.34}$$

where we have used (6.30) and where the summation convention is enforced. In the evaluation of the left-hand side of (6.28), therefore, we should use this formula for each function F_i^s ($i = 0, \ldots, s$). But, according to the expressions of Exercise 6.3, the only nonvanishing results will correspond to the cases $i = 0, 1$, since all the other cases will have a vanishing differential. Thus, from Equation (6.21), we obtain:

$$F_0^{s*}(\sigma^*(\omega)) = f_1(1 - (y^1 + \cdots + y^{s-1}), y^1, \ldots, y^{s-1})\, dy^1 \wedge \cdots \wedge dy^{s-1}, \tag{6.35}$$

and from (6.22):

$$F_1^{s*}(\sigma^*(\omega)) = f_1(0, y^1, \ldots, y^{s-1})\, dy^1 \wedge \cdots \wedge dy^{s-1}. \tag{6.36}$$

Introducing these results into the left-hand side of (6.28) and using (6.12), we obtain:

$$\sum_{i=0}^s (-1)^i \int_{\Delta^{s-1}} F_i^{s-1*} \circ \sigma^*(\omega) \tag{6.37}$$

$$= \int_0^1 \cdots \int_0^{1-(y^2+\cdots+y^{s-1})} \left(f_1(1 - (y^1 + \cdots + y^{s-1}), y^1, \ldots, y^{s-1}) - f_1(0, y^1, \ldots, y^{s-1}) \right) dy^1 \ldots dy^{s-1},$$

which is the same as the right-hand side of (6.33). This completes the proof.   □

EXAMPLE 6.1. **The divergence theorem in** $\mathbb{R}^3$: Let $\mathcal{D}$ be a triangulated domain in $\mathbb{R}^3$ and let:

$$\mathbf{v} = A(x^1,x^2,x^3)\frac{\partial}{\partial x^1} + B(x^1,x^2,x^3)\frac{\partial}{\partial x^2} + C(x^1,x^2,x^3)\frac{\partial}{\partial x^3} \qquad (6.38)$$

be a vector field defined on an open set containing $\mathcal{D}$. By means of the star operator, defined in Section 3.8, we can associate to this vector field the 2-form β given by:

$$\beta = A\,dx^2 \wedge dx^3 - B\,dx^1 \wedge dx^3 + C\,dx^1 \wedge dx^2. \qquad (6.39)$$

The exterior derivative of this form is:

$$d\beta = \left(\frac{\partial A}{\partial x^1} + \frac{\partial B}{\partial x^2} + \frac{\partial C}{\partial x^3}\right) dx^1 \wedge dx^2 \wedge dx^3, \qquad (6.40)$$

which we have already encountered in Example 4.13. This 3-form can in turn be interpreted as a scalar field,[6] namely, the divergence $\nabla\cdot\mathbf{v}$ of the original vector field. In particular, we have:

$$\iiint_{\mathcal{D}} \nabla\cdot\mathbf{v}\, dx^1\,dx^2\,dx^3 = \int_{\mathcal{D}} d\beta. \qquad (6.41)$$

Now let:

$$x^i = x^i(\xi^1,\xi^2) \quad (i=1,2,3) \qquad (6.42)$$

be (at least locally) the equation of a surface $\mathcal{S}$ in $\mathbb{R}^3$ within the domain of definition of β. We have, by direct calculation, from (6.39) and (6.42) and taking (6.12) into consideration:

$$\int_{\mathcal{S}} \beta = \iint_{\mathcal{S}} \begin{vmatrix} A & B & C \\[4pt] \dfrac{\partial x^1}{\partial \xi^1} & \dfrac{\partial x^2}{\partial \xi^1} & \dfrac{\partial x^3}{\partial \xi^1} \\[10pt] \dfrac{\partial x^1}{\partial \xi^2} & \dfrac{\partial x^2}{\partial \xi^2} & \dfrac{\partial x^3}{\partial \xi^2} \end{vmatrix} d\xi^1 d\xi^2. \qquad (6.43)$$

But this expression can also be written in conventional $\mathbb{R}^3$ vector calculus notation as:

$$\int_{\mathcal{S}} \beta = \iint_{\mathcal{S}} \mathbf{v}\cdot\left(\frac{\partial\mathbf{r}}{\partial\xi^1} \times \frac{\partial\mathbf{r}}{\partial\xi^2}\right) d\xi^1 d\xi^2 = \int_{\mathcal{S}} \mathbf{v}\cdot\mathbf{n}\,dS, \qquad (6.44)$$

where $\mathbf{n}$ is the unit normal to the surface and dS is the so-called element of area. Identifying a surface $\mathcal{S}$ successively with each boundary face, and using therein the parametrization induced via the corresponding face map, the unit vector $\mathbf{n}$ becomes (by consistency of orientation) the *exterior* unit normal to the boundary $\partial\mathcal{D}$. Integrating over the boundary chain, we obtain:

$$\int_{\partial\mathcal{D}} \beta = \int_{\partial\mathcal{D}} \mathbf{v}\cdot\mathbf{n}\,dS. \qquad (6.45)$$

---

[6] An interesting subtlety is that, since the star operator has been invoked twice, the result is independent of the orientation of $\mathcal{D}$.

From Equations (6.41), (6.45) and Stokes' theorem, and with $dV = dx^1 dx^2 dx^3$, we finally obtain the divergence theorem in $\mathbb{R}^3$ as:

$$\int_{\mathcal{D}} \nabla \cdot \mathbf{v} \, dV = \int_{\partial \mathcal{D}} \mathbf{v} \cdot \mathbf{n} \, dS. \qquad (6.46)$$

EXERCISE 6.4. **The curl theorem in $\mathbb{R}^3$:** Given an oriented surface bounded by an oriented curve, the classical curl theorem asserts that the flux of the curl of a vector field on the surface equals the circulation of this field over the boundary curve. Show that this theorem is a particular case of Stokes' formula (6.26).

6.3. Integration of Forms on Oriented Manifolds

The theory of integration over chains in manifolds is very elegant and provides a bridge between techniques of combinatorial and point-set topology. It is also very general, as it allows for rather singular domains of integration with arbitrary formal multipliers. It does not require orientability of the underlying set. In principle, the theory of integration over regular domains in a smooth manifold should be obtainable from it, and indeed this is the case. Nevertheless, the interaction between simplexes and charts is a delicate issue that needs to be treated with care. On the other hand, it is possible to derive, more or less independently, a complete theory of integration on regular domains of an oriented manifold, a theory that is usually all that one may need in applications. Charts (not even mentioned in the theory of integration over chains) play a prominent role here from the start. In every chart, things are reasonably clear. The way to glue these local results together to obtain a coherent global result is provided by a technique known as *partition of unity*. This device is very useful in many other unrelated contexts (such as the theory of connections), so that it is reasonable to start with a brief introduction to the topic.

6.3.1. Partitions of Unity

At the outset, it is important to keep in mind the few topological notions introduced in Section 4.2, particularly the following: (open) covering, subcovering, Hausdorff, and compact. We will need a few more definitions.

DEFINITION 6.5. Given two coverings $\mathcal{U} = \{\mathcal{U}_\alpha\}$ and $\mathcal{V} = \{\mathcal{V}_\beta\}$ of a topological space $\mathcal{T}$, $\mathcal{V}$ is said to be a *refinement* of $\mathcal{U}$ if for every member V_β of $\mathcal{V}$ there exists a member U_α of $\mathcal{U}$ such that $V_\beta \subset U_\alpha$.

DEFINITION 6.6. A collection of subsets $\mathcal{U} = \{\mathcal{U}_\alpha\}$ (in particular, a covering) of a topological space $\mathcal{T}$ is *locally finite* if every point $p \in \mathcal{T}$ has a neighbourhood $\mathcal{N}(p)$ such that $\mathcal{N}(p) \cap U_\alpha \neq \emptyset$ for only a finite number of indices α. In other words, $\mathcal{N}(p)$ intersects only a finite number of members of the collection.

DEFINITION 6.7. A topological space is *paracompact* if it is Hausdorff and if every covering has a locally finite refinement.

DEFINITION 6.8. Let $\mathcal{T}$ be a topological space and let $A \subset \mathcal{T}$ be a subset. The *closure* of A, denoted by $\overline{A}$, is the intersection of all the closed sets of $\mathcal{T}$ that contain A. In other words, the closure of A is the smallest closed set that contains A.

DEFINITION 6.9. Let $f : \mathcal{T} \to \mathbb{R}$ be a function defined over the topological space $\mathcal{T}$. The *support* of f, denoted by $supp(f)$, is the smallest closed set of $\mathcal{T}$ outside which $f \equiv 0$. In terms of the previous definition, we can write:

$$supp(f) = \overline{f^{-1}(\mathbb{R} - \{0\})}. \tag{6.47}$$

We are now in a position to define a partition of unity of a manifold $\mathcal{M}$.

DEFINITION 6.10. A *partition of unity* of a manifold $\mathcal{M}$ is a collection $\{\phi\}$ of non-negative smooth functions $\phi_\alpha : \mathcal{M} \to \mathbb{R}$ with compact support (where α may range over an arbitrary set of indices) such that the following conditions are satisfied:

(1) Local finiteness: The collection of supports $\{supp(\phi_\alpha)\}$ is locally finite,
(2) Unit sum: $\sum_\alpha \phi_\alpha(p) = 1$ for all $p \in \mathcal{M}$.

DEFINITION 6.11. A partition of unity is *subordinate to the covering* $\{\mathcal{U}\}$ if for every ϕ_α there exists some $U_\beta \in \{\mathcal{U}\}$ such that $supp(\phi_\alpha) \subset U_\beta$. This definition does not require that the two sets of indices be the same.

A first constructive step towards the establishment of the existence of partitions of unity consists of exhibiting what kind of functions one can expect to encounter in this regard when working in $\mathbb{R}^n$. We start by defining the following function in $\mathbb{R}$:

$$f(t) = \begin{cases} e^{-1/t} & t > 0 \\ 0 & t \leq 0. \end{cases} \tag{6.48}$$

This function, though not analytic, is smooth (C^∞). Its value and the values of its derivatives of all orders are zero at the origin. On the basis of this function, we can construct the smooth *bump* $h(t)$ by:

$$h(t) = \left(\frac{f(t+2)}{f(t+2)+f(-t-1)} \right) \left(\frac{f(2-t)}{f(2-t)+f(t-1)} \right). \tag{6.49}$$

This C^∞ bump function vanishes outside the closed interval $[-2,2]$, is equal to its maximum value 1 in the closed interval $[-1,1]$ and is positive elsewhere. Notice that $supp(h) = [-2,2]$, so that h has compact support.

EXERCISE 6.5. Produce a plot of the bump function (6.49). Where and why does this function fail to be analytic?

Once this bump function has been constructed for $\mathbb{R}$, it can be generalized to $\mathbb{R}^n$ by spherical symmetry, namely:

$$h_n(x^1, \ldots, x^n) := h(|x|), \tag{6.50}$$

where:

$$|x| = \sqrt{(x^1)^2 + \cdots + (x^n)^2}. \tag{6.51}$$

This function has as its support the closed ball in $\mathbb{R}^n$ of radius 2. Alternatively, we can define the bump function:

$$\hat{h}_n(x^1,\ldots,x^n) := h(x^1)h(x^2)\ldots h(x^n), \tag{6.52}$$

whose compact support is the n-dimensional cube of side 4 around the origin.

The bump functions defined so far have a mesa-like appearance with a pronounced constant plateau. Equally admissible functions for a partition of unity can be obtained without this feature. For example, the function $\tilde{h}_{n,a} : \mathbb{R}^n \to \mathbb{R}$ given by the formula:

$$\tilde{h}_{n,a}(x^1,\ldots,x^n) = \begin{cases} e^{-\frac{a^2}{a^2-|x|^2}} & |x| \leq a \\ 0 & |x| > a \end{cases} \tag{6.53}$$

is C^∞, non-negative and has as a support the closed ball with centre at the origin and radius a.

EXERCISE 6.6. Let $\mathcal{U}_a$ be a countable covering of $\mathbb{R}^n$ obtained by open balls of radius a with centres at a countable collection of points P_i (for example, set $a = 2$ and choose as centres all the points with integer coordinates). Show that the functions:

$$\phi_i(X) = \frac{\tilde{h}_{n,a}(X - P_i)}{\sum\limits_{j=1}^{\infty} \tilde{h}_{n,a}(X - P_j)}, \tag{6.54}$$

constitute a partition of unity of $\mathbb{R}^n$ subordinate to the covering $\mathcal{U}_b$, with $b > a$. Notice that, for any given point $X = (x^1,\ldots,x^n)$, the summation in (6.54) involves only a finite number of nonzero terms. The same idea can be used with any of the other bump functions introduced above.

THEOREM 6.2. *For every manifold $\mathcal{M}$ and every atlas in $\mathcal{M}$, there exists a partition of unity subordinate to the atlas.*

PROOF. The proof of the theorem makes use of the fact that, according to the definition, manifolds are second-countable topological spaces. It can then be shown that they are paracompact. What this means is that the construction that we have shown in Exercise 6.6 can be essentially imported to any manifold.[7]   □

6.3.2. Definition of the Integral

In this section (as in so many others) we will purposely avoid a number of important technical issues. For example, the detailed definition of a domain of integration, the fact that we may be dealing with manifolds with boundary, and so on, are issues that we are going to bypass cavalierly, while trying to stick to the bare conceptual bones of the theory.[8]

---

[7] For a complete proof of the theorem see, e.g., Warner or Lee, op. cit.
[8] Once again, the reader is referred to Warner or Lee or Spivak (op. cit.) for a detailed treatment.

Let $\mathcal{M}$ be an oriented manifold and let $(\mathcal{U}, \psi)$ be a (consistently) oriented chart. The integral of an m-form ω over $\mathcal{U}$ is defined as:

$$\int_{\mathcal{U}} \omega := \int_{\psi(\mathcal{U})} \left(\psi^{-1}\right)^* \omega. \tag{6.55}$$

Notice that the right-hand side is a standard Riemann (or Lebesgue) integral of a function in $\mathbb{R}^n$, according to Equation (6.12). Let now ω be assumed to have a compact support in $\mathcal{M}$. We want to define the integral of ω over an n-dimensional submanifold (perhaps with boundary, a detail that we omit) $\mathcal{D} \subset \mathcal{M}$. In particular, we may have $\mathcal{D} = \mathcal{M}$, which we assume in what follows. Let $\{\mathcal{U}_i, \psi_i\}$ $(i = 1, \ldots, p)$ be a finite cover of $supp(\omega)$, where each $\{\mathcal{U}_i, \psi_i\}$ is a consistently oriented chart. We can choose a partition of unity ϕ_i (of the union of the domains of these charts) subordinate to this cover. We define:

$$\int_{\mathcal{M}} \omega := \sum_i \int_{\mathcal{M}} \phi_i \omega, \tag{6.56}$$

where the integrand on the right-hand side is just the product of a function times a differential form. Each integral on the right-hand side is well defined by (6.55), since the integrand has a compact support in a single chart. Notice that when integrating over a submanifold $\mathcal{D}$, one should consider the corresponding intersections.

For the definition implied by Equation (6.56) to make sense, we must prove that the result is independent of the choice of charts and of the choice of partition of unity. This can be done quite straightforwardly by expressing each integral of the right-hand side of one choice in terms of the quantities of the second choice, and vice versa.

6.3.3. Stokes' Theorem

Given an n-dimensional submanifold $\mathcal{D} \subset \mathcal{M}$, its boundary $\partial \mathcal{D}$ can be consistently oriented (by a similar procedure as was done for the faces of a simplex). Thus, given an $(n-1)$-form on $\mathcal{M}$ with compact support, it makes sense to calculate its integral over the (oriented) boundary $\partial \mathcal{D}$, which is an oriented $(n-1)$-dimensional manifold. Stokes' theorem asserts that:

$$\int_{\partial \mathcal{D}} \omega = \int_{\mathcal{D}} d\omega. \tag{6.57}$$

This is formally the same statement as for the case of chains, but requires a fresh proof that we shall omit.

6.4. Fluxes in Continuum Physics

One of the basic notions of Continuum Physics is that of an *extensive property*, a term that describes a property that may be assigned to *subsets* of a given universe, such as the mass of various parts of a material body, the electrical charge enclosed in a certain region of space, and so on. Mathematically speaking, therefore, an extensive property is expressed as a real-valued *set function p*, whose argument ranges over subsets $\mathcal{R}$ of a universe $\mathcal{U}$. It is usually assumed, on physical grounds, that the function p is additive, namely:

$$p(\mathcal{R}_1 \cup \mathcal{R}_2) = p(\mathcal{R}_1) + p(\mathcal{R}_2) \quad \text{whenever } \mathcal{R}_1 \cap \mathcal{R}_2 = \emptyset. \tag{6.58}$$

With proper regularity assumptions, additivity means that, from the mathematical standpoint, p is a *measure* in $\mathcal{U}$.

In the appropriate space-time context, the *balance* of an extensive property expresses a relation between the rate of change of the property in a given region and the causes responsible for that change. Of particular importance is the idea of *flux* of the property through the boundary of a region, which is an expression of the rate of change of the property as a result of interaction with other regions. It is a common assumption that the flux between regions takes place through, and only through, common boundaries. In principle, the flux is a set function on the boundaries of regions. In most physical theories, however, this complicated dependence can be greatly simplified by means of the so-called *Cauchy postulates* and Cauchy's theorem.[9]

6.4.1. Extensive-Property Densities

We will identify the universe $\mathcal{U}$ as an m-dimensional differentiable manifold. Under appropriate continuity assumptions, a set function such as the extensive property p is characterized by a *density*. Physically, this means that the property at hand cannot be concentrated on subsets of dimension lower than m. More specifically, we assume that the density ρ of the extensive property p is a smooth m-form on $\mathcal{U}$ such that:

$$p(\mathcal{R}) = \int\limits_{\mathcal{R}} \rho, \tag{6.59}$$

for any subset $\mathcal{R} \subset \mathcal{U}$ for which the integral is defined. Clearly, the additivity condition (6.58) is satisfied automatically.

We note at this point that, although we have heuristically introduced the domains of integration as subsets of $\mathcal{U}$, in fact we may wish to allow a laxer interpretation that corresponds to the theory of integration over chains, as presented earlier, where simplexes can be multiplied by arbitrary coefficients. There are more advanced theories of integration, such as Whitney's Geometric Integration Theory,[10] whereby the domains of integration are even more general (such as fractal chains). We will review the general lines of this important line of approach in Section 6.5.

---

[9] See Segev R (2000), The geometry of Cauchy's fluxes, *Archive for Rational Mechanics and Analysis* **154**, 183–198. The material presented from here to the end of this chapter is largely based on various publications by Segev.

[10] Whitney H (1957) , *Geometric Integration Theory*, Princeton University Press.

We introduce now the time variable t as if space-time were just a product manifold $\mathbb{R} \times \mathcal{U}$. In fact, this trivialization is observer-dependent, but it will serve for our present purposes. The density ρ of the extensive property p should, accordingly, be conceived as a function $\rho = \rho(t,x)$, where $x \in \mathcal{U}$. Notice that, since for fixed x and variable t, ρ belongs to the same vector space $\Lambda^m \left(T_x^* \mathcal{U} \right)$, it makes sense to take the partial derivative with respect to t to obtain the new m-form:

$$\beta = \frac{\partial \rho}{\partial t}, \tag{6.60}$$

defined on $\mathcal{U}$. For a fixed (i.e., time-independent) region $\mathcal{R}$, we may write:

$$\frac{dp(\mathcal{R})}{dt} = \int_{\mathcal{R}} \beta. \tag{6.61}$$

In other words, the integral of the m-form β over a fixed region measures the rate of change of the content of the property p inside that region.

6.4.2. Balance Laws, Flux Densities and Sources

In the classical setting of Continuum Mechanics it is assumed that the change of the content of a smooth extensive property p within a fixed region $\mathcal{R}$ can be attributed to just two causes: (1) the rate at which the property is produced (or destroyed) within $\mathcal{R}$ by the presence of sources and sinks, and (2) the rate at which the property enters or leaves $\mathcal{R}$ through its boundaries, namely, the *flux* of p. For the sake of definiteness, in this section we adopt the convention that the production rate is positive for sources (rather than sinks) and that the flux is positive when there is an outflow (rather than an inflow) of the property. The *balance equation* for the extensive property p states that *the rate of change of p in a fixed region $\mathcal{R}$ equals the difference between the production rate and the flux*. A good physical example is the balance of internal energy in a rigid body due to volumetric heat sources and heat flux through the boundaries.

Since we have assumed continuity for p as a set function, we will do the same for both the production and the flux. As a result, we postulate the existence of an m-form s, called the *source density* such that the production rate in a region $\mathcal{R}$ is given by the integral:

$$\int_{\mathcal{R}} s. \tag{6.62}$$

Just as ρ itself, the m-form s is defined over all of $\mathcal{U}$ and is independent of $\mathcal{R}$. Thus, from the physical point of view, we are assuming that the phenomena at hand can be described locally (this assumption excludes interesting phenomena, such as internal actions at a distance or surface-tension effects).

As far as the flux term is concerned, we also assume that it is a continuous function of subsets of the boundary ∂R. We postulate the existence, for each region $\mathcal{R}$, of a smooth $(m-1)$-form $\tau_{\mathcal{R}}$, called the *flux density*, such that the flux of p is given by:

$$\int_{\partial\mathcal{R}} \tau_{\mathcal{R}}. \tag{6.63}$$

Thus, the classical balance law of the property p assumes the form:

$$\int_{\mathcal{R}} \beta = \int_{\mathcal{R}} s - \int_{\partial\mathcal{R}} \tau_{\mathcal{R}}. \tag{6.64}$$

An equation of balance is said to be a *conservation law* if both s and $\tau_{\mathcal{R}}$ vanish identically.

6.4.3. Flux Forms and Cauchy's Formula

We note that (beyond the obvious fact that β and s are m-forms, whereas $\tau_{\mathcal{R}}$ is an $(m-1)$-form), there is an essential complication peculiar to the flux densities $\tau_{\mathcal{R}}$. Indeed, in order to specify the flux for the various regions of interest, it seems that one has to specify the form $\tau_{\mathcal{R}}$ for each and every region $\mathcal{R}$. In other words, while the rate of change of the property and the production term are specified by forms whose domain (for each time t) is the entire space $\mathcal{U}$, the flux term must be specified by means of a set function, whose domain is the collection of all regions. We refer to the set function $\mathcal{R} \mapsto \tau_{\mathcal{R}}$ as a *system of flux densities*. Consider, for example, a point $x \in \mathcal{U}$ belonging simultaneously to the boundaries of two different regions. Clearly, we do not expect that the flux density will be the same for both. The example of sun-tanning, used in Appendix A in connection with Equation (A.47), should be sufficiently convincing in this regard. Consider, however, the following particular case. Let the natural *inclusion* map:

$$\iota : \partial\mathcal{R} \longrightarrow \mathcal{U}, \tag{6.65}$$

be defined by:

$$\iota(x) = x \quad \forall x \in \partial\mathcal{R}. \tag{6.66}$$

Notice that this formula makes sense, since $\partial\mathcal{R} \subset \mathcal{U}$. Moreover, the map ι is smooth. It can, therefore, be used to pull back forms of any order on $\mathcal{U}$ to forms of the same order on $\partial\mathcal{R}$. In particular, we can define:

$$\int_{\partial\mathcal{R}} \phi := \int_{\partial\mathcal{R}} \iota^*(\phi), \tag{6.67}$$

for any form ϕ on $\mathcal{U}$. Let us now assume the existence of a globally defined $(m-1)$-*flux form* Φ on $\mathcal{U}$ and let us define the associated system of flux densities by means of the formula:

$$\tau_{\mathcal{R}} := \iota^*_{\partial\mathcal{R}}(\Phi), \tag{6.68}$$

where we use the subscript $\partial\mathcal{R}$ to emphasize the fact that each region requires its own inclusion map. Equation (6.68) is known as *Cauchy's formula*. Clearly, this is a very special system of flux densities (just as a conservative force field is a special vector field derivable from a single scalar field). Nevertheless, it is one of the fundamental results of classical Continuum Mechanics that, under rather general assumptions (known as *Cauchy's postulates*), every system of flux densities can be shown to derive from a unique flux form using Cauchy's formula (6.68). We will omit the general proof of this fact, known as Cauchy's theorem.[11]

In less technical terms, Cauchy's formula is the direct result of assuming that the flux is given by a *single* 2-form defined over the 3-dimensional domain of the body. The fact that one and the same form is to be used for a given location, and integrated over the given boundary, is trivially seen to imply (and generalize) the linear dependence of the flux on the normal to the boundary, as described in the standard treatments (such as Section A.8 of Appendix A). This fact, elementary as it is, doesn't seem to have been generally noticed in most treatments of Continuum Mechanics. It opens the door, however, to the point of view that the most general boundaries that one might want to consider are those over which differential forms can be legitimately defined. This is the point at which Whitney's geometric integration theory is ready to make its contribution, as we shall show in Section 6.5.

EXERCISE 6.7. **Linear dependence:** Show that, when working in $\mathbb{R}^3$, the existence of a single flux 2-form at each point can be reinterpreted as the linear dependence of the flux on the normal to the boundary at that point (via the inner product with a *flux vector*). [Hint: emulate the relevant parts of Exercise 6.1.]

EXERCISE 6.8. **Piola transformation:** Show that the Piola transformation given by Equation (A.51) is automatically contained in the expression of the pull-back of the flux 2-form. [Hint: refer to Exercise 4.20.]

6.4.4. Differential version of the Balance Law

Assuming the existence of a flux form Φ, the general balance law (6.64) can be written as:

$$\int_{\mathcal{R}} \beta = \int_{\mathcal{R}} s - \int_{\partial \mathcal{R}} \iota_{\partial \mathcal{R}}^*(\Phi). \tag{6.69}$$

Using Stokes' theorem (Equation (6.57)), we can rewrite the last term as:

$$\int_{\partial \mathcal{R}} \iota_{\partial \mathcal{R}}^*(\Phi) = \int_{\mathcal{R}} d\Phi, \tag{6.70}$$

where the dependence on $\partial \mathcal{R}$ has evaporated. Using this result, we write (6.69) as:

$$\int_{\mathcal{R}} \beta = \int_{\mathcal{R}} s - \int_{\mathcal{R}} d\Phi. \tag{6.71}$$

---

[11] The interested reader should consult: Segev (op. cit.) and also Segev R, Rodnay G (1999), Cauchy's theorem on manifolds, *Journal of Elasticity* **56**, 129–144.

Since this balance law should be valid for arbitrary $\mathcal{R}$, and since the forms β, s and Φ are defined globally and independently of the region of integration, we obtain:

$$\beta = s - d\Phi. \tag{6.72}$$

This equation is known as the *differential version of the general balance law*. It is the differential-geometric counterpart of Equation (A.54).

6.5. General Bodies and Whitney's Geometric Integration Theory

Given the importance of fluxes in the formulation of the fundamental laws of Continuum Mechanics, the following question should arise naturally: What is the most general subset of $\mathbb{R}^3$ that can sustain fluxes of a physically reasonable kind? If, for some specific definition of this reasonableness, one were able to answer this question, one would have indirectly obtained the answer to an even more important question, namely: What is the most general material body that can sustain the formalism of Continuum Mechanics? We have already pointed out the insufficiency of the standard definition of a body as a 3-dimensional manifold (without boundary) that can be covered with a single chart. Leaving the theoretical aspects aside, consider the everyday phenomenon of a drop of oil advancing in water, or a phase boundary, or the interface between two possibly reacting components of a compound, or the growth of crystals. These interfaces or boundaries tend to evolve in rather sophisticated ways, with finger-like structures being formed spontaneously, or self-replicating fractal patterns emerging naturally. Laws of balance (of momentum, energy) need to be formulated not only within the interior of the nice regions, but also at the interfaces themselves. One of the active areas of research today is precisely the concept of the so-called configurational (or Eshelby) forces responsible for the evolution of these phenomena. Therefore, the question as to how weird a boundary can be and still allow for the calculation of fluxes is of paramount importance. It turns out that Hassler Whitney's geometric integration theory has something important to say about this issue. In the preface to his widely revered but much less widely studied book[12] on the subject, Whitney himself shows that he was aware of this fact. In the opening paragraph of the Preface he states: "In various branches of mathematics and its applications, in particular in differential geometry and in physics, one has often to integrate a quantity over an r-dimensional manifold $\mathcal{M}$ in n-space $\mathbb{E}^n$, for instance, over a surface in ordinary 3-space." He continues: "it is important to know in what manner the integral over $\mathcal{M}$ depends on the position of $\mathcal{M}$ in $\mathbb{E}^n$, assuming the quantity to be integrated is defined throughout a region $\mathcal{R}$ containing $\mathcal{M}$. The main purpose of [geometric integration theory] is to study this function, in a broad geometric and analytic setting."

An essential property of fluxes in field theories is that they should be bounded with respect to both the volume and the area of their domains. It can be shown that the space of flat chains defined by Whitney is the largest possible Banach space

---

[12] Whitney, op. cit.

for which these bounded fluxes are continuous. Geometric integration theory is the mathematical discipline that deals with these issues. The more recent field of geometric measure theory[13] offers an alternative (dual) approach to the subject, but geometric integration theory, by adopting as its point of departure a space of domains of integration, is closer in spirit to Continuum Mechanics, as already noticed by Whitney himself when using examples of fluid flow to illustrate the more intuitive aspects of his theory. The realization of the relevance of these remarks in particular and of geometric integration theory in general to Continuum Mechanics must be credited to Reuven Segev, some of whose ideas we try to summarize in the following sections.

6.5.1. Polyhedral Chains

Although one could develop the theory in terms of simplicial chains, Whitney prefers to work with polyhedral chains. The general setting is the affine space $\mathbb{E}^n$ of dimension n, with supporting space V, an inner-product space. A *closed half-space* of $\mathbb{E}^n$ consists of an affine subspace $\mathbb{P}$ of dimension $n - 1$ together with all the points of $\mathbb{E}^n$ lying to one side of $\mathbb{P}$. A *(convex polyhedral) cell* σ in $\mathbb{E}^n$ is a nonempty bounded (closed) subset of $\mathbb{E}^n$ expressible as the intersection of a finite number of closed half-spaces. The smallest affine subspace containing σ determines (through its own dimension[14]) the dimension of σ. If that dimension is r, we call σ an r-cell. More loosely, an r-cell σ consists of a closed bounded part of an r-plane (i.e., an r-dimensional subspace of $\mathbb{E}^n$) bounded by a finite number of pieces of $(r - 1)$-planes. Note that r can vary between 0 and n. These r-cells are some of the simplest figures on which we want to be able to perform integration. Since an r-plane has precisely two different orientations, an r-cell can (and will) always be oriented (by orienting its plane). We denote by $-\sigma$ the cell consisting of the same points as σ, but with the opposite orientation of its plane.

The boundary $\partial\sigma$ of an r-cell σ consists of a finite number of $(r - 1)$-cells (a 0-cell has no boundary). It is important to realize that each of these cells can be oriented in a manner consistent with the orientation of σ, just as we have done for simplexes. The trick resides in the fact that to choose an orientation of the r-plane containing σ is tantamount to choosing a parity for one (and, therefore, every) ordered set of r linearly independent vectors lying in this plane. This means that, given any one of the cells making up the boundary, all we need to do to define its orientation is to give the parity of any set of $r - 1$ linearly independent vectors lying on its plane. This is easily achieved by adding a first vector to the set in such a way that it points away from the interior of σ and that it lies in its plane.

Having thus defined oriented r-cells and their oriented boundaries, we are in a position to define *polyhedral r-chains* as formal finite linear combinations of r-cells with real coefficients. We need some extra conditions to make this work. First, we

---

[13] See Federer H (1969), *Geometric Measure Theory*, Springer.

[14] Notice that this dimension is not necessarily n, since two tetrahedra, for example, may intersect along an edge rather than a face.

identify 1σ with σ, and -1σ with $-\sigma$ (the oppositely oriented cell). The product 0σ is the empty set. Moreover, if an oriented r-cell σ is the union of the oriented r-cells σ_i with disjoint interiors (namely, if one cuts a cell into pieces), then the original cell σ is identified with the linear combination $\Sigma_i \sigma_i$. Under these conditions, and with the obvious definitions of addition and multiplication by a scalar, we obtain a vector space C_r. Clearly, the collection of all oriented r-cells forms a basis of this vector space. The boundary of an r-chain is the $(r-1)$-chain defined by linearity as:

$$\partial \left(\sum_i a_i \sigma_i \right) = \sum_i a_i \partial \sigma_i, \tag{6.73}$$

We remark that polyhedral r-chains live in $\mathbb{E}^n$ with $r \leq n$ and that, therefore, the r-cells used in forming a linear combination do not necessarily lie in the same r-plane.

The main idea behind Whitney's geometric integration theory is the completion of the various vector spaces C_r of polyhedral chains by means of certain norms satisfying certain optimality conditions. Whitney defines two such norms and calls them flat and sharp.[15] Before giving a precise definition of these norms, it is important to recall that the Euclidean r-volume (remembering that the supporting space V is an inner-product space) can be used to construct the obvious Euclidean norm:

$$\left| \sum_i a_i \sigma_i \right| = \sum_i |a_i| \, |\sigma_i|, \tag{6.74}$$

where $|\sigma|$ denotes the Euclidean volume of the r-cell σ . This norm is called the *mass* of the polyhedral r-chain. Completion with respect to the mass norm would give a very dull result, and many of the physically interesting domains would escape the scope of the theory of integration. In other words, this norm is too large, resulting in the fact that many interesting sequences are not Cauchy sequences. We need smaller norms. In fact, what we need is the smallest possible norms that still guarantee certain continuity conditions for the integral. To an engineering-trained mind, Whitney's achievement in determining these norms is nothing short of prodigious.

6.5.2. The Flat Norm

The flat norm $|A|^\flat$ of the polyhedral r-chain A is defined by the formula:

$$|A|^\flat = \inf\{|A - \partial D| + |D|\}, \tag{6.75}$$

where the *infimum*[16] is to be evaluated considering all $(r+1)$-chains D in $\mathbb{E}^n$. This is undoubtedly a difficult definition. It does not provide a constructive means for

---

[15] It appears that this use of musically inspired terms and symbols in Mathematics is the first of its kind (it is now widespread, for example, to denote certain isomorphisms between vectors and 1-forms). A generalization of Whitney's approach by Jenny Harrison introduces yet another norm, the so-called *natural norm*, also musically inspired and notated. See Harrison J (1993), Stokes' Theorem for Nonsmooth Chains, *Bulletin of the American Mathematical Society* **29**, 235–242.

[16] Recall that the infimum (or greatest lower bound) $\inf(H)$ of a subset $H \in \mathbb{R}$ is the largest real number smaller than or equal to every number in H.

actually calculating the norm, since one has to consider *all* $(r + 1)$-chains D in $\mathbb{E}^n$. Moreover, it is not even clear that this is indeed a norm. Let us check this last point. It is not difficult to verify that $|aA|^\flat \leq |a||A|^\flat$, and that $|A + B|^\flat \leq |A|^\flat + |B|^\flat$. From these inequalities it follows that for the zero chain: $|0|^\flat = 0$ and that $|A|^\flat \geq 0$ for all A. What does not appear to be too obvious is that $|A|^\flat = 0 \Rightarrow A = 0$. In other words, it is easy enough to prove that $|.|^\flat$ is a semi-norm, but the proof that it is a norm takes some work. In fact, Whitney asks the reader to suspend disbelief for a good 20 very dense pages before giving the proof.

EXERCISE 6.9. **The flat norm of a bracket:** The flat norm is notoriously difficult to calculate. Its importance resides in the determination of Cauchy sequences, for which upper bounds usually suffice. On the other hand, it may prove an interesting exercise to try to evaluate the flat norm of at least one nontrivial example. Consider an equal legged, right-angled bracket consisting of two sequentially oriented mutually perpendicular unit segments $p_0 p_1$ and $p_1 p_2$. Show that an upper bound for the flat norm of this 1-chain in $\mathbb{E}^2$ is $1 + \pi/4$. Suggest why it is likely that this is indeed the value of the flat norm. [Hint: consider D as the oriented region enclosed by $p_0 p_1, p_1 p_2$ and a curve joining p_2 with p_0. Formulate the finding of the infimum as a variational problem. Alternatively, if you are only interested in checking the upper bound, choose this curve as the quarter circumference with centre located at the missing corner of the square defined by p_0, p_1, p_2.]

Eventually, accepting that $|.|^\flat$ is indeed a norm, we shall justify why it has been defined in this apparently strange way. Recall that what we intend to do with this norm is to create a Banach space of Cauchy sequences of polyhedral r-chains (we will call the elements of this completion "flat chains"). If we set $D = 0$ we immediately obtain:

$$|A|^\flat \leq |A|, \tag{6.76}$$

for all polyhedral r-chains A. It is also easy to prove that, for a polyhedral r-chain, the flat norm of its boundary is less or equal than the flat norm of the chain itself,[17] namely:

$$|\partial A|^\flat \leq |A|^\flat. \tag{6.77}$$

EXERCISE 6.10. Prove the inequality (6.77). [Hint: See Whitney, op.cit.]

This being the case, given a Cauchy sequence of r-chains, the sequence of their boundaries is also Cauchy, which allows one to define the boundary of any flat chain as the limit of the sequence of boundaries. The boundary operator is continuous in the flat norm. It can be shown that the flat norm is the largest semi-norm in the space of polyhedral chains satisfying the inequalities (6.76) and (6.77).

We will now look at a few simple instructive examples so as to show what kinds of flat chains one actually can get by completing the space of polyhedral chains by means

---

[17] For those concerned with the fact that the dimensional units of a chain are not the same as those of its boundary, Whitney shows that one can develop the theory by means of an equivalent *flat ρ-norm* defined by the formula $\inf\{|A - \partial D| + |D|/\rho\}$, where ρ is a constant having the dimension of length. Equation (6.77) is then replaced by $|\partial A|^\flat \leq |A|^\flat/\rho$.

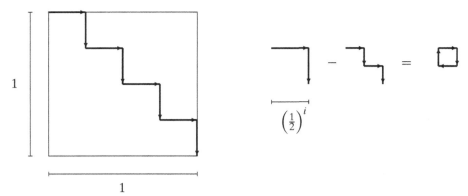

Figure 6.4. The staircase

of the flat norm. Consider the chain A_i (in $\mathbb{E}^2$) made up of two equal, parallel and oppositely oriented segments of length L at a distance d_i. Its mass is $2L$, regardless of the distance. Now let one of the segments be kept fixed while d_i goes to zero, say, as $1/i$. It is obvious that, in the mass norm, the sequence $\{A_1, A_2, \ldots, A_i, \ldots\}$ is not a Cauchy sequence (the norm of the difference of two consecutive terms is constantly equal to $2L$). On the other hand, define the polyhedral 2-chain D_i as the (consistently) oriented rectangle enclosed by A_i. Using the definition of the flat norm we immediately get that $|A_i|^\flat \leq (L+2)d_i$, implying that, in the flat norm, the sequence is a Cauchy sequence (that converges to the zero chain). By taking into consideration the distance between the segments, the flat norm perceives the successive A_i as being closer and closer. The flat norm recognizes more things as being close to each other than the mass norm does.

Another interesting example is provided by the staircase depicted in Figure 6.4. Here A_i is a descending staircase inscribed in the unit square, with rise and run each equal to $1/2^i$. The difference $A_i - A_{i+1}$ is a collection of 2^i closed squares of side $1/2^{i+1}$ oriented clockwise. If we consider as D_i the union of the corresponding 2-dimensional squares, we obtain that $|A_{i+1} - A_i|^\flat \leq 2^{-i-2}$, so that this is a Cauchy sequence[18] in the flat norm (it actually converges to the diagonal). Notice that this sequence would not converge in the mass norm (the mass of the difference of two successive terms is constantly equal to 2).

The examples so far may convey the false impression that flat limits are always polyhedral chains (if that were the case, the vector space of polyhedral chains would already be complete under the flat norm). Apart from the obvious fact that one can approximate curved shapes (even with the mass norm), there exist somewhat unexpected Cauchy sequences in the flat norm. One such example is provided by the Koch curve (or snowflake), obtained by starting from a unit oriented segment and consistently replacing its central third by the correspondingly oriented other two

---

[18] One has to check that the norm of the difference between all elements beyond some N is smaller than any given number. In this example, those norms are bounded by a geometric progression.

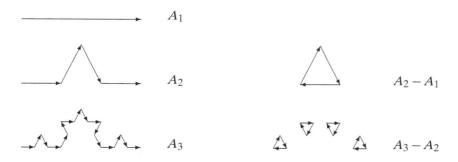

Figure 6.5. The Koch curve

sides of an equilateral triangle. This process is repeated for each of the resulting four segments, and so on ad infinitum. The mass of this sequence tends to infinity, but the flat norm renders it a Cauchy sequence, as suggested in Figure 6.5

6.5.3. Flat Cochains

Completing the space C_r of polyhedral r-chains by means of the flat norm, we have at our disposal the Banach space $C_r^\flat$ of flat r-chains. Its dual space consists of all bounded linear functions defined on $C_r^\flat$. These are called *r-cochains*. In Whitney's vision, these are precisely *integrands*, namely, objects that assign linearly to each domain a number, the integral. The boundedness condition means that there exists a number $N \geq 0$ such that, for a given cochain X, we have

$$|X(A)| \leq N|A|^\flat, \tag{6.78}$$

for all flat r-chains A.

We define the flat norm in the space of cochains by:

$$|X|^\flat := \sup \left(\frac{|X(A)|}{|A|^\flat} \right), \tag{6.79}$$

where the supremum (or least upper bound) is with respect to all nonzero flat chains A. Equivalently:

$$|X|^\flat = \sup |X(A)|, \tag{6.80}$$

over all flat chains A with $|A|^\flat = 1$.

In a similar vein, one can define the *comass* of a cochain by using the mass norm, namely:

$$|X| := \sup \left(\frac{|X(A)|}{|A|} \right), \tag{6.81}$$

over all nonzero (polyhedral) chains A. For this definition to make sense, we must use only polyhedral chains, so that the mass is defined. Notice that $|X| \leq |X|^\flat$ (on account of the fact that the flat norm of a polyhedral chain A is never larger than its mass).

We have seen that any flat r-chain A has a boundary ∂A (this is always a well-defined flat $(r-1)$-chain, since the sequence of the boundaries is bounded in norm by the sequence of the chains, term by term). For a flat r-cochain X, therefore, we can define its *coboundary* dX as the linear function on $(r+1)$-chains obtained as:

$$dX(A) := X(\partial A). \tag{6.82}$$

The boundedness follows from:

$$|dX(A)| = |X(\partial A)| \le |X|^\flat |\partial A|^\flat \le |X|^\flat |A|^\flat. \tag{6.83}$$

It follows that $|dX|^\flat \le |X|^\flat$. More importantly, it can be shown that the flat norm of a cochain X is precisely given by:

$$|X|^\flat = \sup(|X|, |dX|). \tag{6.84}$$

6.5.4. Significance for Continuum Mechanics

Particularizing for 2-dimensional domains in 3-dimensional Euclidean space, we see that cochains are really fluxes. Notice that the additivity with respect to disjoint domains and the skew-symmetry required for interactions ("Newton's third law") are automatically satisfied. Another important property is that the flux is bounded both by the volume and the area of the corresponding domain. In more detail, let us assume that we have defined some linear operation X on all polyhedral r-chains such that there exist two numbers, N_1 and N_2, with the properties $|X(A)| \le N_1|A|$ and $|X(\partial D)| \le N_2|D|$ for all polyhedral r-chains A and all polyhedral $(r+1)$-chains D. We have not yet invented the flat norm. We can write:

$$|X(A)| = |X(A) - X(\partial D) + X(\partial D)| \le |X(A - \partial D)| + |X(\partial D)| \le N_1|A - \partial D| + N_2|D|. \tag{6.85}$$

Since this inequality holds for all D, there exists a least upper bound C such that

$$|X(A)| \le C(|A - \partial D| + |D|). \tag{6.86}$$

In fact, if we let N_1 and N_2 be the smallest possible values, then C is exactly the larger of these two numbers. Letting D be arbitrary, we conclude that

$$|X(A)| \le C\inf(|A - \partial D| + |D|). \tag{6.87}$$

This calculation justifies a posteriori the definition of the flat norm for chains. We see that the two logical boundedness conditions imply that the operation X is bounded (i.e., continuous) with respect to the flat norm. The converse is also true. In fact, it can be shown that the flat norm is optimal in the sense that it is the smallest norm guaranteeing continuity (of fluxes) while satisfying the two boundedness conditions. When we complete the space of polyhedral chains with this norm, we obtain the largest possible Banach space for which the bounded fluxes are continuous! It is in this sense that one can say that flat chains are the most general objects possible, for which (physically) bounded fluxes are continuous.

6.5.5. Cochains and Differential Forms

We have indicated that flat cochains can be regarded as integrands over the corresponding spaces of flat chains. The reasons for this identification are as follows: (i) The action of a cochain on a chain is linear. In particular, reversal of orientation of the chain (i.e., orientation reversal of the domain of integration) changes the sign of the result, just as an integral is expected to behave. Moreover, again by linearity, the action of the cochain on a chain is additive with respect to subdivisions of cells into pieces with disjoint interiors. (ii) The action of a flat r-cochain on polyhedral chains satisfies boundedness and continuity conditions with respect to both the Euclidean "volume" of polyhedral r-chains and the Euclidean "area" of the boundary of polyhedral $(r+1)$-chains. On the other hand, from the consideration in the earlier sections of this chapter, we know that integrands over simplicial complexes are given by differential forms. It is, therefore, legitimate to ask whether or not these two kinds of objects (cochains and differential forms) might be the same. One of the strong results of geometric integration theory is a generally positive answer to this question. Whitney devotes a whole chapter of his book to this issue and proves several theorems beyond the scope of this overview. Nevertheless, except for important technical details, it can be said that to each flat r-cochain X defined in an open set $U \subset \mathbb{E}^n$, there corresponds a unique bounded r-form D_X such that[19]:

$$\int_\sigma D_X = X(\sigma), \tag{6.88}$$

for all r-simplexes σ. The correspondence between cochains and forms is one-to-one.

It is interesting that the definition (6.82) of the coboundary dX of a cochain X, namely:

$$dX(A) := X(\partial A), \tag{6.89}$$

when combined with Stokes' formula (6.26):

$$\int_{\partial A} \omega = \int_A d\omega, \tag{6.90}$$

and with Equation (6.88), is consistent with the identification:

$$D_{dX} = dD_X. \tag{6.91}$$

6.5.6. Continuous Chains

Intuitively speaking, we would like to be able to define flat chains with an infinite number of infinitesimal cells, each of which has been multiplied by a real number that varies continuously from cell to cell. As a first step in this direction, we introduce

---

[19] The form D_X is, technically speaking, a Lebesgue measurable flat form defined uniquely to within a set of measure zero.

the difficult notion of *continuous (flat) chains*. Let:

$$\alpha : \mathbb{E}^n \to \Lambda^r(V) \tag{6.92}$$

be a function on $\mathbb{E}^n$ whose values are r-vectors (defined with respect to the supporting vector space V of $\mathbb{E}^n$). Several technical assumptions need to be made as to the nature of this function, such as measurability and cellwise continuity. The identity:

$$X(\tilde{\alpha}) = \int_{\mathbb{E}^n} \langle D_X | \alpha \rangle, \tag{6.93}$$

for all flat r-cochains X, can be shown to uniquely define the flat chain $\tilde{\alpha}$, called the continuous chain associated with the function α.

We now introduce the idea of multiplication of functions by chains. Let ϕ be a real-valued sharp function[20] defined on $\mathbb{E}^n$, and let σ be an oriented r-simplex belonging to some oriented r-dimensional subspace W of $\mathbb{E}^n$. We denote by α the r-vector corresponding to a unit oriented r-volume in W. On the basis of this r-vector and the given function, we now define the following r-vector valued function β on the subspace W:

$$\beta(x) = \begin{cases} \phi(x)\,\alpha & \text{if} \quad x \in \sigma \\ 0 & \text{if} \quad x \in W - \sigma \end{cases} \tag{6.94}$$

To this function, we can assign uniquely a continuous r-chain $\tilde{\beta}$ according to Equation (6.93) applied within the Euclidean space W. The product of the function ϕ by the oriented r-simplex σ is defined (in W) as:

$$\phi\sigma := \tilde{\beta}. \tag{6.95}$$

This is a well-defined r-chain in W and hence in $\mathbb{E}^n$. This definition can now be extended by linearity to any (finite) r-chain $A = \sum a_i \sigma_i$ in $\mathbb{E}^n$ as follows:

$$\phi A := \sum a_i \phi \sigma_i. \tag{6.96}$$

For any flat r-chain (i.e., a Cauchy sequence of finite chains A_j), we set:

$$\phi A := \lim_{j \to \infty}{}^\flat \phi A_j. \tag{6.97}$$

It can be shown that this limit exists and is unique.

It is not difficult to see how the concept of multiplying a chain by a function may be useful in terms of regarding a continuous field over a domain as a flat chain. A potential limitation of this approach, however, is that the function must be defined over (an open subset of) $\mathbb{E}^n$. In the description of fractal bodies, for instance, this may be too severe a limitation, as we already pointed out when dealing with the induced differential structure in Section 4.18. For this reason, we will exhibit below in Section 6.5.9 a possible alternative approach.

---

[20] A sharp function is a bounded function satisfying a Lipschitz condition.

6.5.7. Balance Laws and Virtual Work in Terms of Flat Chains

In Section 6.4.2, we considered an extensive property p, whose content (under certain smoothness assumptions) in a region $\mathcal{R} \subset \mathbb{E}^n$ is obtained as the integral of a smooth n-form ρ over $\mathcal{R}$. We now substitute for this form ρ a flat n-cochain P. The total content of p in an oriented n-cell σ is given by the evaluation $P(\sigma)$. The balance equation (6.64) will thus acquire a form such as:

$$\frac{dP(\mathcal{R})}{dt} = S(\mathcal{R}) + T(\partial \mathcal{R}), \qquad (6.98)$$

where $\mathcal{R}$ is now the flat n-chain representing the body, S is a production flat n-cochain and T is a flux flat $(n-1)$-cochain. By definition of coboundary, we can write:

$$T(\partial \mathcal{R}) = dT(\mathcal{R}), \qquad (6.99)$$

which, together with (6.98) and the arbitrariness of the domain of integration yields the generic balance equation:

$$\frac{dP}{dt} = S + dT. \qquad (6.100)$$

This should be clear enough, but now we can change our perspective slightly and consider that P represents a quantity that performs "work" on another field. For example, P may represent the component of a force in a given global direction. To obtain the work, we should also provide the value of the corresponding displacement component, ϕ say. With this interpretation, the question naturally arises as to where is this last element (the "displacement") going to reside. Given what we have learned in the previous section, the answer presents itself to us naturally: the intensity of the displacement field is encoded by constructing the product chain $\phi\sigma$. The total work on the cell is thus given by the simple expression $P(\phi\sigma)$. This idea can be extended by linearity to an arbitrary flat n-chain B (representing, for example, a material body). The total work is given by the evaluation $P(\phi B)$, where ϕ is the intensity of the displacement field over B. In a similar way, we can conceive of a flux $(n-1)$-cochain T representing a component of the surface traction. Its work over the boundary is given by the evaluation $T(\phi\partial B)$. In this way, we have given a clear physical meaning to the action of a cochain over any chain. If P represents a component of the body force, and T the corresponding component of the surface traction flux, the total virtual work is given by:

$$VW = P(\phi B) + T(\phi\partial B). \qquad (6.101)$$

By definition of coboundary, we can also write:

$$VW = (P + dT)(\phi B). \qquad (6.102)$$

When expressed in terms of more conventional differential forms, the flux term can be seen to split into the virtual work of the divergence of the stress plus the so-called internal virtual work.

6.5.8. The Sharp Norm

Apart from the two "logical" (physically motivated) continuity conditions that we have just discussed, one may want to consider a further similar condition that will result in restricting the space of cochains, and simultaneously enlarging the space of chains. What we want is that if a cell σ is translated (without rotation) by any vectorial amount $\mathbf{v}$ to a new cell $T_v\sigma$, then the evaluation $X(\sigma)$ should be close to $X(T_v\sigma)$. In other words, there exists a number N_3 such that for all oriented cells σ the following inequality holds:

$$|X(T_v\sigma) - X(\sigma)| \leq N_3 |\sigma| \|\mathbf{v}\|. \tag{6.103}$$

It turns out that the smallest norm in the space of polyhedral chains for which the operation X satisfying the three boundedness conditions is continuous, is the sharp norm, which we will presently define. Since the sharp norm is never larger than the flat norm, chains that are relatively far apart in the flat norm are felt as being closer together under the sharp norm. This means that there are now more Cauchy sequences at our disposal or, in other words, that even more irregular domains of integration have become available. The sharp norm in the space of polyhedral chains is defined, for a given r-chain $A = \sum a_i\sigma_i$, by:

$$|A|^\sharp = \inf\left(\frac{(\sum |a_i| \|\sigma_i\| \|\mathbf{v}_i\|)}{(r+1)} + \left|\sum a_i T_{v_i}\sigma_i\right|^\flat\right), \tag{6.104}$$

where the infimum is to be evaluated over all vectors $\mathbf{v}_i$ and all representations of the polyhedral chain (it may be that the expression is independent of representation, but the proof is not available). Taking in particular $\mathbf{v}_i = \mathbf{0}$, it follows that:

$$|A|^\sharp \leq |A|^\flat. \tag{6.105}$$

Completing the space of polyhedral chains with this norm, we obtain the Banach space of *sharp chains*. We note that in this space the boundary of a (sharp) chain cannot be defined in general (as a sharp chain). To appreciate the kind of limits which are now admitted but were not admitted before, consider the "strainer" obtained by eliminating from the staircase the vertical segments, as shown in Figure 6.6.

It can be shown that the sequence of successive strainers does not converge in the flat norm, but it does in the sharp norm. The importance of this example is that it acts as a Venetian blind in the sense that the Venetian blind is roughly globally parallel to the windowpane, but the flux is biased in a different direction. This means that we have some kind of a line with a global slope different from its local slope. An interesting example of this behaviour, not necessarily related to the present context, is given by the mechanics of two intermeshing combs or brushes,[21] a useful model to explain the stability of muscle fibres in the apparently unstable so-called *descending limb* of the force-length relation. If differential equations can be written in terms of functions with graphs given by such sharp chains one may rigorously establish the Lyapunov stability of such dynamical systems.

---

[21] See: Epstein M, Herzog W (1998), *Theoretical Models of Skeletal Muscle*, John Wiley.

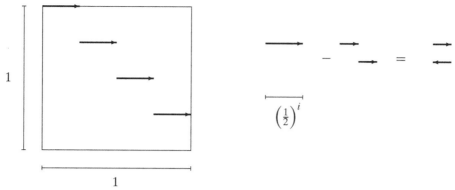

Figure 6.6. The strainer (or Venetian blind)

6.5.9. Fields on Chains as Chains

In Section 6.5.6 we mentioned an alternative approach to the introduction of fields on chains as products of functions times chains. In cases such as fractal bodies, the use of continuous chains as introduced above may be too restrictive to capture the inherent fractality of the virtual displacements. For this reason, we will here propose a nonrigorous alternative that must be considered for heuristical purposes only.

Let a material body $\mathcal{B}$ be modelled as a flat r-chain in $\mathbb{E}^n$, with $r \leq n$. Recall that flat r-chains are elements of the Banach space obtained by completing the space of polyhedral chains by means of the flat norm. In other words, every flat r-chain A can be represented as an infinite Cauchy sequence of polyhedral chains, namely, $A = \{A_i\}$. This representation, however, is not unique, a fact that needs to be taken into consideration if using one particular representation to define a field over the chain, as we will soon see.

Each polyhedral chain A_i in the sequence is itself decomposed as a (finite) linear combination of (nonoverlapping) polyhedral cells σ_{ij} , namely:

$$A_i = \sum_{j=1}^{n_i} \beta_{ij}\, \sigma_{ij}, \quad \text{(no sum on } i\text{)}. \tag{6.106}$$

The n_i real numbers β_{ij} may be construed as equal to 1 (or perhaps -1) for the case of a material body, but this is not an essential restriction of the theory. More importantly, though, the decomposition (6.106) is not unique since, for example, any cell can be partitioned into a sum of nonoverlapping cells. Therefore, let, a particular representation of the r-chain A as a Cauchy sequence $\{A_i\}$ and a particular decomposition of each polyhedral chain A_i as a finite sum (6.106) be given. For each value of i, let $\alpha_{ij}(j = 1, \ldots, n_i)$ be real numbers, and define the polyhedral chain A_i^α associated with these numbers as:

$$A_i^\alpha = \sum_{j=1}^{n_i} \alpha_{ij}\, \beta_{ij}\, \sigma_{ij}. \tag{6.107}$$

We now form the sequence $\{A_i^\alpha\}$. If this sequence happens to be a Cauchy sequence with respect to the flat norm, we denote it by A^α and call it a *(zero-grade) field* on the flat chain A. It is worthwhile recalling that this field is defined relative to a particular Cauchy sequence representing A and a particular decomposition of each member of this sequence. Let A^α and A^γ be two fields of the same flat chain A, each of them relative to a particular sequence and a particular decomposition of each of its members. We say that the two fields coincide if, and only if, $A^\alpha = A^\gamma$ as flat chains. Thus, we judge a field, as we should, by the limit, rather than by the way in which this limit has been achieved.

REMARK 6.3. For simplicity of the exposition, we have considered scalar-valued coefficients α_{ij}. To represent actual fields in a mechanical context (such as virtual displacements or virtual strains), we should consider vector-valued coefficients.

If the chain of departure, modelling the material body $\mathcal{B}$, happens to be "smooth", that is, an r-polyhedral chain or, more generally, an r-dimensional embedded submanifold of $\mathbb{E}^n$, the preceding construction is reminiscent of the finite-element method. Experience with this method, however, teaches us that other types of fields ("of higher grade") may be more suitable for the eventual formulation of variational principles. Following this lead, we proceed to generalize our definition. We note, however, that the case of arbitrary flat chains (such as fractals) does not fall directly under the purview of the finite element method, whereas it is covered by our definition.

REMARK 6.4. For the case of smooth bodies, it is not difficult to prove that, if the coefficients α_{ij} are systematically chosen from the values of a fixed sharp function over the body, the resulting sequence is always Cauchy in the flat norm. Thus, a (classical) field defined over a smooth body can be regarded as a continuous chain in the sense of Whitney or as the multiplication of the body-chain by a sharp function. In the case of general flat chains, however, a policy of confining fields to Lipschitz (or smooth) mappings over an open set of $\mathbb{E}^n$ containing the chain, may result in a crucial loss of flexibility of the theory. This is, for example, the case of the theory based on the concept of differential spaces of Sikorski if one uses the induced differential structure (see Section 4.18). The concept of field presented here allows for less regular transformations.

We now proceed to define *first-grade fields* by specifying for each member A_i of the generating sequence of the chain A a particular decomposition into a finite sum of r-simplexes s_{ij} ($j = 1, \ldots, n_i$):

$$A_i = \sum_{j=1}^{n_i} \beta_{ij}\, s_{ij}. \tag{6.108}$$

To every A_i there corresponds a particular finite 0-chain, namely, the collection of its vertices, say N_{ij} in number. For each value of i, let α_{ij} ($j = 1, \ldots, N_i$) be real numbers, which we may call *nodal parameters* associated with the polyhedral chain A_i. Within each simplex of this chain, the nodal parameters define uniquely a linear function ω_{ij}^α, $j = 1, \ldots, n_i$. Each of the products $\omega_{ij}^\alpha \beta_{ij} s_{ij}$ is a valid flat chain (supported by a simplex). We can, therefore, define the *first-grade field* over the flat chain A_i^α associated with the given set of nodal numbers as:

$$A_i^\alpha = \sum_{j=1}^{n_i} \omega_{ij}^\alpha \, \beta_{ij} \, s_{ij}. \tag{6.109}$$

We note that once multiplied by the linear functions over each simplex, the resulting chain is no longer polyhedral. Just as in the case of the zero-grade fields, we now form the sequence $\{A_i^\alpha\}$. If this sequence happens to be a Cauchy sequence with respect to the flat norm, we denote it by A^α and call it a *first-grade field* of the flat chain A. As before, we identify those decompositions and sequences of nodal displacements that give rise to the same field.

REMARK 6.5. If two fields on a chain A are defined over two different representations, $\{A_i\}$ and $\{A_i'\}$ say, it is possible to redefine them over a common representation by considering, at each step i, a refined partition (such as a common simplicial refinement). By this means, it appears that the set of fields over a given chain has the natural structure of an infinite-dimensional affine space. Adopting the original chain as origin, the differences would form a vector space.

An interesting observation is that, if at each step of the generation of a first-grade field we consider the differential of the linear function over each simplex, we obtain a zero-grade (perhaps vector-valued) field over each simplex. In particular, if the first-grade field represents one component of a virtual displacement, then its derivatives within each simplex can be seen as virtual strains. Even if the first-grade field happens to be properly defined (in other words, the first-grade chains converge in the flat norm), it may so happen that the corresponding zero-grade derived field does not converge in the flat norm. Intuitively, though, it is possible that it converges in the sharp norm. This fact may be relevant for the implementation of the principle of virtual work, in which the surface tractions need to be defined as a flat chain, but the so-called internal virtual work does not.

EXAMPLE 6.2. Consider the staircase chain on which a first-grade field is defined in the following way: The nodal parameters are chosen so that the interpolated linear fields ω_{ij}^α have the same slope k in all the horizontal segments, while a zero slope in the vertical segments. If we consider the associated zero-grade field obtained by taking the derivative of the given field at each stage of the generation, we obtain, in the limit, the strainer multiplied by the number k. Thus, a zero-grade field over a flat chain turned out to be a sharp chain.

EXAMPLE 6.3. A composite bar of a constant cross section A is generated by a systematic subdivision of the length L into 2^I equal segments. Each segment is assigned an elastic modulus E_1 or E_2 in alternating fashion, that is in the pattern $E_1, E_2, E_1, E_2, E_1, \ldots, E_2$. We assume that the left end of the segment is fixed to a

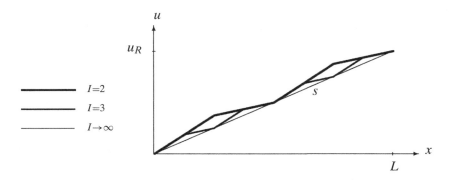

Figure 6.7. Displacement function

rigid wall, and then apply a longitudinal force F at the free (right) end. We now ask the question: What is the equilibrium configuration of this bar (consisting of an infinite number of segments)? Since, at each stage I of the generation, half of the bar is occupied by each of the two materials, the displacement of the right end is given by:

$$u_R = \frac{FL\left(1 + \frac{E_2}{E_1}\right)}{2E_2 A}. \tag{6.110}$$

As far as the internal points are concerned, the graph of the displacement as a function of the running coordinate x along the bar axis is a jagged line that, as the generation advances, gets closer and closer to the straight line segment s joining the origin with the point (L, u_R), as illustrated in Figure 6.7.

Thus, the displacement field converges to s in the flat norm. But can we use this line as a basis for the calculation of stresses within the bar? Clearly not, since such a configuration would not be in equilibrium with the applied force. What has happened in this case is easily explained by the fact that the strain field is discontinuous. The graph of the displacement illustrated in Figure 6.7 is reminiscent of the *staircase* flat chain. The graph of its derivatives, on the other hand, is reminiscent of the *strainer*, which is a sharp chain (but not a flat one).

In concluding this section, we mention once again that, unlike the rigorous theory of continuous chains presented before, this alternative formulation is not supported by any reliable theory and that it has been included here for heuristic purposes only.

PART THREE

FURTHER TOPICS

Fibre Bundles

We have already emphasized the central role played in Continuum Mechanics by the concept of differentiable manifold, and we will continue to do so. In this chapter, we will consider a class of differentiable manifolds endowed with extra structure, manifolds whose points have, as it were, a double loyalty, or a double sense of belonging: not only to the manifold itself but also to a smaller entity, which will be called a fibre. The manifold itself will be called a differentiable fibre bundle, once this extra structure has been rigorously defined. We will discover that the identification of this extra internal structure will serve, as one might expect, to represent materials with microstructure, such as granular media, liquid crystals and the like. But, more interestingly, even without that particular application in mind, we will find that fibre bundles are ubiquitous as convenient, sometimes indispensable, tools for completing the geometric picture of Continuum Mechanics that we have developed so far.

7.1. Product Bundles

The simplest instance of a fibre bundle is demonstrated by a *product bundle*, that is, a manifold $\mathcal{C}$ obtained as the Cartesian product of two given manifolds, $\mathcal{B}$ and $\mathcal{F}$, namely:

$$\mathcal{C} = \mathcal{B} \times \mathcal{F}, \tag{7.1}$$

where $\mathcal{B}$ is called the *base manifold* and $\mathcal{F}$ is the *fibre*.[1] We already know that a product of two manifolds is itself a manifold, whose dimension is the sum of the dimensions of the factors and whose topology is the product topology. But here we want to emphasize that $\mathcal{C}$ is endowed with two natural projection maps, namely:

$$pr_1 : \mathcal{C} \longrightarrow \mathcal{B}, \tag{7.2}$$

---

[1] Strictly speaking, fibre bundles are defined using topological spaces, rather than manifolds. When the topological spaces involved are manifolds and certain smoothness conditions are satisfied (as will be the case in our treatment), one should use the terminology of (smooth) differentiable fibre bundles. For the sake of economy, however, we will continue using the term "fibre bundle." Unless otherwise stated, all manifolds in this chapter will be finite-dimensional.

and

$$pr_2 : C \longrightarrow F,\tag{7.3}$$

which assign to any given pair $(b,f) \in C$ its first and second components, b and f, respectively. It is clear that these two maps, in addition to being surjective, are continuous in the product topology. Indeed, let U be an open subset of B. We have, by definition of pr_1, that $pr_1^{-1}(U) = U \times F$, which is open. A similar reasoning applies to pr_2.

For each $b \in B$ the set $pr_1^{-1}(b)$ is called the *fibre at b*, denoted as C_b, which in this case is simply a copy of F. From this definition it follows that fibres at different points are disjoint sets and that each point $c \in C$ necessarily belongs to a fibre, namely, to $C_{pr_1(c)}$. The fibres can, therefore, also be seen as the equivalence classes corresponding to the equivalence relation of "having the same first projection."

Given an atlas in B and an atlas in F, the product of these atlases is an atlas of the product bundle C. Naturally, as a manifold, C may be endowed with other atlases. Nevertheless, we will always restrict attention to *product atlases*, namely, those that emphasize the product nature of the bundle C. In a product chart, the projections acquire a particularly simple form. Indeed, let $x^i, i = 1,\ldots,m = \dim(B)$ and $y^\alpha, \alpha = 1,\ldots,n = \dim(F)$ be coordinate systems for some charts in the base and the fibre, respectively. Then we have:

$$pr_1 : C \longrightarrow B$$
$$(x^i, y^\alpha) \mapsto (x^i),\tag{7.4}$$

and

$$pr_2 : C \longrightarrow F$$
$$(x^i, y^\alpha) \mapsto (y^\alpha).\tag{7.5}$$

It follows from this coordinate representation that the projections are not just continuous, but also C^∞ maps. Technically, they are surjective submersions.

EXAMPLE 7.1. **Shells:** One of the many different ways to describe a *shell* in structural engineering is to regard it as the product bundle of an oriented two-dimensional manifold (with or without boundary) B times the open (or sometimes closed) segment $F = (-1,1) \in \mathbb{R}$. The base manifold is known as the *middle surface* while the fibre conveys the idea of thickness, eventually responsible for the bending stiffness of the shell. The fact that this is a product bundle means that one can in a natural way identify corresponding locations throughout the thickness at different points of the middle surface. Thus, two points of the shell standing on different points of the middle surface can be said to correspond to each other if they have the same value of the second projection. This fact can be interpreted as being on the same side of the middle surface and at the same fraction of the respective thicknesses.

7.2. Trivial Bundles

The next degree of complexity towards the most general notion of fibre bundle, is the so-called *trivial* (or *simple*, or *globally trivializable*) bundle. In essence, what we want to achieve is the loss of the second projection (pr_2), while preserving the first (pr_1). A clear example of the convenience of such a generalization is provided by the concept of *space-time*, already encountered in Chapter 1. The ancients (particularly, but not only, Aristotle) seemed to have thought that space and time were absolute entities existing, as it were, independently of each other. In our terminology, therefore, we would say that for them space-time was a product bundle with, say, time as the base manifold and space as the fibre. The physical meaning of pr_1 would thus be that of providing the time of occurrence of an event, such as a collision. Accordingly, two events p and q are *simultaneous* if $pr_1(p) = pr_1(q)$. Similarly, since space is absolute in this vision of the world, the second projection provides information about the location of an event in absolute space. Thus, two events p and q can be said to have occurred at the same place (regardless of their times of occurrence) if $pr_2(p) = pr_2(q)$. The principle of Galilean relativity can be said to have demolished the second projection. Indeed, unless two events happen to be simultaneous (a concept only questioned much later by Einstein's relativity principle), it is impossible, according to Galileo and Newton, to compare in an absolute way the places at which they occurred. While it is true that a given observer can make such a judgement, a different observer will in general legitimately disagree. Herein lies the clue to our generalization, namely, that although the total manifold C is no longer a product it still looks like a product (albeit a different one) to each observer. This intuitive idea leads to the following, somewhat more formal, provisional definition.

Let two manifolds, B and F, be given as before. A trivial bundle with base B and *typical fibre* F consists of a manifold C (the *total manifold*) and a smooth surjective map

$$\pi : C \longrightarrow B, \tag{7.6}$$

called the *bundle projection*, such that there exists a diffeomorphism (called a *global trivialization*), namely, a map

$$\psi : C \longrightarrow B \times F, \tag{7.7}$$

with the property

$$\pi = pr_1 \circ \psi. \tag{7.8}$$

This last property can be expressed by means of the following commutative diagram:

$$
\begin{array}{ccc}
C & \xrightarrow{\ \psi\ } & B \times F \\
 & \searrow{\scriptstyle \pi} & \big\downarrow{\scriptstyle pr_1} \\
 & & B
\end{array}
\tag{7.9}
$$

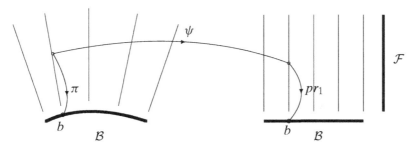

Figure 7.1. A trivial bundle

Figure (7.1) provides a graphical representation. Equation (7.8) guarantees that the inverse image by π of each point $b \in \mathcal{B}$ is diffeomorphic to the typical fibre $\mathcal{F}$ (but not necessarily equal to it, as in the case of a product bundle).

As before, we call $\mathcal{C}_b = \pi^{-1}(b)$ the *fibre at b*. Each point $c \in \mathcal{C}$ belongs to one and only one fibre, namely, that at $\pi(c)$. By being diffeomorphic to the typical fibre, all fibres are mutually diffeomorphic. But these diffeomorphisms are not natural (or canonical), depending as they do on the trivialization. This remark serves to indicate that a particular observer in Classical Mechanics may simply correspond to a particular trivialization of space-time.

EXAMPLE 7.2. **Bodies with microstructure:** Physically speaking, we say that a body is endowed with *microstructure* if, in addition to the usual kinematic degrees of freedom afforded by the deformation of a material matrix, or macromedium, a complete kinematic description must also include some information regarding extra (internal) degrees of freedom. An example of this situation is provided by an everyday material such as concrete, which is formed by embedding in a cement matrix an aggregate consisting of stones whose size is relatively large when compared with the grains of cement. Each of these stones can then be considered as a *micromedium*. In a continuum model we expect to have these micromedia smoothly assigned to each point of the matrix, thus generating a fibre bundle, whose typical fibre is the micromedium.[2] In contradistinction with the case of the shell, discussed in Example 7.1, there is no canonical correspondence between points belonging to micromedia attached at different points of the macromedium.

As defined so far, if we find any other diffeomorphism $\phi : \mathcal{C} \to \mathcal{B} \times \mathcal{F}$ satisfying Equation (7.8), will it define the same trivial bundle? To decide how to answer this question, let us observe that every trivialization ψ, by definition, results in a pair which, with an obvious notation, we can write $\psi(c) = (\pi(c), \tilde{\psi}(c))$, where $\tilde{\psi}(c)$ belongs to $\mathcal{F}$. If we now fix $\pi(c) = b$ letting c vary (namely, if we move along a single fibre), we obtain the restricted maps $\tilde{\psi}_b$ and $\tilde{\phi}_b$, which are diffeomorphisms

---

[2] This example has been chosen for its graphical clarity. In the actual practice of Civil Engineering it is rare to find that concrete is treated in such a degree of detail. Instead, the contribution of the micromedium is averaged or *homogenized* into a supposedly equivalent ordinary macromedium

between the fibre at b and the typical fibre. The map $\tilde{\phi}_b \circ \tilde{\psi}_b^{-1} : \mathcal{F} \to \mathcal{F}$ is, therefore, a diffeomorphism of the typical fibre. We call this a *transition map* at b. As part and parcel of the correct definition of a fibre bundle we will impose the restriction that the allowed transition maps must belong to a given subgroup $\mathcal{G}$ of the group of diffeomorphisms of the typical fibre. This group is called the *structure group* of the bundle. For technical reasons, it is customary to assume that $\mathcal{G}$ has been given the structure of a finite-dimensional Lie group and that the transition maps depend smoothly on position along $\mathcal{B}$.

An equivalent way to introduce the structure group is to consider it as a Lie group of transformations of $\mathcal{F}$ whose (left) action is *effective*.[3] This way of defining the structure group is more appealing, since it allows to have two different fibre bundles with different typical fibres, but with the same structure group. However the structure group is defined, we shall say that two global trivializations define the same fibre bundle if the corresponding transition maps belong to the structure group and if the dependence of the assigned element of $\mathcal{G}$ is smooth in $\mathcal{B}$.

Revisiting now the definition of a product bundle, we realize that for that particular case the structure group must consist just of the identity transformation of $\mathcal{F}$. A globally trivializable bundle whose group is just the identity is *equivalent* to a product bundle, in a precise sense that we will explain later. At any rate, we see that in the case in which the structure group is the identity alone, there exists a canonical diffeomorphism between fibres at different points. In the case of a globally trivializable bundle, this is true too for any discrete structure group, as can be understood from the smoothness of ϕ. For general (not necessarily globally trivializable) fibre bundles even this last statement is no longer true.

EXAMPLE 7.3. **A disjointed cylinder:** Consider a trivializable fibre bundle whose base manifold is $\mathbb{R}$ and whose typical fibre is the unit circle. If we adopt as the structure group the identity alone, this bundle represents a cylinder, thought of as a collection of infinitesimally thin discs glued together in a definite fashion. Let us now enlarge the structure group to the group consisting of the identity and the 180°-rotation. Due to the continuity of any global trivialization, we obviously obtain the same cylinder as before (since we cannot jump continuously from one element of the group to the other). But let us now enlarge the structure group further to the whole group of rotations of the circle. We now have an infinite choice of trivializations out of any initially given one. Two different trivializations differ in the way the discs are attached to each other. So, the discs can no longer be thought of as glued together, but rather as allowing for arbitrary amounts of continuously distributed twist, the amount of the twist being in the eyes of the beholder (namely, the trivialization). It is clear how this example is similar to the more complicated case of classical space-time, in which the structure group is the collection of affine maps of $\mathbb{E}^3$.

---

[3] We recall that the action of a group on a manifold is effective if, except for the group identity, there is no group element whose action transforms every point of the manifold into itself.

7.3. General Fibre Bundles

The need for more general fibre bundles is foreshadowed by the classical example of the Moebius band, although many other examples arise in practical applications in Continuum Mechanics, as we shall see in later chapters. For the sake of fostering intuition, we will now briefly discuss the Moebius band.

A Moebius band is the surface obtained by taking a long rectangular strip, twisting one end by 180°, and then gluing together the two ends, rendering the strip ringlike. If the twist had not been effected, we could have conceived of the resulting cylindrical surface as the product of a circumference (the base) by a segment (the fibre), namely, the width of the strip. Now, however, this is no longer the case. In fact, the product of two orientable manifolds is orientable, whereas the Moebius band is not (one can continuously pass from what appears to be one side to the other!), so they cannot be diffeomorphic. It is obvious, on the other hand, that for any point of the base, there exists an open neighbourhood such that the corresponding part of the Moebius band is isomorphic to a product. We may, therefore, cover the band with (at least two) pieces, erected upon (two) open sets of the base. On each piece, we can define a diffeomorphism with a product of the corresponding open set and the typical fibre. Wherever these open sets intersect, we have to explain how these diffeomorphisms are to be related to each other along fibres. The structure group tells us precisely the freedom permitted in this gluing process of the fibres. It is clear in this example that this structure group must include at least one nonorientation preserving fibre transformation. In fact, the group consisting of the identity and the reflection about the mid-point of the fibre will do the job.

With this classical example in mind, we define a general fibre bundle (Figure 7.2) with base $\mathcal{B}$, typical fibre $\mathcal{F}$ and structure group $\mathcal{G}$, as a manifold $\mathcal{C}$ and a smooth surjective bundle-projection map $\pi : \mathcal{C} \to \mathcal{B}$ such that there exists an open covering $\mathcal{U}_\alpha$ of $\mathcal{B}$ and respective *local trivializations*:

$$\psi_\alpha : \pi^{-1}(\mathcal{U}_\alpha) \longrightarrow \mathcal{U}_\alpha \times \mathcal{F}, \tag{7.10}$$

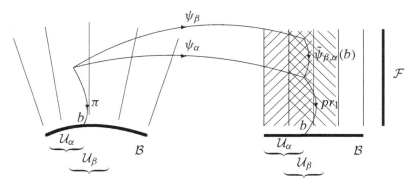

Figure 7.2. A general fibre bundle

with the property $\pi = pr_1 \circ \psi_\alpha$, as illustrated in the following commutative diagram:

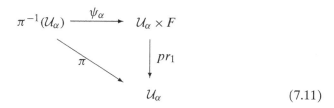

$$(7.11)$$

Moreover, whenever $b \in \mathcal{U}_\alpha \cap \mathcal{U}_\beta \neq \emptyset$, the transition maps $\tilde{\psi}_{\beta,\alpha}(b) := \tilde{\psi}_{\alpha,b} \circ \tilde{\psi}_{\beta,b}^{-1}$ belong to the structure group $\mathcal{G}$ and depend smoothly on position throughout the intersection.

Consider now, for the same $\mathcal{C}$, $\mathcal{B}$, $\mathcal{F}$, π and $\mathcal{G}$, a different open covering $\mathcal{V}_\beta$ with local trivializations ϕ_β. We say that it defines the same fibre bundle as before if, on nonvanishing intersections, the transition maps $\tilde{\psi}_{\alpha,b} \circ \tilde{\phi}_{\beta,b}^{-1}$ belong to the structure group $\mathcal{G}$ and depend smoothly on position b throughout the intersection. The two trivializations are said to be compatible. In this sense, we can say that the union of the two trivialization coverings becomes itself a new trivialization covering of the fibre bundle.[4] When there is no room for confusion, a fibre bundle is denoted as a pair $(\mathcal{C}, \pi)$ indicating just the total space and the projection. An alternative notation is $\pi : \mathcal{C} \to \mathcal{B}$. A more complete notation would be $(\mathcal{C}, \pi, \mathcal{B}, \mathcal{F}, \mathcal{G})$.

Consider further two fibre bundles, $(\mathcal{C}, \pi)$ and $(\mathcal{C}', \pi')$, having the same base, typical fibre and structure group, but defined through incompatible trivializations $\mathcal{U}_\alpha, \psi_\alpha$, and $\mathcal{V}_\beta, \phi_\beta$. If it is possible to fix this incompatibility by just rearranging each fibre by means of a fibre-diffeomorphism that depends smoothly on position throughout the base space, we say that the two bundles are *equivalent*. More precisely, we are looking for a smooth map $F : \mathcal{C} \to \mathcal{C}'$ between the two bundles such that: (i) $\pi' \circ F = \pi$, (ii) for each fixed $b \in \mathcal{B}$ the restriction $F|_{\pi^{-1}(b)} : \pi^{-1}(b) \to (\pi')^{-1}(b)$ is a diffeomorphism, (iii) the maps $\tilde{\phi}_\beta \circ F \circ \tilde{\psi}_\alpha^{-1}$, defined on the appropriate intersections, belong to the structure group $\mathcal{G}$ and depend smoothly on position. If such a (fibre-preserving, base-identity-inducing) map exists, the two fibre bundles are said to be equivalent. This type of map is a particular case of a *fibre-bundle morphism*, which we shall later discuss in some more detail. As an example of equivalent fibre-bundles, consider the case in which the structure group is just the identity. Each global trivialization gives rise perforce to a different fibre bundle. But all these bundles are equivalent to each other and to the product bundle of the base manifold times the typical fibre.

7.3.1. Adapted Coordinate Systems

The total space $\mathcal{C}$ of a smooth fibre bundle $(\mathcal{C}, \pi, \mathcal{B}, \mathcal{F}, \mathcal{G})$ is a differentiable manifold in its own right and, as such, can sustain a whole class of equivalent atlases. Among these atlases, however, there are some that enjoy the property of being in some

---

[4] One could, therefore, define a fibre bundle by means of a maximal covering, in a similar way as a manifold differentiable structure is defined by means of a maximal atlas.

sense adapted to the fibred structure of C seen as a fibre bundle. When working in coordinates, these *adapted atlases* or, more particularly, the corresponding *adapted coordinate charts* are used almost exclusively. The construction of these special charts mimics the construction of charts in a product manifold, by taking advantage of the local triviality of fibre bundles. Thus, let p be a point in C and let $\{\mathcal{U}, \psi\}$ be a local trivialization, namely a diffeomorphism $\psi : \pi^{-1}(\mathcal{U}) \to \mathcal{U} \times \mathcal{F}$, such that $\pi(p) \in \mathcal{U}$. Without loss of generality, we may assume that the trivialization is subordinate to a chart in the base manifold $\mathcal{B}$ in the sense that $\mathcal{U}$ is the domain of a chart in $\mathcal{B}$ with coordinates x^i $(i = 1, \ldots, m)$. Moreover, since the typical fibre $\mathcal{F}$ is itself a differentiable manifold of dimension n, the point $f = pr_2(\psi(p))$ belongs to some chart of $\mathcal{F}$ with domain $\mathcal{V} \subset \mathcal{F}$ and coordinates u^α $(\alpha = 1, \ldots, n)$. The induced adapted coordinate chart in C is the map $u_\psi : \psi^{-1}(\mathcal{U} \times \mathcal{V}) \to \mathbb{R}^{m+n}$ whose value at p is $(x^1, \ldots, x^m, u^1, \ldots, u^n)$. The proof that in this way an atlas can be constructed in C is straightforward. It is implicitly assumed that the differential structure of C is compatible with the one induced by the adapted atlases.

7.4. The Fundamental Existence Theorem

The *fundamental existence theorem* of fibre bundles states that given the manifolds $\mathcal{B}$ and $\mathcal{F}$ and a Lie group $\mathcal{G}$ acting effectively to the left on $\mathcal{F}$, and given, moreover, an open covering $\mathcal{U}_\alpha$ of $\mathcal{B}$ and a smooth assignment of an element of $\mathcal{G}$ to each point in every nonvanishing intersection of the covering, then there exists a fibre bundle (C, π) with local trivializations based upon that covering and with the assigned elements of $\mathcal{G}$ as transition maps. Furthermore, any two bundles with this property are equivalent.[5]

An important application of the fundamental existence theorem is that given a bundle $(C, \pi, \mathcal{B}, \mathcal{F}, \mathcal{G})$, we can associate to it other bundles with the same base manifold and the same structure group, but with different typical fibre $\mathcal{F}'$, in a precise way. Indeed, we can choose a trivialization covering of the given bundle, calculate the transition maps, and then define the *associated bundle* $(C', \pi', \mathcal{B}, \mathcal{F}', \mathcal{G})$, modulo an equivalence, by means of the assertion of the fundamental theorem. A case of particular interest is that in which the new fibre is identified with the structure group. This gives rise to the so-called *associated principal bundle*, which we shall discuss later in more detail.

From the physical point of view, the typical fibre is likely to elicit the intuitive picture pertaining to the meaning of the extra structure afforded by the fibre bundle (such as the thickness of a shell, or the presence of a micromedium). From the mathematical point of view, however, the fibre itself is less important than the specification of the structure group and the transition maps. In some sense, the fibre plays a secondary role and, in many cases, can be replaced by other typical fibres without altering the ability of the model to describe the physical phenomenon at hand. In fact, the replacement of the typical fibre by the structure group itself (thus rendering the associated principal bundle) often results in a clearer mathematical picture.

---

[5] See Steenrod N (1951), *The Topology of Fibre Bundles*, Princeton University Press.

7.5. The Tangent and Cotangent Bundles

Given an n-dimensional differentiable manifold $\mathcal{B}$, we have seen how at each point $b \in \mathcal{B}$ one can define the tangent space $T_b\mathcal{B}$. It is logical, therefore, to investigate whether the intuitive idea of attaching to each point its tangent space can lead to a rigorous definition of a fibre bundle $T\mathcal{B}$ which can appropriately be called the *tangent bundle of $\mathcal{B}$*. The total space of this bundle will consist, as just described, of all the ordered pairs (b, v_b), where $b \in \mathcal{B}$ and $v_b \in T_b\mathcal{B}$. But it is important to bear in mind that this is not a Cartesian product, since the tangent spaces at different points are not canonically isomorphic. The projection map π is defined by $\pi(b, v_b) = b$. Since all the tangent spaces are n-dimensional vector spaces, they are all isomorphic (albeit not canonically) to $\mathbb{R}^n$, and so it is reasonable to declare $\mathbb{R}^n$ as the typical fibre. To complete the definition, we need to give a trivialization and define the structure group. We will require that every atlas covering of the base manifold $\mathcal{B}$ must be admissible as a trivialization covering of the bundle. Let $(\mathcal{U}, \bar{\psi})$ be a chart of $\mathcal{B}$, and denote by x^i ($i = 1, \ldots, n$) the corresponding chart coordinates. As we know, the vectors $\frac{\partial}{\partial x^i}\big|_b$ constitute a basis of $T_b\mathcal{B}$, and thus establish an isomorphism between $T_b\mathcal{B}$ and $\mathbb{R}^n$. More explicitly, a vector $v_b \in T\mathcal{B}$ has a unique expression $v_b = v^i \frac{\partial}{\partial x^i}\big|_b$, thus assigning to v_b the n-tuple $(v^1, \ldots, v^n)$ in $\mathbb{R}^n$, and vice versa. If $(\mathcal{V}, \bar{\phi})$ is another chart with $b \in \mathcal{U} \cap \mathcal{V}$, and if we denote by y^i ($i = 1, \ldots, n$) the coordinates in the new chart, then the vectors $\frac{\partial}{\partial y^i}\big|_b$ will establish a new isomorphism between $T_b\mathcal{B}$ and $\mathbb{R}^n$. If $v_b = w^i \frac{\partial}{\partial y^i}\big|_b$, we obtain that $w^i = \frac{\partial y^i}{\partial x^j}\big|_b v^j$, which defines an automorphism of $\mathbb{R}^n$. For this construction to be valid no matter what is the atlas of departure, since all diffeomorphims are permitted in the definition of the transition functions at the base manifold, we have to admit that the matrix $\{a^i_j\} = \left\{\frac{\partial y^i}{\partial x^j}\big|_b\right\}$ can be any element of the general linear group $GL(n; \mathbb{R})$, which thus becomes the structure group of $T\mathcal{B}$. It is not difficult to verify that the required smoothness conditions, both for the projection and for the bundle transition maps, are satisfied. It is also easy to show (by explicitly exhibiting an atlas) that the total space is a differentiable manifold. This completes the definition of the tangent bundle.

EXAMPLE 7.4. Classical Lagrangian Mechanics: The fundamental idea of analytical mechanics, as conceived by Lagrange, is that of the *configuration space* of a mechanical system. In modern terminology, the configuration space $\mathcal{Q}$ is a (finite-dimensional) differentiable manifold whose dimension measures the number of degrees of freedom of the system at hand. An example is provided by the plane double-pendulum, consisting of two point-masses, the first one attached by an inextensible link to a fixed point, and the second attached to the first in the same manner. The configuration space of this system is a torus, since the kinematical state of the system is completely specified by the location of the first mass on a circle around the fixed support, plus the location of the second mass on a circle around the present position of the first mass. A velocity of the system consists of a vector tangent to the torus at the point defining its present state, since velocities are by definition vectors tangent to parametrized trajectories. Thus, the tangent bundle TQ is the repository of all possible states of the system together with all possible velocities. The Lagrangian density $\mathcal{L}$, measuring the difference between kinetic and potential energies, is, therefore a function:

$$\mathcal{L} : TQ \longrightarrow \mathbb{R}. \tag{7.12}$$

Given a coordinate system q^i in the base manifold Q, it is customary to denote by $\dot{q}^i$ the components of tangent vectors (velocities) in the natural basis $\frac{\partial}{\partial q^i}$.

EXERCISE 7.1. **Orientability:** Show that a tangent bundle is always orientable. [Hint: given an atlas in the base manifold, use the standard isomorphisms described above to construct an atlas of the tangent bundle, and show that the determinant of the transition functions is always positive.]

In a similar way, by attaching the cotangent (or dual) spaces $T_b^*\mathcal{B}$ to the corresponding points in $\mathcal{B}$, we can construct the *cotangent bundle* $T^*\mathcal{B}$. Its structure group is again $GL(n; \mathbb{R})$ and it can be shown that, in fact, $T\mathcal{B}$ and $T^*\mathcal{B}$ are equivalent fibre bundles. Observe, however, that if a covector $\omega \in T^*\mathcal{B}$ is represented in two different coordinate systems as $\omega = \omega_i dx^i = \rho_i dy^i$, we have that

$$\rho_j = \left. \frac{\partial x^i}{\partial y^j} \right|_b \omega_i. \tag{7.13}$$

We can then say that the group $GL(n; \mathbb{R})$ acts on the typical fibre by direct left multiplication $a^i_j r^j$ in the case of $T\mathcal{B}$, and by multiplication through the inverse-transpose $(a^{-1})^j_i r_j$ in the case of $T^*\mathcal{B}$. In a more formal approach, these two different (but equivalent) left actions could have been taken as a starting point for the definition of these bundles.

EXAMPLE 7.5. **Classical Hamiltonian Mechanics:** The fundamental notion in Hamiltonian Mechanics is the *phase space*, namely the space of positions and momenta of the system. The momenta are obtained (in a local chart of TQ) by differentiating the Lagrangian density with respect to the velocity coordinates, namely:

$$p_i = \frac{\partial \mathcal{L}}{\partial \dot{q}^i}. \tag{7.14}$$

It is not difficult to verify that, fixing a point q in the base manifold Q, these quantities p_i are components of a covector $p \in T_q^*Q$. It follows that the arena of Hamiltonian Mechanics is the cotangent bundle T^*Q. A Hamiltonian density $\mathcal{H}$ is a map:

$$\mathcal{H} : T^*Q \longrightarrow \mathbb{R}. \tag{7.15}$$

For further results, see Example 7.14.   □

The tangent and cotangent bundles are two instances of *vector bundles*, namely, bundles whose typical fibre is a vector space. In a vector bundle one assumes, moreover, that the fibre-wise admissible charts are consistent with the vector-space structure of the typical fibre. This means that the transition maps are at most elements of $GL(n;\mathbb{R})$, where n is the dimension of the fibre.

In the same way as for the tangent and cotangent bundles, one can define the bundle of tensors of type (r,s). The typical fibre is $\mathbb{R}^{n^{r+s}}$, and the (left) action of $GL(n;\mathbb{R})$ is given by:

$$(aT)^{k_1...k_r}_{h_1...h_s} = a^{k_1}_{i_1}...a^{k_r}_{i_r}\,(a^{-1})^{j_1}_{h_1}...(a^{-1})^{j_s}_{h_s}\,T^{i_1...i_r}_{j_1...j_s}, \qquad (7.16)$$

where $a \in GL(n;\mathbb{R})$. The tangent and cotangent bundles are the particular cases of type $(1,0)$ and $(0,1)$, respectively. All these bundles differ just by the typical fibre, and so they are associated bundles, as defined above.

EXERCISE 7.2. **Left action:** Show that the action of the general linear group shown in Equation (7.16) is indeed a left action.

7.6. The Bundle of Linear Frames

The *bundle of linear frames*, or *linear frame bundle*, $F\mathcal{B}$ of a base n-dimensional manifold $\mathcal{B}$ can be defined constructively in the following way. At each point $b \in \mathcal{B}$ we form the set $F_b\mathcal{B}$ of all ordered n-tuples $\{e\}_b = (e_1,...,e_n)$ of linearly independent vectors e_i in $T_b\mathcal{B}$, namely, the set of all bases of $T_b\mathcal{B}$. Our total space will consist of all ordered pairs of the form $(b,\{e\}_b)$ with the obvious projection onto $\mathcal{B}$. The pair $(b,\{e\}_b)$ is called *a linear frame at b*. Following a procedure identical to the one used for the tangent bundle, we obtain that each basis $\{e\}_b$ is expressible uniquely as:

$$e_j = p^i{}_j\,\frac{\partial}{\partial x^i} \qquad (7.17)$$

in a coordinate system x^i, where $\{p^i{}_j\}$ is a nonsingular matrix. We conclude that the typical fibre in this case is $GL(n;\mathbb{R})$. But so is the structure group. Indeed, in another coordinate system, y^i, we have:

$$e_j = q^i{}_j\,\frac{\partial}{\partial y^i}, \qquad (7.18)$$

where

$$q^i{}_j = \frac{\partial y^i}{\partial x^m}\,p^m{}_j = a^i{}_m\,p^m{}_j. \qquad (7.19)$$

This is an instance of a *principal fibre bundle*, namely, a fibre bundle whose typical fibre and structure group coincide. The action of the group on the typical fibre is the natural left action of the group on itself. One of the interesting features of a principal bundle is that the structure group has also a natural *right* action *on the bundle itself*, and this property can be used to provide an alternative definition of principal bundles, which we shall pursue later. In the case of $F\mathcal{B}$, for example, the right action is defined, in a given coordinate system x^i, by

$$R_a\{e\} = p_i^k a_j^i \frac{\partial}{\partial x^k}, \quad j = 1, \ldots, n, \tag{7.20}$$

which sends the basis (7.17) at b to another basis at b, that is, the action is fibre preserving. One can verify that this definition of the action is independent of the system of coordinates adopted (see Box 7.1). The principal bundle of linear frames of a manifold is associated to all the tensor bundles, including the tangent and the cotangent bundles, of the same manifold. By a direct application of the fundamental existence theorem, we know that the associated principal bundle is defined uniquely up to an equivalence. We will later see that many properties of bundles can be better understood by working first on the associated principal bundle.

EXAMPLE 7.6. **Bodies with linear microstructure:** The idea of endowing bodies with a microstructure represented by affine deformations of micromedia, or grains, embedded in a matrix goes back to the pioneering work of the Cosserat brothers. Clearly, this is a particular case of bodies with arbitrary microstructure (see Example 7.2), where no specific limitation is imposed on the microdeformations.[6] If each of the grains is permitted to undergo just affine deformations (namely, deformations with a constant deformation gradient), it is clear that the extra kinematics can be described in terms of a linear map of any basis attached to the grain. Knowing how one basis deforms is enough to determine how all other bases at the same point deform. The choice of basis remaining arbitrary, we are naturally led to the conclusion that the appropriate geometric counterpart of a body $\mathcal{B}$ with linear microstructure (a *general Cosserat body*) is the bundle of linear frames $F\mathcal{B}$.

BOX 7.1. Left and Right Actions

Why can't one define a canonical (i.e., trivialization independent) right action of the structure group on the fibre bundle in the general case? Why can one do it in the case of a principal fibre bundle, such as the bundle of linear frames? Consider a fibre $\pi^{-1}(b)$ in a general fibre bundle with typical fibre $\mathcal{F}$ and structure group $\mathcal{G}$. A trivialization results in a diffeomorphism:

$$\tilde{\psi}_b : \pi^{-1}(b) \longrightarrow \mathcal{F}. \tag{7.21}$$

The only available action of $\mathcal{G}$ on $\mathcal{F}$ is the (left) action L_g involved in the definition of the fibre bundle. Accordingly, given another trivialization, the corresponding map $\tilde{\phi}_b : \pi^{-1}(b) \to \mathcal{F}$ must satisfy:

$$\tilde{\psi}_b \circ \tilde{\phi}_b^{-1} = L_a, \tag{7.22}$$

---

[6] The terminology *Cosserat medium* is often used in the literature to designate the particular case in which the grains can undergo rotations only. For this reason, we use here the longer and more descriptive title. An alternative terminology distinguishes between micropolar and micromorphic media.

for some $a \in \mathcal{G}$. Let us now assume that we want to define an action, $\tilde{L}_g$ say, of $\mathcal{G}$ directly on the fibre $\pi^{-1}(b)$. The only way to do it is to choose one of the trivializations to map the fibre in question to the typical fibre, apply the (only available, left) action and then return to the fibre by the inverse map, namely:

$$\tilde{L}_g = \tilde{\psi}_b^{-1} \circ L_g \circ \tilde{\psi}_b. \tag{7.23}$$

There is nothing else we can do which will involve just g (except possibly use $L_{g^{-1}}$, which does not make any substantial difference to the argument). But repeating the process using the second trivialization, we would obtain:

$$\tilde{L}_g' = \tilde{\phi}_b^{-1} \circ L_g \circ \tilde{\phi}_b, \tag{7.24}$$

which, when combined with (7.22), yields:

$$\tilde{L}_g' = \tilde{\psi}_b^{-1} \circ L_a \circ L_g \circ L_{a^{-1}} \circ \tilde{\psi}_b \neq \tilde{L}_g, \tag{7.25}$$

so the action is not canonical!

The reason we get away with a canonical action in the case of a principal bundle is that a group always has *two* canonical actions on itself, a left one and a right one. These actions are commutative. The trick is then to take advantage of the right action to define the following *right* action on the fibre $\pi^{-1}(b)$:

$$\tilde{R}_g = \tilde{\psi}_b^{-1} \circ R_g \circ \tilde{\psi}_b. \tag{7.26}$$

With a similar notation (a prime to indicate the second trivialization), we now obtain:

$$\tilde{R}_g' = \tilde{\phi}_b^{-1} \circ R_g \circ \tilde{\phi}_b = \tilde{\psi}_b^{-1} \circ L_a \circ R_g \circ L_{a^{-1}} \circ \tilde{\psi}_b = \tilde{\psi}_b^{-1} \circ R_g \circ L_a \circ L_{a^{-1}} \circ \tilde{\psi}_b = \tilde{R}_g, \tag{7.27}$$

which shows that this (right) action is independent of the trivialization and, hence, canonical.

7.7. Principal Bundles

Given a fibre bundle $(\mathcal{C}, \pi, \mathcal{B}, \mathcal{F}, \mathcal{G})$, we have defined its associated principal bundle $(\mathcal{P}, \pi_P, \mathcal{B}, \mathcal{G}, \mathcal{G})$ by invoking the fundamental existence theorem to construct (uniquely up to an equivalence) a fibre bundle with the same base manifold, structure group, trivialization covering, and transition maps, but with a typical fibre equal to the structure group itself. As an example, we have shown that the principal bundle associated with all the tensor bundles of $\mathcal{B}$ is the bundle of linear frames $F\mathcal{B}$. We also have shown how the existence of a canonical right action of $\mathcal{G}$ on itself (in addition to the canonical left action that $\mathcal{G}$ must have on the typical fibre) allows us to define a canonical right action of $\mathcal{G}$ on the principal bundle itself. It is not difficult to verify that this action is *free*.

EXERCISE 7.3. Prove that the canonical right action of the structure group on a principal bundle is free.

The existence of a free right action is strong enough to provide an independent definition of a principal fibre bundle which, although equivalent to the one already given, has the merit of being independent of the notion of transition maps. Moreover, once this more elegant and constructive definition has been secured, a subsidiary definition of the associated (nonprincipal) bundles becomes available, again without an explicit mention of the transition maps. Finally, this more abstract definition brings out intrinsically the nature and meaning of the associated bundles.

Let $\mathcal{P}$ be a differentiable manifold (the *total space*) and $\mathcal{G}$ a Lie group (the *structure group*), and let $\mathcal{G}$ act freely to the right on $\mathcal{P}$. This means that there exists a smooth map:

$$R_g : \mathcal{P} \times \mathcal{G} \longrightarrow \mathcal{P}$$
$$(p,g) \mapsto R_g p = pg, \tag{7.28}$$

such that, for all $p \in \mathcal{P}$ and all $g, h \in \mathcal{G}$, we have:

$$R_{gh} p = R_h R_g p = pgh,$$
$$R_e p = p, \tag{7.29}$$

where e is the group identity. The fact that the action is free means that if, for some $p \in \mathcal{P}$ and some $g \in \mathcal{G}$, $R_g p = p$, then necessarily $g = e$. Define now the quotient $\mathcal{B} = \mathcal{P}/\mathcal{G}$ and check that $\mathcal{B}$ is a differentiable manifold[7] and that the canonical projection $\pi_P : \mathcal{P} \to \mathcal{P}/\mathcal{G}$ is differentiable. The set $\pi_P^{-1}(b)$ is called the fibre over $b \in \mathcal{B}$.

REMARK 7.1. Recall that an element of the quotient $\mathcal{B} = \mathcal{P}/\mathcal{G}$ is, by definition of quotient, an equivalence class in $\mathcal{P}$ by the action of the group $\mathcal{G}$. In other words, each element b of the quotient (namely, of the base manifold $\mathcal{B}$) can be regarded as representing an orbit. The projection map assigns to each element of $\mathcal{P}$ the orbit to which it belongs. The fibre over b consists of all the elements of $\mathcal{P}$ that belong to the specific orbit represented by b.

To complete the definition of a principal bundle, we need only to add the condition that $\mathcal{P}$ be locally trivial, namely, that for each $b \in \mathcal{B}$, there exists a neighbourhood $\mathcal{U} \subset \mathcal{P}$ such that $\pi_P^{-1}(\mathcal{U})$ is isomorphic to the product $\mathcal{U} \times \mathcal{G}$. More precisely, there exists a fibre-preserving diffeomorphism:

$$\psi : \pi_P^{-1}(\mathcal{U}) \longrightarrow \mathcal{U} \times \mathcal{G}$$
$$p \mapsto (b, \tilde{\psi}_b), \tag{7.30}$$

---

[7] A sufficient condition for the quotient to be a differentiable manifold is that the group $\mathcal{G}$ be *properly discontinuous*. Roughly speaking, this condition guarantees that fibres at different points of the base manifold don't get arbitrarily close to each other.

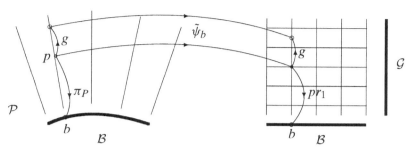

Figure 7.3. The group consistency condition

where $b = \pi_P(p)$, with the additional property that it must be *consistent with the group action*, namely (see Figure 7.3):

$$\tilde{\psi}_b(pg) = \tilde{\psi}_b(p)g \quad \forall p \in \pi_P^{-1}(\mathcal{U}), g \in \mathcal{G}. \tag{7.31}$$

This completes the definition of the principal bundle. The right action is fibre preserving and every fibre is diffeomorphic to $\mathcal{G}$. Moreover, every fibre is a closed submanifold of $\mathcal{P}$ and coincides with an orbit of the right action of $\mathcal{G}$.

BOX 7.2. **Why the Group-Consistency Condition is Crucial**

We have introduced an alternative definition of a principal fibre bundle through the free right action of a group $\mathcal{G}$ on a total manifold $\mathcal{P}$. This definition does not make specific mention of the transition maps, which are an essential part of the standard definition of a fibre bundle. It behooves us now to show that this new definition is equivalent to the standard one. To do so, we consider a principal fibre bundle $\mathcal{P}$, with structure group $\mathcal{G}$, defined in the new way. Consider two local trivializations, $(\mathcal{U}, \psi)$ and $(\mathcal{V}, \phi)$, and let $b \in \mathcal{U} \bigcap \mathcal{V}$. We want to show that for each point $b \in \mathcal{B}$ the restricted map $\tilde{\phi}_b \circ \tilde{\psi}_b^{-1} : \mathcal{G} \to \mathcal{G}$ is in fact equivalent to the left action of a fixed element $a \in \mathcal{G}$. Let $g \in \mathcal{G}$. Since $\tilde{\phi}_b \circ \tilde{\psi}_b^{-1}(g)$ belongs to $\mathcal{G}$ and since the left action of $\mathcal{G}$ on itself is transitive, there exists a unique $a \in \mathcal{G}$ such that:

$$\tilde{\phi}_b \circ \tilde{\psi}_b^{-1}(g) = L_a g = ag. \tag{7.32}$$

We need to prove that this $a \in \mathcal{G}$ is independent of g. Therefore, let, $h \in \mathcal{G}$ be some other point in the typical fibre. By transitivity of the right action there exists a unique $c \in \mathcal{G}$ such that $h = gc$. Therefore,

$$\tilde{\phi}_b \circ \tilde{\psi}_b^{-1}(h) = \tilde{\phi}_b \circ \tilde{\psi}_b^{-1}(gc) = \tilde{\phi}_b[R_c \tilde{\psi}_b^{-1}(g)]$$
$$= R_c[\tilde{\phi}_b \circ \tilde{\psi}_b^{-1}(g)] = R_c L_a(g) = L_a R_c g = L_a h. \tag{7.33}$$

> The crucial step in this chain is the commutation of the trivializations $\tilde{\phi}_b$ and $\tilde{\psi}_b$ with the right action R_c, namely, the consistency condition. The last result clearly shows that transition maps can indeed be defined and that they belong to the structure group, as required by the standard definition. The equivalence of both definitions follows now from the fundamental existence theorem.

EXAMPLE 7.7. **A relativistic material body-time.** In the context of General Relativity, it is convenient to define a 4-dimensional body time $\mathcal{C}$. This notion corresponds roughly to the following physical picture: an ordinary 3-dimensional material body $\mathcal{B}$ each of whose points is equipped with a clock. All clocks are identical (they measure proper time), but they cannot be canonically synchronized. If we, accordingly, adopt as the structure group the group of orientation-preserving isometries of $\mathbb{R}$ (namely, the group of translations of the real line),[8] we can define $\mathcal{C}$ as a 4-dimensional manifold on which this group acts. For consistency with classical nonrelativistic Continuum Mechanics, we may impose the extra condition that the quotient manifold be coverable with a single coordinate chart.

EXERCISE 7.4. **Transitivity:** The right and left natural actions of a group $\mathcal{G}$ on itself are not only free but also *transitive*, namely, for every $g, h \in \mathcal{G}$ there exist unique $a, b \in \mathcal{G}$ such that $R_a g = h$ and $L_b g = h$. Use the group consistency condition of a principal fibre bundle as just defined to show that, on the fibre $\pi_P^{-1}(b)$ at a point $b \in \mathcal{B}$, the right action of the structure group $\mathcal{G}$ is transitive. Moreover, if $g, a \in \mathcal{G}$ and $b \in \mathcal{U} \subset \mathcal{B}$, show that $\tilde{\psi}_b^{-1}(ga) = R_a \tilde{\psi}_b^{-1}(g)$, where ψ is a local trivialization defined on $\mathcal{U}$.

7.8. Associated Bundles

The concept of associated bundle has already been defined and used to introduce the notion of the principal bundle associated with any given fibre bundle. On the other hand, in the preceding section we have introduced an independent definition of principal bundles by means of the idea of a right action of a group on a given total manifold. We want now to show that this line of thought can be pursued to obtain another view of the collection of all (nonprincipal) fibre bundles associated with a given principal bundle.

As a more or less intuitive motivation for this procedure, it is convenient to think of the example of the principal bundle of linear frames $F\mathcal{B}$ of a manifold $\mathcal{B}$. We already know that this bundle is associated to the tangent bundle $T\mathcal{B}$. Consider now a pair (f, v), where $f \in F\mathcal{B}$ and $v \in T\mathcal{B}$, such that $\pi_P(f) = \pi(v) = b$. In other words, f and v represent, respectively, a basis and a vector of the tangent space at some point $b \in \mathcal{B}$. We can, therefore, identify v with its components on the linear frame f,

---

[8] The reason to choose isometries is that all the clocks are supposed to be endowed with the same time scale. See Epstein M, Burton D, Tucker R (2006), Relativistic anelasticity, *Classical and Quantum Gravity* **23**, 3545–3571.

namely, with an element of the typical fibre ($\mathbb{R}^n$) of $T\mathcal{B}$. If we consider now a pair $(\hat{f}, v)$, where v is the same as before but $\hat{f}$ is a new linear frame at b, the corresponding element of the typical fibre representing the *same* vector v changes. More explicitly, with an obvious notation, if $\hat{f}_j = a^i_j f_i$, then $v^i = a^i_j \hat{v}^j$ or $\hat{v}^i = (a^{-1})^i_j v^j$. We conclude that to represent the *same object* under a change of frame, there needs to be some kind of compensatory action in the change of the components. The object itself (in this case, the tangent vector) can be identified with the collection (or equivalence class) of all pairs made up of a frame and a set of components related in this compensatory way. In terms of the group actions on the typical fibres, if $\hat{f} = R_a f$, then the representative r of the vector v in $\mathbb{R}^n$ changes according to $\hat{r} = L_{a^{-1}} r$. We may, therefore, think of a vector as an equivalence class of elements of the Cartesian product $\mathcal{G} \times R^n$, corresponding to the following equivalence relation: $(g, r) \sim (\hat{g}, \hat{r})$ if, and only if, there exists $a \in \mathcal{G}$ such that $\hat{g} = ga$ and $\hat{r} = L_{a^{-1}} r$.[9]

REMARK 7.2. The cotangent bundle: A similar point of view can be applied to the cotangent bundle and to all tensor bundles. In the case of the cotangent bundle, for example, the representation of a covector ω changes, upon a change of basis, according to the formula $\hat{\omega}_i = a^j_i \omega_j$. At first sight, this transformation law seems to contradict the criterion $\hat{r} = L_{a^{-1}} r$ (for the representatives in $\mathbb{R}^n$) given above. But one has to remember that for the cotangent bundle the left action of the general linear group on the typical fibre is given through left multiplication by the inverse-transpose. Therefore, in this case, the expression $L_{a^{-1}}$ provides the desired left multiplication by the transpose.

REMARK 7.3. Full circle: Note how, in a not-so-subtle way, the previous remarks reveal that the notion of associated bundle is a sophisticated throwback to the traditional definition of vectors and tensors by a declaration of the way in which their components change with a change of coordinates, as described in Exercise 4.5.

With the above motivation in mind, the following construction of a fibre bundle associated to a given principal bundle will seem less artificial than it otherwise would. We start from the principal bundle $(\mathcal{P}, \pi_P, \mathcal{B}, \mathcal{G}, \mathcal{G})$ and a manifold $\mathcal{F}$, which we want to construe as the typical fibre of a new fibre bundle $(\mathcal{C}, \pi, \mathcal{B}, \mathcal{F}, \mathcal{G})$ associated with $\mathcal{P}$. For this to be possible, we need to have an effective left action of $\mathcal{G}$ on $\mathcal{F}$, which we assume to have been given. To start off, we form the Cartesian product $\mathcal{P} \times \mathcal{F}$ and notice that the structure group $\mathcal{G}$ acts on it with a right action induced by its right action on $\mathcal{P}$ and its left action on $\mathcal{F}$. To describe this new right action, we will keep abusing the notation in the sense that we will use the same symbols for all the actions in sight, since the context should make clear which action is being used in each particular expression. Let (p, f) be an element of the product $\mathcal{P} \times \mathcal{F}$, and let $a \in \mathcal{G}$. We define:

$$R_a(p, f) = (R_a p, L_{a^{-1}} f). \tag{7.34}$$

---

[9] By transitivity, it is enough to check that $\hat{r} = L_{\hat{g}^{-1} g} r$.

EXERCISE 7.5. **Right action on the product:** Verify that the formula above constitutes an effective right action.

The next step towards the construction of the associated bundle with typical fibre $\mathcal{F}$ consists of taking the quotient space $\mathcal{C}$ generated by this action. In other words, we want to deal with a set whose elements are equivalence classes in $\mathcal{C} \times \mathcal{F}$ by the equivalence relation: "$(p_1, f_1) \sim (p_2, f_2)$ if, and only if, there exists $a \in \mathcal{G}$ such that $(p_2, f_2) = R_a(p_1, f_1)$." The motivation for this line of attack should be clear from the introductory remarks to this section. Recalling that the right action of $\mathcal{G}$ on $\mathcal{P}$ is fibre preserving, it becomes obvious that all the pairs (p, f) in a given equivalence class have first components p with the same projection $\pi_P(p)$ on $\mathcal{B}$. This means that we have a perfectly well-defined projection π in the quotient space $\mathcal{C}$, namely, $\pi : \mathcal{C} \to \mathcal{B}$ is a map that assigns to each equivalence class the common value of the projection of the first component of all its constituent pairs.

Having a projection, we can now define the fibre of $\mathcal{C}$ over $b \in \mathcal{B}$ naturally as $\pi^{-1}(b)$. We need to show now that each such fibre is diffeomorphic to the putative typical fibre $\mathcal{F}$. More precisely, we want to show that for each local trivialization $(\mathcal{U}, \psi)$ of the original principal bundle $\mathcal{P}$, we can also construct a local trivialization of $\pi^{-1}(\mathcal{U})$, namely, a diffeomorphism $\rho : \pi^{-1}(\mathcal{U}) \to \mathcal{U} \times \mathcal{F}$. To understand how this works, let us fix a point $b \in \mathcal{U}$ and recall that, given the local trivialization $(\mathcal{U}, \psi)$, the map $\tilde{\psi}_b$ provides us with a diffeomorphism of the fibre $\pi_P^{-1}(b)$ with $\mathcal{G}$. We now form the product map of $\tilde{\psi}_b$ with the identity map of $\mathcal{F}$, namely, $(\tilde{\psi}_b, id_\mathcal{F}) : \pi_P^{-1}(b) \times \mathcal{F} \to \mathcal{G} \times \mathcal{F}$. Each equivalence class by the right action (7.34) is mapped by the product map $(\tilde{\psi}_b, id_\mathcal{F})$ into a submanifold (an orbit), as shown in Figure 7.4.

These orbits do not intersect with each other. Moreover, they can be seen as graphs of single-valued $\mathcal{F}$-valued functions of $\mathcal{G}$ (Why? See Exercise 7.6.) Therefore, choosing any particular value $g \in \mathcal{G}$, we see that these orbits can be parametrized by $\mathcal{F}$. This provides the desired one-to-one and onto relation between the fibre $\pi^{-1}(b)$ and the manifold $\mathcal{F}$, which can now legitimately be called the typical fibre of $\mathcal{C}$. To complete the construction of the desired fibre bundle, we need to guarantee that the fibre-wise isomorphism that we have just constructed depends differentiably on b, a requirement that we assume fulfilled.

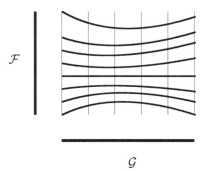

Figure 7.4. Images of equivalence classes

EXERCISE 7.6. **Revealing the typical fibre:** Show that the orbits generated by the right action (7.34) on $\mathcal{G} \times \mathcal{F}$ do not intersect and look like single-valued functions of $\mathcal{G}$ with values in $\mathcal{F}$. [Hint: the right action of $\mathcal{G}$ on itself is free and transitive, while its left action on $\mathcal{F}$ is only effective. See also the next exercise.]

EXERCISE 7.7. **The case of the tangent bundle:** Consider the specific case of the tangent bundle of a 2-dimensional manifold to illustrate the nature of the above orbits. The typical fibre $\mathcal{F}$ is identified with $\mathbb{R}^2$. How does the orbit corresponding to the zero vector look? Show that no matter which vector is considered, its corresponding orbit will attain any given value of $\mathcal{F}$ an infinite number of times. [Hint: consider changes of basis that are given by matrices with a fixed eigenvector whose eigenvalue is equal to 1.] Conclude, therefore, that the orbits cannot possibly be parametrized by $\mathcal{G}$.

7.9. Fibre-Bundle Morphisms

We are interested in considering (smooth) mappings between any pair of fibre bundles, say $(\mathcal{C}, \mathcal{B}, \pi, \mathcal{F}, \mathcal{G})$ and $(\mathcal{C}', \mathcal{B}', \pi', \mathcal{F}', \mathcal{G}')$. Naturally, considering the total spaces $\mathcal{C}$ and $\mathcal{C}'$ as manifolds, we have at our disposal the collection of all smooth mappings $K : \mathcal{C} \to \mathcal{C}'$. Nevertheless, for obvious reasons, our interest lies in those maps that preserve fibres, namely, those maps that carry entire fibres into fibres, thus respecting the fundamental nature of the bundles. We shall call these special maps *fibre-bundle morphisms*. The fibre-preserving restriction can be expressed by the condition that, for each $b \in \mathcal{B}$, the set $\pi' \circ K \circ \pi^{-1}(b)$ must be a singleton, that is, it must consist of a single point b' of $\mathcal{B}'$. If this is the case, we obtain a perfectly well-defined subsidiary map k at the base-manifold level, namely, $k : \mathcal{B} \to \mathcal{B}'$. We can, therefore, define a fibre-bundle morphism as a pair of (smooth) maps, $K : \mathcal{C} \to \mathcal{C}'$ and $k : \mathcal{B} \to \mathcal{B}'$ such that $\pi' \circ K = k \circ \pi$, as represented in the following commutative diagram:

$$
\begin{array}{ccc}
\mathcal{C} & \xrightarrow{\ K\ } & \mathcal{C}' \\
{\scriptstyle \pi}\big\downarrow & & \big\downarrow{\scriptstyle \pi'} \\
\mathcal{B} & \xrightarrow{\ k\ } & \mathcal{B}'
\end{array}
\qquad (7.35)
$$

When there is no room for confusion, one indicates a fibre-bundle morphism by means of the "upper" map alone, understanding (as we have done) that a cognate symbol will be used, whenever needed, for the subsidiary ("lower") map between the base manifolds.

EXAMPLE 7.8. **The tangent map:** We have seen in Chapter 4 that, given a smooth map $k : \mathcal{B} \to \mathcal{B}'$ between two manifolds, $\mathcal{B}$ and $\mathcal{B}'$, the differential $k_*(b)$ of k at a point $b \in \mathcal{B}$ is a well-defined linear map between the tangent spaces $T_b\mathcal{B}$ and $T_{k(b)}\mathcal{B}'$. Since, by construction, the differential map takes fibres to fibres, we see that it induces a fibre-bundle morphism between the respective tangent bundles. This morphism is called the *tangent map of k* and is denoted by $Tk : T\mathcal{B} \to T\mathcal{B}'$. Naturally, the subsidiary map of Tk is k itself. This is a special instance in which a fibre-bundle morphism is constructed from the bottom up.

Given the fibre-bundle morphisms $K_1 : \mathcal{C} \to \mathcal{C}'$ and $K_2 : \mathcal{C}' \to \mathcal{C}''$, the composition $K_2 \circ K_1 : \mathcal{C} \to \mathcal{C}''$ is also a fibre-bundle morphism. Moreover, the composition of fibre-bundle morphisms (whenever the domains permit it) is associative, namely, $K_1 \circ (K_2 \circ K_3) = (K_1 \circ K_2) \circ K_3$. Finally, the identity map $id_{\mathcal{C}} : \mathcal{C} \to \mathcal{C}$ is a well-defined fibre-bundle morphism that can be composed with other fibre-bundle morphisms (to the right or to the left) leaving the result unaltered. These properties endow the collection of all fibre-bundles with the structure of a *category* (see Box 7.3).

BOX 7.3. **Categories**

The conceptual framework of category theory plays in contemporary mathematics a unifying role similar to that inaugurated by set theory at the turn of the twentieth century. And, much as in the case of set Theory, one can derive a considerable benefit from the mere use of the language of category theory, without a need to be familiar with its intricacies and subtleties.

A *category* C is a class of *objects* to each ordered pair (A, B) of which there corresponds a set denoted by $Hom(A, B)$. The elements of these sets are called *morphisms*. Moreover, for each ordered triple (A, B, C) of objects in C there exists a function, called *composition*:

$$Hom(A, B) \times Hom(B, C) \longrightarrow Hom(A, C)$$

$$(f, g) \mapsto g \circ f. \tag{7.36}$$

To constitute a category these basic elements (objects, morphisms, and compositions) must satisfy the following two properties:

(1) Associativity: If $f \in Hom(A, B)$, $g \in Hom(B, C)$ and $h \in Hom(C, D)$, then

$$h \circ (g \circ f) = (h \circ g) \circ f. \tag{7.37}$$

(2) Existence of identity: For each object A there exists a (unique) *identity morphism* $id_A \in Hom(A, A)$ such that for any morphism $f \in Hom(A, B)$ we have that $id_B \circ f = f \circ id_A = f$.

The usual (but by no means only) application of this concept consists of identifying the objects in the category with sets endowed with some structure, and the morphisms with maps that preserve that structure. When this is the case, it is customary to use the notation $f : A \to B$ interchangeably with $f \in Hom(A, B)$. In this case, moreover, the compositions and the associativity of morphisms acquire a natural meaning, as does the existence of the identity map. Examples of categories of this type are sets (functions), topological spaces (continuous maps), vector spaces (linear maps), smooth manifolds (smooth maps), where we have indicated the morphisms in parentheses.

A morphism $f \in Hom(A, B)$ is called an *isomorphism* if there exists a morphism $g \in Hom(B, A)$ such that $f \circ g = id_B$ and $g \circ f = id_A$.

By restricting the objects and/or the admissible morphisms to a subclass (always satisfying the above properties), we may obtain useful subcategories of fibre bundles. Two of these subcategories of bundles are worthy of mention. The first one consists of all fibre bundles having the same typical fibre $\mathcal{F}$ and the same structure group $\mathcal{G}$. For this restricted collection, we will further restrict the admissible morphisms K in the following way: Since the typical fibres and structure groups are the same, we can, and will, require that, for local trivializations $(\mathcal{U}, \psi)$ and $(\mathcal{V}, \phi)$ (of either bundle) such that $\mathcal{U} \bigcap f^{-1}(\mathcal{V}) \neq \emptyset$, the map $\phi \circ K \circ \psi^{-1} : \mathcal{U} \times \mathcal{F} \to \mathcal{V} \times \mathcal{F}$ must consist of left actions of $\mathcal{G}$ on $\mathcal{F}$ depending smoothly on position throughout this intersection. Two bundles having the same base space $\mathcal{B}$, typical fibre $\mathcal{F}$ and structure group $\mathcal{G}$ are, therefore, equivalent (see Section 7.3) if, and only if, there exists a fibre-bundle morphism K (of the type just described) with $k = id_\mathcal{B}$.

The second subcategory of fibre bundles that deserves special mention is that consisting of principal bundles. The corresponding morphisms, called *principal-bundle morphisms*, are defined as follows: Given two principal bundles, $(\mathcal{P}, \pi_P, \mathcal{B}, \mathcal{G}, \mathcal{G})$ and $(\mathcal{P}', \pi'_P, \mathcal{B}', \mathcal{G}', \mathcal{G}')$, a principal-bundle morphism consists of a (smooth) map $K : \mathcal{P} \to \mathcal{P}'$ and a Lie-group homomorphism[10] $\hat{K} : \mathcal{G} \to \mathcal{G}'$ such that:

$$K(R_g p) = R'_{g'} K(p), \quad \forall p \in \mathcal{P}, g \in \mathcal{G}, \tag{7.38}$$

where $g' = \hat{K}(g)$. We have denoted the right actions of $\mathcal{G}$ and $\mathcal{G}'$ on $\mathcal{P}$ and $\mathcal{P}'$ by appropriately subscripted R and R', respectively. If, instead, we want to indicate the right actions by simple apposition to the right, we may rewrite the last equation more elegantly as: $K(pg) = K(p)\hat{K}(g)$.

EXERCISE 7.8. **Principal-bundle morphism:** Show that with the above definition a principal-bundle morphism is indeed a fibre-bundle morphism (namely, a fibre-preserving map). Reason on a drawing similar to Figure 7.3.

EXAMPLE 7.9. **Configurations of general Cosserat bodies:** In Example 7.6 we introduced the notion of a body with linear microstructure (or general Cosserat body), and we concluded that such a body can be identified with the principal bundle of linear frames $F\mathcal{B}$ of an ordinary body $\mathcal{B}$ representing the macromedium. A configuration of the macromedium, as we know, can be described as an embedding of $\mathcal{B}$ into classical space, which is an affine 3-dimensional Euclidean space $\mathbb{E}^3$. The linear frame bundle $F\mathbb{E}^3$ is, therefore, the natural setting for the configurations of $F\mathcal{B}$. Each configuration is a principal-bundle morphism $K : F\mathcal{B} \to F\mathbb{E}^3$, with the group homomorphism $\hat{K}$ being just the identity. The subsidiary map $\kappa : \mathcal{B} \to \mathbb{E}^3$ is

---

[10] Recall that a homomorphism of groups is a mapping that respects the group operation.

an ordinary configuration of the macromedium. Note that $F\mathbb{E}^3$ is a principal bundle whose structure group is the whole general linear group $GL(3;\mathbb{R})$, rather than just the (nonprincipal) product bundle $\mathbb{E}^3 \times GL(3;\mathbb{R})$ (whose structure group is the identity). Nevertheless, one often uses the product terminology to refer to a globally trivializable bundle, if the structure group is clear from the context.

7.10. Cross Sections

A *cross section* σ of a fibre bundle $(\mathcal{C}, \pi, \mathcal{B}, \mathcal{F}, \mathcal{G})$ is a (smooth) map:

$$\sigma : \mathcal{B} \longrightarrow \mathcal{C}, \tag{7.39}$$

such that $\pi \circ \sigma = id_{\mathcal{B}}$, as shown in the following commutative diagram:

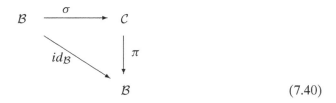

$$(7.40)$$

A cross section is thus nothing but a smooth assignment, to each point b in the base manifold, of an element of its fibre $\mathcal{C}_b$.

EXAMPLE 7.10. **Vector fields:** A vector field over a manifold $\mathcal{B}$ is defined as a smooth assignment of a tangent vector at each point of the manifold. In the terminology of fibre bundles, therefore, a vector field over a manifold $\mathcal{B}$ is a cross section of the tangent bundle $T\mathcal{B}$.

EXAMPLE 7.11. **One-forms:** A 1-form over a manifold $\mathcal{B}$ is, by definition, a smooth assignment of a covector at each point. In other words, a 1-form over a manifold $\mathcal{B}$ is a cross section of the cotangent bundle $T^*\mathcal{B}$. In a similar way one can identify tensor fields over a manifold with cross sections of the corresponding tensor bundle.

EXAMPLE 7.12. **Moving frames:** A moving frame (or *repère mobile of Cartan*) is a smooth assignment of a basis for tangent vectors at each point of a manifold. It can be seen, therefore, as a cross section of the principal bundle of linear frames.

REMARK 7.4. **Existence of cross sections**: It is important to realize that not all fibre bundles admit (smooth, or even continuous) cross sections. Globally trivializable fibre bundles, on the other hand, always admit cross sections. Indeed, let $(\mathcal{B}, \psi)$ be a global trivialization of a globally trivializable fibre bundle and let a denote a fixed element of the typical fibre $\mathcal{F}$. Then, the function defined as $\sigma(b) = \tilde{\psi}_b^{-1}(a)$ is a cross section. One should not think, however, that this sufficient condition is also necessary for the existence of a cross section. As an intuitive example to the contrary, one can take the Moebius band (see Section 7.3). Any smooth curve drawn on the original strip such that its end points are symmetrically arranged will do the job as a cross

Figure 7.5. Cross sections of a Moebius band

section, as shown in Figure 7.5. (Notice, however, that any two such cross sections will end up having at least one point in common!)

The previous remark notwithstanding, we can prove that in the case of principal bundles trivializability is equivalent to the existence of a cross section. We need to prove just the "only if" part, since the "if" part is always true, as we have shown in the previous remark. Assume, therefore, that a given principal bundle $(\mathcal{P}, \pi, \mathcal{B}, \mathcal{G}, \mathcal{G})$ admits a cross section $\sigma : \mathcal{B} \to \mathcal{P}$. The proof follows from the transitivity of the right action of $\mathcal{G}$ on $\mathcal{P}$. For, let $p \in \mathcal{P}$. By the transitivity of the right action on each fibre, there exists a unique $g(p) \in \mathcal{G}$ such that $p = R_{g(p)} \circ \sigma \circ \pi_P(p)$. Define the map:

$$\psi : \mathcal{P} \longrightarrow \mathcal{B} \times \mathcal{G}$$

$$p \mapsto (\pi_P(p), g(p)). \tag{7.41}$$

It is not difficult to verify that this map is a global trivialization of the principal bundle, satisfying the group consistency condition. A nice way to picture this situation (Figure 7.6) is to imagine that the given cross section σ is translated by the right action of the group to give rise to a family of cross sections $R_g \circ \sigma$. Since the action is effective, no two such cross sections will intersect (provided just g is not the group identity). Moreover, since the action is transitive on fibres, every point of the fibre bundle will belong to one (and only one) translated cross section. In other words, we have a family of cross sections, parametrized by the structure group, that completely spans the total space. This is, naturally, tantamount to a global trivialization.

An interesting corollary of this theorem is that, since the trivialization coverings of all associated bundles are the same, a fibre bundle is trivializable if, and only if, its associated principal bundle admits a cross section. This result is just one illustration

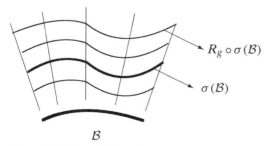

Figure 7.6. Translation of a cross section by the right action

of the assertion we made in Section 7.4 to the effect that working with the associated principal bundle often simplifies and helps in understanding the mathematical picture.

EXERCISE 7.9. **The Moebius band revisited:** Describe the principal bundle associated with the Moebius band (see Section 7.3). Notice that the associated bundle must be constructed upon the same trivialization covering and transition maps as the band itself. Thus, if we cover the base manifold (a circumference) by means of two arcs, each of extent $3\pi/2$, overlapping on two disjoint arcs, each of extent $\pi/2$, on one of these overlaps the transition map is the identity and on the other the inversion. The structure group is, therefore, the group of permutations of two symbols, also called the symmetric group of order 2. The associated principal bundle will have this group as its typical fibre. But, since this fibre is discrete,[11] it is not possible to pass continuously from one element to the other (as required by one of the transition maps). It follows that the associated principal bundle, in contradistinction with the Moebius band itself, does not admit a cross section. We conclude, therefore, that the Moebius band is not a trivial bundle. To what do you attribute the existence of cross sections in the Moebius band?

EXAMPLE 7.13. **The canonical cross section of a vector bundle:** A vector bundle, whether globally trivializable or not, always possesses a canonical cross section, namely, the map that assigns to each point of the base manifold the zero vector of its tangent space. It is sometimes called the *zero section*.

7.11. Iterated Fibre Bundles

Given a manifold $\mathcal{B}$, we have shown how to define and construct its tangent bundle $T\mathcal{B}$. As a by-product of this construction we obtained a differentiable structure on the total space (of pairs (b, v_b), where $b \in \mathcal{B}$ and $v_b \in T_b\mathcal{B}$). In other words, the total space is a differentiable manifold in its own right, regardless of the special atlases (compatible with the fibred structure) that were used to define it. It makes sense, therefore, to calculate its tangent bundle, which will be denoted as $TT\mathcal{B}$. Put differently, we can consider T as an operator that produces manifolds out of manifolds and can, therefore, be applied repeatedly. A similar reasoning applies to the cotangent (T^*) and linear frame bundle (F) operators as well as to all the associated tensor bundles. These operators can be applied in succession ($TT^*\mathcal{B}, T^*T^*\mathcal{B}, TF\mathcal{B}$, and so on). So far, nothing remarkable has been said. The interesting feature of an *iterated bundle* is that it can also be regarded as a fibre bundle *over the underlying original base manifold* $\mathcal{B}$, and not just as a fibre bundle over the total space of the first bundle. The case of the tangent operator is particularly striking in that its repeated

---

[11] It consists of just two elements. With the discrete topology, it can be regarded as a topological group. Somewhat imprecisely, we may say that the typical fibre in this case is a manifold of dimension zero.

application provides a third canonical bundle structure, as we shall see. The importance of iterated bundles is multifaceted. From the purely mathematical point of view, they appear as an essential ingredient in the theory of connections. In Classical Mechanics, they play an important role in both the Lagrangian and the Hamiltonian formulations. In Continuum Mechanics, they arise naturally in the various theories of continuous distributions of dislocations and in the formulation of theories of materials with internal structure.

7.11.1. The Tangent Bundle of a Fibre Bundle

Before embarking into the issue of iterated bundles just outlined, it will be useful to consider the tangent bundle TC of the total space C of an arbitrary fibre bundle (C, π, B, F, G). Let us denote by τ_C its natural projection:

$$\tau_C : TC \longrightarrow C. \tag{7.42}$$

If the dimensions of B and F are, respectively, m and n, the dimension of C is $m+n$, and the typical fibre of (TC, τ_C) is $\mathbb{R}^{m+n}$, with structure group $GL(m+n; \mathbb{R})$. We remark that (TC, τ_C) is, in fact, a vector bundle.

Although TC cannot in general be considered as a fibre bundle over the original base manifold B (see Box 7.4), there exists another canonical fibre-bundle structure of TC over TB. Indeed, consider the following commutative diagram obtained by the mere definition of the tangent map π_* of the original projection π:

$$
\begin{array}{ccc}
TC & \xrightarrow{\ \pi_* \ } & TB \\
{\scriptstyle \tau_C}\big\downarrow & & \big\downarrow{\scriptstyle \tau_B} \\
C & \xrightarrow{\ \pi \ } & B
\end{array}
\tag{7.43}
$$

We want to show that the projection $\pi_* : TC \to TB$ endows TC with the structure of a fibre bundle over TB. To this end, consider a trivialization (U_α, ψ_α) of (C, π). The maps:

$$(\psi_\alpha)_* : T(\pi^{-1}(U_\alpha)) = \pi_*^{-1}(TU_\alpha) \longrightarrow T(U_\alpha \times F) \approx TU_\alpha \times TF, \tag{7.44}$$

provide the desired trivialization. We conclude that the typical fibre of the new bundle is TF. As far as the structure group is concerned, the trivialization just constructed can be allowed to be further composed with any vector-bundle automorphism of TF consistent with G, namely, with a fibre-wise-linear fibre-bundle morphism that induces in the base F an admissible transformation (i.e., a left action of an element of G). We conclude that the structure group of the new bundle (TC, π_*) is the product $G \times GL(n; \mathbb{R})$. Note that the new bundle $(TC, \pi_*, TB, TF, G \times GL(n; \mathbb{R}))$ is not, in general, a vector bundle.

BOX 7.4. **Can Fibre Bundles be Composed?**

Let $(\mathcal{C}, \pi, \mathcal{B}, \mathcal{F}, \mathcal{G})$ and $(\mathcal{D}, \tau, \mathcal{C}, \mathcal{K}, \mathcal{H})$ be two fibre bundles related (as shown by the notation) by the mere fact that the base manifold of the second fibre bundle happens to be the total space of the first. The composition:

$$\rho = \pi \circ \tau : \mathcal{D} \longrightarrow \mathcal{B} \tag{7.45}$$

is a smooth map onto $\mathcal{B}$. Can it be used unconditionally to define a new fibre bundle? The answer in general is negative. Naively, we might think that, letting the point b vary throughout $\mathcal{B}$, the sets $\rho^{-1}(b)$ may turn out to be diffeomorphic to the product $\mathcal{F} \times \mathcal{K}$ (on which the product group $\mathcal{G} \times \mathcal{H}$ acts on the left) and thus serve to define $\mathcal{F} \times \mathcal{K}$ as the typical fibre of a fibre bundle with total space $\mathcal{D}$ over the base manifold $\mathcal{B}$. What may go wrong, however, is the possibility of defining local trivializations. Why so? Because the trivializing coverings of $\mathcal{C}$ (as the base of the second fibre bundle) may have nothing to do with the fibred nature of $\mathcal{C}$ (as the total space of the first fibre bundle). This situation is shown schematically in Figure 7.7, where $\mathcal{V}$ is a trivializing open set for the second fibre bundle.

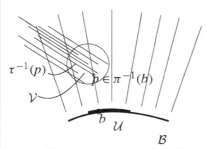

Figure 7.7. A bundle over a bundle

A sufficient condition for the composition to make sense as a fibre bundle is, therefore, the following: there exists a trivializing covering $(\mathcal{U}_\alpha, \psi_\alpha)$ of the bundle $(\mathcal{C}, \pi, \mathcal{B}, \mathcal{F}, \mathcal{G})$ such that the sets $\mathcal{V}_\alpha = \pi^{-1}(\mathcal{U}_\alpha)$ (together with some functions ϕ_α) provide a trivializing covering $(\mathcal{V}_\alpha, \phi_\alpha)$ of the bundle $(\mathcal{D}, \tau, \mathcal{C}, \mathcal{K}, \mathcal{H})$. This will, of course, always be the case if the latter fibre bundle is trivial. It also happens to be the case for iterated tangent, cotangent, tensor, and frame bundles in general.

EXERCISE 7.10. **The case of a vector bundle:** Show that the tangent bundle of a vector bundle over a base manifold $\mathcal{B}$ can always be seen as a fibre bundle on $\mathcal{B}$. Is this a vector bundle? Why not? Extend your result to the case in which the original fibre bundle is not necessarily a vector bundle, but has a trivial typical fibre.

Consider now the kernel of the projection map π_*. This is, by definition, the set of all elements $p \in TC$ that are projected by π_* onto the zero vector of $T_x \mathcal{B}$, where $x = \tau_B \circ \pi_*(p) = \pi \circ \tau_C(p)$. This set stands, as it were, over the zero section of $T\mathcal{B}$ and it can be seen to constitute thereat a fibre bundle with typical fibre $T\mathcal{F}$. By projecting back the zero section onto $\mathcal{B}$, we have a fibre bundle over $\mathcal{B}$ with typical fibre $T\mathcal{F}$ and structure group $\mathcal{G} \times GL(n; \mathbb{R})$. This fibre bundle is called the *vertical bundle* corresponding to the fibre bundle $\mathcal{C}$.

REMARK 7.5. **Coordinate expressions**: Let $x^i (i = 1, \dots, m)$ and $y^\alpha (\alpha = 1, \dots, n)$ be coordinate charts in $\mathcal{B}$ and $\mathcal{F}$, respectively. Their Cartesian product constitutes a (local) coordinate chart x^i, y^α for the fibre bundle $\mathcal{C}$. A tangent vector to $\mathcal{C}$ will have a natural coordinate representation as $v^i \frac{\partial}{\partial x^i} + w^\alpha \frac{\partial}{\partial y^\alpha}$, where the natural base vectors are evaluated at the corresponding point of $\mathcal{C}$. It follows that the induced natural chart for (TC, τ_C) is of the form $x^i, y^\alpha, v^i, w^\alpha$. The projection π has the coordinate expression $(x^i, y^\alpha) \mapsto (x^i)$. Consequently, its differential π_* is expressed as: $(x^i, y^\alpha, v^i, w^\alpha) \mapsto (x^i, v^i)$, as can be easily verified by definition of differential map. It follows that an element of the fibre bundle (TC, π_*) has the coordinate expression $x^i, v^i, y^\alpha, w^\alpha$, the first two entries representing the base point and the last two the position on the fibre. We see that the passage from (TC, τ_C) to (TC, π_*) has the effect of *switching* around the two central entries of the coordinate charts. As for the vertical bundle corresponding to $\mathcal{C}$, its coordinate expression is x^i, y^α, w^α, obtained by eliminating the zero entries (v^i) of the definition.

7.11.2. The Iterated Tangent Bundle

From the previous section, if we identify $\mathcal{C}$ with the tangent bundle $T\mathcal{B}$, it follows that the resulting bundle $TT\mathcal{B}$ has two distinct fibre bundle structures, but now both fibre bundles are defined over the same base manifold $T\mathcal{B}$, and are equivalent. The "switching" operation in this case is called the *canonical involution*. The following diagram clarifies the situation:

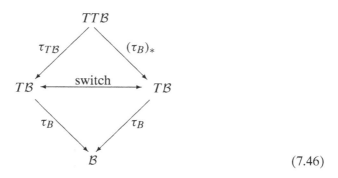

$$(7.46)$$

Given a manifold chart in $\mathcal{B}$, we can lift it, by differentiating twice, to a local trivialization of $TT\mathcal{B}$ as a fibre bundle over $\mathcal{B}$, as can be easily checked. The typical fibre of this fibre bundle is $\mathbb{R}^m \times \mathbb{R}^{2m}$, and the structure group is $GL(m; \mathbb{R}) \times GL(2m; \mathbb{R})$.

BOX 7.5. **Working with Coordinates**

Consider two manifolds, $\mathcal{M}$ and $\mathcal{N}$ say, of dimensions m and n, respectively. Let $x^i (i = 1, \ldots, m)$ and $y^\alpha (\alpha = 1, \ldots, n)$ be coordinates in some local charts. The coordinate expression of a map $f : \mathcal{M} \to \mathcal{N}$ will be in the form of n functions of m variables:

$$y^\alpha = y^\alpha(x^i), \quad (\alpha = 1, \ldots, n \; i = 1, \ldots m), \tag{7.47}$$

where we adopt the common, but slightly confusing, notation that calls the function by the same name as the dependent variable. The differential of this map, namely, $f_* : T\mathcal{M} \to T\mathcal{N}$, is expressed in coordinates as:

$$(x^i, v^i) \mapsto \left(y^\alpha, u^\alpha = \frac{\partial y^\alpha}{\partial x^i} v^i \right), \tag{7.48}$$

where, for simplicity, we avoid indicating the point at which the functions and derivatives are evaluated. We recall that in the previous expression v^i are the components of a vector in the natural basis $\frac{\partial}{\partial x^i}$, and similarly for its image in $T\mathcal{N}$. Identifying $\mathcal{N}$ with $\mathcal{M}$, and assuming that we have in each a different coordinate system, say x^i and $\hat{x}^i$, if we consider the identity map, we obtain the following law of transformation of components of elements of the tangent bundle:

$$(x^i, v^i) \mapsto \left(\hat{x}^j, \hat{v}^j = \frac{\partial \hat{x}^j}{\partial x^i} v^i \right). \tag{7.49}$$

We now apply the same routine to the iterated tangent bundle. An element of $TT\mathcal{M}$ has the coordinate expression $((x^i, v^i), (w^i, z^i))$, where the last two entries represent the components of a tangent vector to $T\mathcal{M}$ at the point with coordinates x^i, v^i. This vector (in a $2m$-dimensional tangent space) is given explicitly by $w^i \frac{\partial}{\partial x^i} + z^i \frac{\partial}{\partial v^i}$. Under a change of coordinates, following the same technique as before, to calculate the new components of the same vector we need to consider the Jacobian matrix $\left\{ \frac{\partial(\hat{x}^i, \hat{v}^i)}{\partial(x^j, v^j)} \right\}$. We obtain:

$$\begin{pmatrix} \hat{w}^i \\ \hat{z}^i \end{pmatrix} = \begin{pmatrix} \dfrac{\partial \hat{x}^i}{\partial x^j} & 0 \\ \dfrac{\partial^2 \hat{x}^i}{\partial x^j \partial x^k} v^k & \dfrac{\partial \hat{x}^i}{\partial x^j} \end{pmatrix} \begin{pmatrix} w^j \\ z^j \end{pmatrix}. \tag{7.50}$$

If we were to demand that just this type of chart be admissible for the fibre bundle (namely, *holonomic* charts obtained by lifting charts of the base manifold), we would obtain a fibre bundle whose structure group consists of the product of $GL(m; \mathbb{R})$ times the subgroup of $GL(2m; \mathbb{R})$ consisting of matrices with four square $(m \times m)$ blocks as follows: $\begin{pmatrix} A & 0 \\ B & A \end{pmatrix}$. No such demand is placed in the definition of $TT\mathcal{M}$, but for some applications it may be convenient to consider such a reduced bundle.

EXAMPLE 7.14. **The canonical 1-form and Hamiltonian Mechanics:** Other iterated bundles are formed in a similar way as the iterated tangent bundle. Although it does not possess the "switching" feature, the iterated cotangent bundle is endowed as a bonus with a canonical cross section. A cross section of the iterated cotangent bundle is, by definition, a 1-form on the cotangent bundle. In Mechanics, the cotangent bundle of the configuration space $\mathcal{Q}$ is the so-called phase space. In short, we conclude that the phase space of Classical Mechanics is endowed with a natural (not identically vanishing) 1-form $\theta : T^*\mathcal{Q} \to T^*T^*\mathcal{Q}$. We will define this 1-form without use of coordinates, to guarantee that its definition is canonical. Let $p \in T^*\mathcal{Q}$ and $u \in T_p T^*\mathcal{Q}$. In other words, u is a vector tangent to $T^*\mathcal{Q}$ at the point p. The *canonical 1-form* will be completely defined if we specified the value of the linear operation $\langle \theta(p), u \rangle$ for all p and for all vectors u at each p. We set:

$$\langle \theta(p), u \rangle := \langle p, \pi_* u \rangle, \tag{7.51}$$

where $\pi : T^*\mathcal{Q} \to \mathcal{Q}$ is the projection map of the cotangent bundle. The reason this definition makes sense is that the point p at which u sits happens to have the meaning of a covector in $\mathcal{Q}$, and so it can be applied linearly to vectors tangent to $\mathcal{Q}$. The exterior derivative $\omega = d\theta$ of the canonical 1-form provides the canonical 2-form which, being nondegenerate, results in a canonical symplectic structure of phase space, a concept central to Hamiltonian Mechanics.

EXERCISE 7.11. **The canonical 1-form in components:** Find the component expression of the canonical 1-form in a coordinate chart induced by a coordinate chart in $\mathcal{Q}$.

Solution: Let q^i be a local system of coordinates in $\mathcal{Q}$. Since an arbitrary 1-form σ on $\mathcal{Q}$ can be expressed uniquely as $p = p_i dq^i$, we obtain a natural chart of $T^*\mathcal{Q}$ as (q^i, p_i). The notation is suggestive of coordinates and momenta of Classical Mechanics. An element of the tangent bundle $T(T^*\mathcal{Q})$ will have the form q^i, p_i, w^i, z_i, since a tangent vector can be uniquely written as $w^i \frac{\partial}{\partial x^i} + z_i \frac{\partial}{\partial p_i}$. The projection π has the coordinate expression $(q^i, p_i) \mapsto (q^i)$, whence its derivative π_* results in $(q^i, p_i, w^i, z_i) \mapsto (q^i, w^i)$. We obtain $\langle \theta(p), u \rangle = \langle p, \pi_* u \rangle = p_i w^i = \langle p_i dq^i, u \rangle$. Since this is true for arbitrary u, we conclude that $\theta = p_i dq^i$. It follows that the canonical 2-form is given by $\omega = dp_i \wedge dq^i$.

8 Inhomogeneity Theory

Rather than continuing with the geometric theory of fibre bundles for its own sake, in this chapter we will introduce the rudiments of inhomogeneity theory, one of whose applications is the modelling of continuous distributions of dislocations and other defects in materials. The origins of this theory can be traced back to the works of Kondo, Bilby, Eshelby, Kroener, and other researchers in the second half of the twentieth century. Our presentation will follow the line of attack of Noll[1] and Wang,[2] which mimics more closely the style of modern differential geometry. This formulation will, therefore, strengthen the understanding of the concepts already introduced and serve as a motivation for what is to come, particularly the notions of connection, G-structure and Lie groupoid.

8.1. Material Uniformity

8.1.1. Material Response

Let B and $\mathbb{E}^3$ denote an ordinary body manifold and the (3-dimensional) Euclidean space of classical physics, respectively, and let Q be the set of (smooth) embeddings of B into $\mathbb{E}^3$. The elements $\kappa \in Q$ are the *configurations* of the body B. One may extend the formulation to include more general bodies (bodies with internal structure, etc.), but the main ideas of the theory of material inhomogeneity are essentially independent of the complexity of the body at hand.

The first concept to be introduced is that of *material response functional*. Since the geometrical structure we are interested in exposing will emerge already in the context of elasticity, we will confine our attention to a purely elastic response. Thus, time-rate effects (such as those appearing in the presence of viscosity or materials with fading memory) and temperature effects (such as those present in thermal conductors) will not be considered. Without loss of geometrical generality, we will also assume

---

[1] Noll W (1967), Materially uniform bodies with inhomogenenities, *Archive for Rational Mechanics and Analysis* **27**, 1-32.

[2] Wang C-C (1967), On the geometric structure of simple bodies, a mathematical foundation for the theory of continuous distributions of dislocations, *Archive for Rational Mechanics and Analysis* **27**, 33–94.

that the (purely mechanical) material response is entirely characterized by a single scalar-valued functional:

$$W : \mathcal{Q} \longrightarrow \mathbb{R}, \tag{8.1}$$

whose physical meaning is the elastic energy stored in the body at a given configuration. In a more technical terminology, the body is said to be *hyperelastic*. As given, the *constitutive law* (8.1) is truly *global*, in the sense that the stored energy may depend arbitrarily on the whole configuration.

REMARK 8.1. Under certain conditions, the set $\mathcal{Q}$ can be given the structure of an infinite dimensional differentiable manifold (a manifold of maps) and a comprehensive treatment can be pursued along elegant lines similar to those of classical particle mechanics, as already suggested in Chapters 1 and 4.

8.1.2. Germ Locality

The notion of *locality* of the material response is so natural that it appears to have been taken for granted by the founders of elasticity theory (e.g., Cauchy), and so useful that most practitioners still take it for granted today. In fact, in spite of the partial success of certain nonlocal formulations to resolve important problems such as singularities of stress fields in cracks, the spectacular achievements of purely local theories in the last three centuries leave no doubt that they will remain in the spotlight for years to come.

In terms of the hyperelastic context that we have established, locality results from the imposition of two restrictions upon the energy functional W:

(1) The first restriction consists of assuming that W is actually obtained as the result of integrating a 3-form ω defined on $\mathcal{B}$, namely:

$$W = \int_{\mathcal{B}} \omega, \tag{8.2}$$

where $\omega = \omega(\kappa)$ is a (3-form valued) functional of the configuration $\kappa \in \mathcal{Q}$. In other words, this assumption recognizes a definite way in which each point of the body contributes to the total energy, but each of these contributions may still depend on the entire configuration. We will often indicate the 3-form ω in the more familiar format $\omega = \psi \, dV$, which is indeed the way it looks in any coordinate system, where dV stands for a volume element and ψ is a scalar functional depending on the configuration:

$$W = \int_{\mathcal{B}} \psi(\kappa; X) \, dV, \tag{8.3}$$

where X denotes the variable point in $\mathcal{B}$.

(2) The second restriction assumes that the first one has already been imposed and drastically reduces the possible dependence of $\psi(\kappa; X)$ on κ. Indeed,

it prescribes that if two configurations, κ_1 and κ_2, happen to coincide on a neighbourhood of a point $X \in \mathcal{B}$, then the value of ψ at the point X is the same for both configurations, namely, $\psi(\kappa_1, X) = \psi(\kappa_2, X)$. Notice that the neighbourhood on which the configurations coincide may be arbitrarily small. Mathematically, an equivalence class of smooth functions that coincide on a neighbourhood of a point of their domain is called a *germ*, so this type of locality is also called *germ locality*. We will soon introduce less restrictive types of locality. A germ at a point X is denoted as $[f]_X$, where f is any function in the equivalence class. Germs of maps between manifolds can be composed by composing, whenever possible, a representative of each class and then taking the germ of the result.

8.1.3. Jet Locality

The concept of *germ locality* just introduced is too restrictive for most applications and far too general to be of practical use. Intuitively speaking, to require that to have the same effect on a given point two configurations must coincide on a whole neighbourhood of the point, no matter how small a neighbourhood, it is simply too much to ask. Moreover, the mathematically correct treatment of germ locality is fraught with technical difficulties of the type expected in dealing with infinite-dimensional manifolds. Germ locality can be said to be the least local of all possible local material behaviours. At the other end of the spectrum, we have materials whose local response is sensitive just to the first derivative of the deformation, which we have already encountered and identified as the deformation gradient $\mathbf{F}$. In a molecular picture, this type of locality would correspond to interatomic interactions with the nearest neighbours only. These are called *simple materials* and are the most widely used. Farther reaching interactions can be arguably modelled by means of the inclusion of higher derivatives of the deformation. Materials whose local response depends both on $\mathbf{F}$ and *its* gradient $\nabla \mathbf{F}$, so-called *second-grade materials*, have found specialized applications too.

Adopting coordinate charts in the body and in space, a configuration κ boils down, in a small neighbourhood of a given point $X \in \mathcal{B}$, to three real functions of three variables, namely:

$$x^i = \kappa^i(X^1, X^2, X^3) \qquad i = 1, 2, 3, \tag{8.4}$$

where X^I ($I = 1, 2, 3$) and x^i ($i = 1, 2, 3$) are the body and space coordinates, respectively.

DEFINITION 8.1. Two configurations, κ_1 and κ_2, are *r-jet equivalent at $X \in \mathcal{B}$* if, in some coordinate charts, the values of these functions as well as all partial derivatives up to and including the order r, coincide at X, that is:

$$\kappa_1^i(X) = \kappa_2^i(X), \qquad i = 1, 2, 3, \tag{8.5}$$

and

$$\left.\frac{\partial^s \kappa_1^i}{\partial X^{J_1} \dots \partial X^{J_s}}\right|_X = \left.\frac{\partial^s \kappa_2^i}{\partial X^{J_1} \dots \partial X^{J_s}}\right|_X, \quad i, J_k = 1, 2, 3, \ 1 \le s \le r. \qquad (8.6)$$

EXERCISE 8.1. **Independence of chart:** Show that two configurations that satisfy Equations (8.5) and (8.6) in one pair of charts, also satisfy them in any other pair of compatible charts.

BOX 8.1. **Jets of Maps**

Given two smooth manifolds, $\mathcal{M}$ and $\mathcal{N}$ (of dimensions m and n, respectively), we denote by $C^\infty(\mathcal{M}, \mathcal{N})$ the collection of all C^∞-maps (i.e., smooth maps) $f : \mathcal{M} \to \mathcal{N}$. We say that two maps $f, g \in C^\infty(\mathcal{M}, \mathcal{N})$ have the same k-jet at a point $X \in \mathcal{M}$ if (i) f(X)=g(X); (ii) in a coordinate chart in $\mathcal{M}$ containing X and a coordinate chart in $\mathcal{N}$ containing the image $f(X)$, all the partial derivatives of f and g up to and including the order k are respectively equal.

Although the above definition is formulated in terms of charts, it is not difficult to show by direct computation that the property of having the same derivatives up to and including order k is in fact independent of the coordinate systems used in either manifold. Notice that, in order for this to work, it is imperative to equate *all* the lower-order derivatives. If, for example, we were to equate just the second derivatives, without regard to the first, the equality of the second derivatives would not be preserved under arbitrary coordinate transformations.

REMARK 8.2. Two functions have the same k-jet at X if, and only if, they map every smooth curve through X to curves having a k-th order contact at the (common) image point. This property can be used as an alternative (coordinate independent) definition.

The property of having the same k-jet at a point is an equivalence relation. The corresponding equivalence classes are called *k-jets at X*. Any function in a given k-jet is then called a *representative* of the k-jet. The k-jet at X of which a given function $f : \mathcal{M} \to \mathcal{N}$ is a representative is denoted by $j_X^k f$. The collection of all k-jets at $X \in \mathcal{M}$ is denoted by $J_X^k(\mathcal{M}, \mathcal{N})$. The point X is called the *source* of $j_X^k f$ and the image point $f(X)$ is called its *target*.

From an intuitive point of view, we can view a k-jet at a point as the result of placing at that point a myopic observer, who does not distinguish between functions as long as they are sufficiently close to each other in a small neighbourhood of the point. The worst-case scenario is $k = 0$, which corresponds to identifying all functions that just have the same value at the point. Next (for $k = 1$) comes the identification of all functions with the same tangent map at the point. With better prescription glasses, the range of vision (and the corresponding k) can be increased.

The relation of r-jet equivalence at a point X is easily seen to be an equivalence relation. The corresponding equivalence classes, denoted by $j_X^r(\kappa)$, are called *r-jets of configurations at X*. The appearance of the symbol κ serves the purpose of identifying a representative map of the equivalence class. One often refers to $j_X^r(\kappa)$ as the *r-jet of κ at X*. The definition of r-jets applies to maps between any two differentiable manifolds (see Box 8.1) and, the same as for germs, the composition of r-jets is accomplished by first composing (whenever possible) representatives of the two equivalence classes and then evaluating the r-jet (see Box 8.2). In more prosaic terms, the composition of jets is the jet of the composition.

BOX 8.2. **Composition of Jets**

Jets of the same order can be composed whenever the target of one jet coincides with the source of the next. Let $\mathcal{M}$, $\mathcal{N}$, and $\mathcal{P}$ be smooth manifolds of dimension m, n, and p, respectively, and let $f \in C^\infty(\mathcal{M},\mathcal{N})$ and $g \in C^\infty(\mathcal{N},\mathcal{P})$. For any point $X \in \mathcal{M}$, the composition of the jets $j_X^k f$ and $j_{f(X)}^k g$ is defined by:

$$j_{f(X)}^k g \circ j_X^k f = j_X^k (g \circ f). \tag{8.7}$$

This definition is independent of the representatives f and g of each jet. Thus, to compose two jets, we choose any representative of the first jet and compose it with any representative of the second, thus obtaining a map from the first manifold ($\mathcal{M}$) to the third ($\mathcal{P}$). We then calculate the jet of this composite function and obtain, by definition, the composition of the given jets.

To illustrate this operation, consider the case $k = 2$. Let coordinate systems X^I ($I = 1,\ldots,m$), x^i ($i = 1,\ldots,n$), and z^α ($\alpha = 1,\ldots,p$) be chosen, respectively, around the points $X \in \mathcal{M}$, $f(X) \in \mathcal{N}$, and $g(f(X)) \in \mathcal{P}$. The functions $f : \mathcal{M} \to \mathcal{N}$ and $g : \mathcal{N} \to \mathcal{P}$ are then given locally in coordinates by:

$$x^i = x^i(X^1,\ldots,X^m) = x^i\big|_X, \quad z^\alpha = z^\alpha(x^1,\ldots,x^n) = z^\alpha\big|_x, \tag{8.8}$$

where the indices take values in the appropriate ranges. The jet $j_X^2 f$ is then given by the following coordinate expressions:

$$x^i\big|_X, \quad \frac{\partial x^i}{\partial X^I}\bigg|_X, \quad \frac{\partial^2 x^i}{\partial X^J \partial X^I}\bigg|_X, \tag{8.9}$$

a total of $n + mn + m^2 n$ numbers. Similarly, the jet $j_{f(X)}^2 g$ is given in coordinates by the following $p + np + n^2 p$ numbers:

$$z^\alpha\big|_x, \quad \frac{\partial z^\alpha}{\partial x^i}\bigg|_{f(X)}, \quad \frac{\partial^2 z^\alpha}{\partial x^j \partial x^i}\bigg|_{f(X)}. \tag{8.10}$$

To obtain the components of the composite jet $j_{f(X)}^2 g \circ j_X^2 f$, we consider the composite functions $z^\alpha(x^i(x^I))$ and take their first and second derivatives. The result consists of the following $p + mp + m^2p$ numbers:

$$z^\alpha\big|_{x|X}, \quad \frac{\partial z^\alpha}{\partial x^i}\bigg|_{f(X)} \frac{\partial x^i}{\partial X^I}\bigg|_X,$$

$$\frac{\partial^2 z^\alpha}{\partial x^j \partial x^i}\bigg|_{f(X)} \frac{\partial x^j}{\partial X^J}\bigg|_X \frac{\partial x^i}{\partial X^I}\bigg|_X + \frac{\partial z^\alpha}{\partial x^i}\bigg|_{f(X)} \frac{\partial^2 x^i}{\partial X^J \partial X^I}\bigg|_X. \tag{8.11}$$

It is this peculiar composition formula for second (and indeed higher) derivatives that confirms our previous observation as to the necessity of including all lower-order derivatives in the definition of jets.

A body-point $X \in \mathcal{B}$ with the constitutive function $\psi\left(\kappa, ; X\right)$ is *r-jet local* if:

$$\psi\left(\kappa; X\right) = \psi\left(j_X^r(\kappa); X\right). \tag{8.12}$$

The response of the material at the point X is, therefore, dictated only by all the derivatives of the deformation (evaluated at that point) up to and including the r-th order. Note that if a material is germ local, then it is automatically r-jet local for all r. Also, r-jet locality implies s-jet locality for all $s < r$.

8.1.4. First- and Second-Grade Materials

A local hyperelastic *first-grade* (or *simple*) body is described by a constitutive law of the form:

$$\psi = \psi(j_X^1 \kappa), \tag{8.13}$$

where we have economized the indication of the explicit dependence on X since it is already included in the definition of the jet. The coordinate version of this equation is:

$$\psi = \psi(F_I^i, X^I), \tag{8.14}$$

where, with the notation of Appendix A:

$$F_I^i := \frac{\partial x^i}{\partial X^I}. \tag{8.15}$$

Notice that in the coordinate expression (8.14) we are not including the explicit spatial argument x^i (which is certainly included in the definition of jet) because of the requirement of translation invariance of the energy. Under a coordinate transformation in the body (change of reference configuration) of the form:

$$Y^M = Y^M(X^I), \quad (I, M = 1, 2, 3), \tag{8.16}$$

Equation (8.14) transforms according to:

$$\psi = \hat{\psi}(F_M^i, Y^M) = \psi(F_M^i \frac{\partial Y^M}{\partial X^I}, Y^M(X^I)). \tag{8.17}$$

BOX 8.3. **Jets and Frames**
In the previous chapter we have introduced the notion of a linear frame at a point b of an n-dimensional manifold $\mathcal{B}$ as a basis $\{e\}_b = (e_1,\ldots,e_n)$ of the tangent space $T_b\mathcal{B}$. An equivalent definition consists of identifying such a basis with a linear isomorphism $f: \mathbb{R}^n \to T_b\mathcal{B}$. The relation between these two definitions is as follows. Let $(r_1,\ldots,r_n)$ denote the *canonical basis* of $\mathbb{R}^n$, namely, $r_1 = (1,0,0,\ldots,0)$, $r_2 = (0,1,0,\ldots 0)$, etc. Then e_i is simply the image of r_i by the linear isomorphism f, that is $e_i = f(r_i)$. The theory of jets provides still a third way to define and interpret a linear frame. Consider a (local) diffeomorphism ϕ of an open neighbourhood of the origin 0 in $\mathbb{R}^n$ onto an open neighbourhood of b in $\mathcal{B}$. Then, the 1-jet $j_0^1\phi$ (which can be identified with the differential of ϕ at 0) is a linear map between $T_0\mathbb{R}^n$ and $T_b\mathcal{B}$. Since $T_0\mathcal{R}^n$ can be canonically identified with $\mathbb{R}^n$ itself, we have a linear frame (according to our second definition above). The advantage of regarding a linear frame as the jet of a local diffeomorphism is that the notion of frame can thus be extended to more general situations ("nonlinear frames"). A case of particular interest for the theory of materials with internal structure is that of second-order frames.

A local hyperelastic *second-grade body* is characterized by a constitutive law of the form:

$$\psi = \psi(j_X^2 \kappa). \tag{8.18}$$

In coordinates, this equation reads as follows:

$$\psi = \psi(F^i_{\ I}, F^i_{\ IJ}, X^I), \tag{8.19}$$

where we have used the notation:

$$F^i_{\ IJ} := \frac{\partial F^i_{\ I}}{\partial X^J} = \frac{\partial^2 x^i}{\partial X^J \partial X^I} = F^i_{\ JI}. \tag{8.20}$$

Under a coordinate transformation (8.16), Equation (8.19) transforms according to:

$$\psi = \hat{\psi}(F^i_{\ M}, F^i_{\ MN}, Y^M)$$

$$\tag{8.21}$$

$$= \psi(F^i_{\ M} \frac{\partial Y^M}{\partial X^I}, F^i_{\ MN} \frac{\partial Y^N}{\partial X^J} \frac{\partial Y^M}{\partial X^I} + F^i_{\ M} \frac{\partial 2Y^M}{\partial X^J \partial X^I}, Y^M(X^I)).$$

8.1.5. Material Isomorphism

Assuming the material response to be local, the following intuitive question can be legitimately asked: Are two body points, X and Y, made of the same material? The subtlety of this question resides in the fact that germs and jets at two different points of a manifold are not directly comparable. In other words, it doesn't make any sense to affirm that, for example, the 1-jets of a configuration at two different points are equal, since the numerical values of the derivatives are coordinate dependent. This being the case, it appears that the material responses at two different points cannot be compared at all. To appreciate the way out of this impasse, consider first a diffeomorphism $\phi : \mathcal{B} \to \mathcal{B}$ of the body, namely, a smooth and smoothly invertible transformation of the body onto itself. This transformation most certainly does not affect the material the points are made of, whatever one may understand by this expression. Nevertheless, the material response of the image $\phi(\mathcal{B})$ differs in general from that of the original $\mathcal{B}$. Specifically, the two material responses are related by the identity

$$\omega'(\kappa) = \omega(\kappa \circ \phi), \tag{8.22}$$

where a prime has been used to denote the response relative to the transformed body. More specifically, for a body with germ locality we have at a fixed point X:

$$\omega'\left([\kappa]_{\phi(X)}; \phi(X)\right) = \omega([\kappa \circ \phi]_X; X)$$
$$= \omega\left([\kappa]_{\phi(X)} \circ [\phi]_X; X\right). \tag{8.23}$$

We have been careful to express this identity in terms of the 3-form ω rather than the scalar function ψ, because of obvious reasons that have to do with the choice of dV. In practice, however, one often works with so-called *reference configurations* and a corresponding expression is obtained for ψ (see Box 8.4). Be that as it may, our point is that a body diffeomorphism (or a "change of reference configuration") affects the material response of each point in a very definite way, while the point retains its material identity. It is precisely this extra degree of freedom what permits us to compare the responses at two different points.

BOX 8.4. Working with Reference Configurations

The main actors in Continuum Mechanics are the material body $\mathcal{B}$ and the physical space-time $\mathcal{S}$, which are differentiable manifolds of dimensions 3 and 4, respectively. Whereas $\mathcal{S}$ is endowed with extra structure, $\mathcal{B}$ is not. As we saw in Chapter 1, the extra structure possessed by $\mathcal{S}$ can be summarized by saying that $\mathcal{S}$ is an *affine bundle* whose typical fibre is $\mathbb{E}^3$. Choosing an origin and an orthonormal basis in $\mathbb{E}^3$, we obtain a global coordinate chart called a *spatial Cartesian coordinate system*. With this choice, $\mathbb{E}^3$ can be identified with $\mathbb{R}^3$.

The body $\mathcal{B}$ is a 3-dimensional manifold that can be covered with a single chart. Now, a chart is a map $\phi : \mathcal{B} \to \mathbb{R}^3$, with a smooth inverse defined on its range. A configuration, on the other hand, is an embedding $\kappa : \mathcal{B} \to \mathbb{E}^3$. This means that a chart in $\mathcal{B}$ looks a lot like a configuration, although it is something completely different. To identify a chart with a configuration, one can make the body inherit the metric structure of one of its charts by declaring, for example, that particular chart to be Cartesian. One refers to this chart as a *reference configuration*. So, we can say that a reference configuration induces a metric (Riemannian) structure on $\mathcal{B}$, just as an ordinary configuration does.

Continuum Mechanics does not need reference configurations to be properly formulated. Naturally, since the actual physical objects and concepts (e.g., a constitutive law) need to be expressed in practice in terms of charts, it has become customary to use expressions such as: "the body will be identified once and for all with one of its configurations, which will be called a reference configuration." Rules are then given for the change of mathematical expressions of the same physical object (such as a constitutive law) under changes of reference configuration. Equation (8.22), for example, can be reinterpreted as the change of form of the constitutive 3-form ω with a change ϕ of reference configuration. If we want to express this change in terms of the scalar ψ of Equation (8.3), the formula will depend on the particular volume form dV chosen. The most physically meaningful choice, but one that is seldom made, is obtained by pulling back the standard volume form of $\mathbb{E}^3$ to the body manifold by means of the present configuration. A second choice derives from assuming that a mass-density 3-form has been specified in the body. Finally, one can adopt the volume form induced by a given reference configuration, in which case the formula for the change of the constitutive 3-form ω (Equation 8.22) will be expressed in terms of the corresponding scalar function ψ as:

$$\psi'(\kappa) = J^{-1}\psi(\kappa \circ \phi), \tag{8.24}$$

where J is the Jacobian determinant of ϕ (now seen as a change of reference configuration) evaluated at X. If one chooses to use the energy density per unit spatial volume or per unit mass, assuming that mass is preserved regardless of the reference configuration used, the determinant J does not appear.

DEFINITION 8.2. Two points, X and Y of a germ-local hyperelastic body $\mathcal{B}$ are *materially isomorphic* (read: made of the same material) if there exists a diffeomorphism $\phi_{XY} : \mathcal{B} \to \mathcal{B}$ such that $Y = \phi_{XY}(X)$ and the equation:

$$\omega([\kappa]_Y; Y) = \omega([\kappa]_Y \circ [\phi_{XY}]_X; X) \tag{8.25}$$

is satisfied identically for all $\kappa \in \mathcal{Q}$. The germ of such a diffeomorphism is called a *material isomorphism* between X and Y.

To simplify the notation, we shall denote a material isomorphism by p_{XY}, as shown in Figure 8.1. Accordingly, we will write Equation (8.25) as follows:

$$\omega([\kappa]_Y; Y) = \omega([\kappa]_Y \circ p_{XY}; X). \tag{8.26}$$

The reason for using the germ (p_{XY}) of ϕ_{XY} rather than ϕ_{XY} itself to define the material isomorphism is obvious: By virtue of the assumption of germ locality, it is only the germ of the diffeomorphism that enters into the material response, and so we identify all the transformations ϕ_{XY} having the same germ p_{XY} at X.

REMARK 8.3. **A surgical perspective**: A material isomorphism can be imagined as a skin graft. A small neighbourhood of a point in, say, the skin of the leg is to be

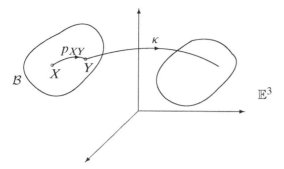

Figure 8.1. A material isomorphism

cut off and transplanted to a damaged neighbourhood of a point in, say, the face. Because the skin at different parts of the body is naturally stretched into different degrees, the surgeon may have to apply a slight stretch (or perhaps a distension or even a shear strain) to the piece before implanting it in the new position. Only then will the graft be perfect. This transplant operation is a rather faithful pictorial representation of a material isomorphism.

REMARK 8.4. **What happened to the mass density?** The definition of material isomorphism is completely independent of any concept of mass density. Nevertheless, assume that a mass density has been given. Technically, the mass density $\tilde{\rho}$ is a 3-form on $\mathcal{B}$ whose integral over any subbody delivers its mass. A material isomorphism p_{XY} (with a representative diffeomorphism ϕ_{XY}, say) is *mass consistent* if it pulls back the density at Y to that at X. From the definition of pull-back of a form, this means that if we take three linearly independent vectors at X ("an elementary parallelepiped"), the derivative map $(\phi_{XY})_*$ maps them into three vectors at Y in such a way that, if we evaluate the mass-density 3-form on them (namely, if we calculate the mass "enclosed" by the deformed parallelepiped), we get the same result as when evaluating the 3-form at X on the original parallelepiped. Unless otherwise stated, we will assume that our material isomorphisms are mass-consistent. Nevertheless, when one is interested in studying material isomorphisms that change with time, the mass inconsistency of the material isomorphisms as time goes on can be exploited to model phenomena of volumetric growth, which are very important in biomedical applications. When using the notion of reference configuration, the mass-consistency condition results in the cancellation of the Jacobian determinant (see Box 8.4) in the energy density per unit mass.

Although the definition of material diffeomorphism was phrased in terms of germ locality, it can be applied equally well, mutatis mutandis, to r-jet locality. In the case of materials of grade r, we will again use the notation p_{XY} to denote material isomorphisms, namely, r-jets of diffeomorphisms satisfying:

$$\omega\left(j_Y^r(\kappa); Y\right) = \omega\left(j_Y^r(\kappa) \circ p_{XY}; X\right) \tag{8.27}$$

for all configurations κ.

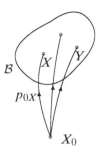

Figure 8.2. Implants of an archetype

For the case of a first-grade (simple) material, for instance, a material isomorphism from point X to point Y consists of a linear nonsingular map:

$$\mathbf{P}_{XY} : T_X\mathcal{B} \longrightarrow T_Y\mathcal{B} \tag{8.28}$$

between their tangent spaces. Equation (8.25) becomes:

$$\psi(\mathbf{F}; Y) = J_{P_{XY}}^{-1} \, \psi(\mathbf{FP}_{XY}; X), \tag{8.29}$$

where ψ is measured per unit volume of the reference configuration and J denotes the determinant.

A body is said to be *materially uniform* if all of its points are mutually materially isomorphic. A body is said to be *smoothly materially uniform* if it is materially uniform and if the material isomorphisms can be chosen smoothly throughout the body (as functions of X and Y).[3] From now on, we will use liberally the terminology "uniform body" to mean "smoothly materially uniform body."

We note that material isomorphism is an equivalence relation. Indeed, every point is materially isomorphic to itself (via the identity map), and the relationship is reflexive (via the inverse map), and transitive (via the composition of maps). This means that a uniform body falls entirely within an equivalence class of this relation. Moreover, by transitivity, all the points of a uniform body are isomorphic to any single point of the body. This last property can be used to isolate a neighbourhood of an *archetypal material point* X_0, typifying the material response, and rather than consider each pair of points, define an *implant operation* of the archetype into each point (Figure 8.2). We will see that each of these two avenues (the consideration of all pairs, on the one hand, or the implantation of the archetype at each point, on the other hand) has a well-defined associated geometric object.

---

[3] It is precisely at this point where the use of germ locality becomes technically difficult, and so, from now on, we will refer exclusively to jet locality. For r-jet locality, smoothness is simply ascertained by checking that in some chart the jets (consisting, after all, of a finite set of numbers) depend smoothly on position.

For a first-grade material, if we denote the implant from the archetype to the point X by $\mathbf{P}(X)$, the uniformity condition (8.29) can be written as:

$$\psi(\mathbf{F};X) = J_{P(X)}^{-1} \, \bar{\psi}(\mathbf{F}\mathbf{P}(X)), \tag{8.30}$$

for all possible deformation gradients $\mathbf{F}$, where $\bar{\psi}$ denotes the response function of the archetype per unit volume induced by the choice of a frame.

EXAMPLE 8.1. **The Eshelby stress in elasticity:** As an example of the application of (8.30), we investigate the change in the internal energy density ψ brought about by a change in the material implant at a point. The physical context for such a calculation is given by various theories of *anelastic evolution*, such as plasticity and remodelling, whereby the passage of time may entail material rearrangements. Taking the derivative of Equation (8.30) with respect to the material implant, we obtain:

$$\frac{\partial \psi}{\partial \mathbf{P}} = -\psi \mathbf{P}^{-T} + J_P^{-1} \, \mathbf{F}^T \left(\frac{\partial \bar{\psi}}{\partial(\mathbf{F}\mathbf{P})} \right), \tag{8.31}$$

where the formula for the derivative of a determinant $(\partial J_P/\partial \mathbf{P} = J_P \, \mathbf{P}^{-T})$ has been used. Recalling (see Equation (A.111)) that in Elasticity the Piola stress is related to the free-energy density by:

$$\mathbf{T} = \frac{\partial \psi}{\partial \mathbf{F}} = J_P^{-1} \frac{\partial \bar{\psi}}{\partial(\mathbf{F}\mathbf{P})} \, \mathbf{P}^T, \tag{8.32}$$

we can rewrite Equation (8.31) as:

$$\frac{\partial \psi}{\partial \mathbf{P}} \, \mathbf{P}^T = -\psi \mathbf{I} + \mathbf{F}^T \mathbf{T} = -\mathbf{b}, \tag{8.33}$$

where $\mathbf{I}$ is the identity in the reference configuration. The tensor $\mathbf{b} = \psi \mathbf{I} - \mathbf{F}^T \mathbf{T}$ is known as the *Eshelby stress*. It is an example of a *configurational force*.

8.1.6. Material Symmetries and the Nonuniqueness of Material Isomorphisms

A *material symmetry* at a point X of a locally hyperelastic body is a material automorphism, that is, a material isomorphism p_{XX} between X and itself, namely:

$$\omega([\kappa]_X;X) = \omega([\kappa]_X \circ p_{XX};X) \tag{8.34}$$

identically for all $\kappa \in \mathcal{Q}$. By the mass consistency condition (which we assume fulfilled),[4] the same expression will apply for the scalar ψ in a reference configuration, since mass consistency is tantamount to demanding that the determinant J of any

---

[4] Naturally, having assumed that material isomorphisms are mass consistent, we are tacitly agreeing that material symmetries are too. The condition of mass consistency is mentioned more often in connection with material symmetries than with material isomorphisms, although whatever physical reasons apply to the former must also apply to the latter. On the other hand, it is clear that the condition of mass consistency for symmetries is independent of the particular mass density prescribed, whereas this is not the case for mass consistency of material isomorphisms between different points.

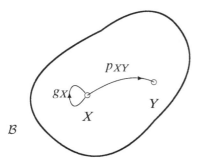

Figure 8.3. Material isomorphisms and symmetries

material symmetry be equal to 1. Notice that the above formula can be written in terms of germs or jets, as the case may be.

It is not difficult to check that all material symmetries at a point form a group $\mathcal{G}_X$, which is called the *material symmetry group at X* relative to the given constitutive law. In the case of r-jet locality, the group identity is given by the r-jet of the identity map, and the group multiplication is the composition of r-jets. The elements (p_{XX}) of the symmetry group at X will also be denoted by g_X. Let X and Y be two different materially isomorphic points in a body, and let p_{XY} be a material isomorphism. Let, moreover, g_X and g_Y be elements of the symmetry groups at X and Y, respectively. Then, it follows that the compositions $p_{XY} \circ g_X$ and $g_Y \circ p_{XY}$ are also material isomorphisms between these points. Conversely, let p_{XY} and p'_{XY} be two material isomorphisms between X and Y. Then, the compositions $p'^{-1}_{XY} \circ p_{XY}$ and $p_{XY} \circ p'^{-1}_{XY}$ belong, respectively, to the symmetry groups $\mathcal{G}_X$ and $\mathcal{G}_Y$ (Figure 8.3). The proof of these assertions is left to the reader. This means that the totality of material isomorphisms between X and Y is a set $\mathcal{P}_{XY}$ that can be obtained from any one of its elements p_{XY} as:

$$\mathcal{P}_{XY} = p_{XY} \circ \mathcal{G}_X = \mathcal{G}_Y \circ p_{XY}. \tag{8.35}$$

Given a material symmetry g_X at X, the composition $p_{XY} \circ g_X \circ p^{-1}_{XY}$ is a material symmetry at Y. In other words, the groups G_X and G_Y are conjugate groups,[5] where the conjugation can be effected with any member of the set $\mathcal{P}_{XY}$. When the archetype concept is adopted, namely, when a point X_0 is singled out as a standard of comparison, it is useful to refer to its symmetry group $\mathcal{G}_0$ as the *archetypal symmetry group* or the body.

It follows from these remarks that the material isomorphisms will be unique if, and only if, each of the symmetry groups (equivalently, the archetypal symmetry group) consists of just the identity. In this case the material is sometimes said to have no material symmetries. The material symmetry group, which may be discrete or continuous, plays a pivotal role in inhomogeneity theory.

---

[5] Two subgroups, A and B, of a group G are *conjugate* if there exists $g \in G$ such that $B = gAg^{-1}$. We extend the terminology to encompass two groups (not necessarily subgroups of a common group) thus related by a map g. In a reference configuration, all the symmetry groups can be regarded as subgroups of a common group.

8.2. The material Lie groupoid

8.2.1. Introduction

Let us recapitulate and see where things stand now. We have discovered that a materially uniform body $\mathcal{B}$ is endowed with a strong geometrical structure that can be summarized as follows: for each pair of points, X and Y, there exists a set $\mathcal{P}_{XY}$ which points to both a "source" (the first subscript) and a "target" (the second subscript).[6] The elements of $\mathcal{P}_{XY}$ and $\mathcal{P}_{YZ}$ can be composed to produce (all of the) elements of $\mathcal{P}_{XZ}$. Moreover, for each $X \in \mathcal{B}$, the set $\mathcal{G}_X = \mathcal{P}_{XX}$ is a group. There is a mathematical object, called a *groupoid*, which mirrors this situation exactly. In fact, a groupoid is a somewhat more general object and we will learn that the property of material uniformity corresponds to a precise particular case. In general terms, one may say that, in much the same way as a group represents the *internal*, or local, symmetries of a geometric entity, a groupoid can be said to also represent the *external*, or distant, symmetries of the same object. A classical example is the tiling of a bathroom floor. Each square tile has a symmetry group consisting of certain rotations and reflections. But it is also intuitively recognized that the floor as a whole has a repetitive pattern and, therefore, some extra symmetry. Because the floor is not infinite, however, we cannot describe all of these extra symmetries by means of a group of global transformations of the plane, such as translations. We will see that the notion of groupoid circumvents this problem. In the case of a material body, the fact that two distant points are made of the same material should be understood as an extra degree of symmetry that the body possesses, just as in the case of the bathroom floor, where distant tiles happen to have the same shape. This analogy should not be pushed too far, but it serves to trigger a useful picture to understand the unifying role that the concept of groupoid plays in terms of encompassing all types of symmetries.

8.2.2. Groupoids

The abstract notion of a *groupoid* emerges as the common structure underlying many constructions that arise naturally in a variety of apparently disconnected applications in algebra, topology, geometry, differential equations, and practically every branch of mathematics. In a restricted way, it can be seen as a generalization of the notion of group, but it is better to understand it as an important mathematical concept in its own right.[7]

A groupoid consists of a total set $\mathcal{Z}$, a base set $\mathcal{B}$, two ("projection") surjective submersions:

$$\alpha : \mathcal{Z} \longrightarrow \mathcal{B} \quad \text{and} \quad \beta : \mathcal{Z} \longrightarrow \mathcal{B} \tag{8.36}$$

[6] This situation is reminiscent of the concept of a fibre bundle, except that now each "fibre" projects not just to one but to two points in the base manifold.

[7] For a thorough treatment of groupoids see MacKenzie K (1987), *Lie Groupoids and Lie Algebroids in Differential Geometry*, London Mathematical Society Lecture Note Series **124**, Cambridge University Press. An informal and illuminating explanation can be found in Weinstein A (2000), Groupoids: Unifying Internal and External Symmetry. A tour through Examples, *Notices of the American Mathematical Society* **43**, 744–752.

called, respectively, the *source* and the *target* maps, and a binary operation ("composition") defined only for those ordered pairs $(y,z) \in \mathcal{Z} \times \mathcal{Z}$ such that:

$$\alpha(z) = \beta(y). \qquad (8.37)$$

This operation (usually indicated just by reverse apposition of the operands) must satisfy the following properties:

(1) Associativity:

$$(xy)z = x(yz), \qquad (8.38)$$

whenever either product is defined;

(2) Existence of identities: for each $b \in \mathcal{B}$ there exists an element $id_b \in \mathcal{Z}$, called the *identity at* b, such that $z\, id_b = z$ whenever $\alpha(z) = b$, and $id_b\, z = z$ whenever $\beta(z) = b$;

(3) Existence of inverse: for each $z \in \mathcal{Z}$ there exists a (unique) *inverse* z^{-1} such that

$$zz^{-1} = id_{\beta(z)} \quad \text{and} \quad z^{-1}z = id_{\alpha(z)}. \qquad (8.39)$$

It follows from this definition that to each ordered pair (a,b) of elements of $\mathcal{B}$ one can associate a definite subset $\mathcal{Z}_{ab}$ of $\mathcal{Z}$, namely, the subset $\{z \in \mathcal{Z} \mid \beta(z) = b,\ \alpha(z) = a\}$. It is clear that these sets (some of which may be empty) are disjoint and that their union is equal to $\mathcal{Z}$. It is also clear that the various identities are elements of subsets of the form $\mathcal{Z}_{bb}$.

EXERCISE 8.2. **The groups inside the groupoid:** Show that each set of the form $\mathcal{Z}_{bb}$ is a group.

A useful way to think of a groupoid is as a collection of symbols $(a,b,c,\ldots)$ and arrows $(x,y,z,\ldots)$ connecting some of them. The symbols correspond to the elements of the base set $\mathcal{B}$, while the arrows correspond to the elements of the total set $\mathcal{Z}$. The tail and tip of an arrow z correspond to the source $\alpha(z)$ and the target $\beta(z)$, respectively. Two arrows, z and y, can be composed if, and only if, the tip of the first ends where the tail of the second begins. The result is an arrow yz whose tail is the tail of z and whose tip is the tip of y:

For this picture to correspond more or less exactly to the more formal definition of a groupoid, however, we have to add the proviso that for each arrow z connecting point a to point b, there exists an "inverse" arrow z^{-1} connecting point b with point a.

It is also very important to bear in mind that there is no need for a given pair of points to be connected by one or more arrows. Some may be connected and some may not. In fact, an extreme case can occur whereby no two (different) points are thus connected. In this extreme case, the set $\mathcal{Z}$ becomes simply the disjoint union of the groups $\mathcal{Z}_{bb}$. In our tiled-floor analogy, we have a case of a floor patched up with tiles all of which are of different shapes and sizes.

EXERCISE 8.3. **Conjugation:** Prove that if $\mathcal{Z}_{ab} \neq \emptyset$, then the groups $\mathcal{Z}_{aa}$ and $\mathcal{Z}_{bb}$ are conjugate, and the conjugation between them is achieved by any element of $\mathcal{Z}_{ab}$. Show, moreover, that the set $\mathcal{Z}_{ab}$ is spanned completely by composing any one of its elements with $\mathcal{Z}_{aa}$ or with $\mathcal{Z}_{bb}$ (to the right or to the left, of course).

EXERCISE 8.4. **Groupoids as categories:** If you enjoyed the idea and language of categories (partially introduced in the previous chapter), you may like to verify that a groupoid is nothing but a category in which every morphism is an isomorphism. (Note: a slight difference remains in that a category need not be a set, but just a class.)

A groupoid is said to be *transitive* if for each pair of points $a, b \in \mathcal{B}$ there exists at least one element of the total set with a and b as the source and target points, respectively. In other words, a groupoid is transitive if, and only if, $\mathcal{Z}_{ab} \neq \emptyset \ \forall (a,b) \in \mathcal{B} \times \mathcal{B}$. In a transitive groupoid all the local groups $\mathcal{Z}_{bb}$ are mutually conjugate (recall Exercise 8.3).

EXERCISE 8.5. **Transitive groupoids:** For each of the examples of groupoids in Box 8.5, determine whether or not it is a transitive groupoid.

A groupoid is a *Lie groupoid* if the total set $\mathcal{Z}$ and the base set $\mathcal{B}$ are differentiable manifolds, the projections α and β are smooth, and so are the operations of composition and of inverse. It follows from the definition that each of the sets $\mathcal{Z}_{bb}$ is a Lie group.

Putting together the various mechanical and geometrical notions introduced thus far, we conclude that *a materially uniform body is endowed with the natural structure of a transitive Lie groupoid*. We call this structure the *material Lie groupoid*. At this point it is worthwhile noticing that if the assumption of smoothness were to be dropped from the definition of material uniformity, then the material groupoid would still exist and would still be transitive, but it would no longer be a Lie groupoid. Such a situation may certainly arise in practice.

EXAMPLE 8.2. **Material discontinuities:** A not very imaginative example of a nonsmooth material groupoid is provided by gluing together two identical pieces of wood, but without respecting the continuity of the grain. A somewhat more sophisticated example could be the presence of an isolated edge dislocation in a uniform body.

Consider now a nonuniform body. The material groupoid is still properly definable, except that it loses its transitivity. It may still preserve its smoothness (namely, it may still be a Lie groupoid).

BOX 8.5. **Some Examples of Groupoids**

To show the versatility of the concept of groupoid, we list a few examples drawn from different areas of mathematics.

(1) The product groupoid: Given a set $\mathcal{B}$, the Cartesian product $\mathcal{B} \times \mathcal{B}$ is a groupoid with $\alpha = pr_1$ and $\beta = pr_2$. You should be able to complete this trivial, but important, example.

(2) The general linear groupoid $GL(\mathbb{R})$: Take as the total set the collection of all nonsingular square matrices of all orders. The base space will be taken as the set $\mathbb{N}$ of natural numbers. The binary operation is matrix multiplication. We can see that this groupoid is nothing but the disjoint union of all the general linear groups $GL(n;\mathbb{R})$.

(3) The fundamental groupoid: Let $\mathcal{T}$ be a topological space. For each pair of points $a, b \in \mathcal{T}$ we consider the collection of all continuous curves starting at a and ending at b. We partition this set into equivalence classes, two curves being considered equivalent if they are homotopic,[a] and we define $\mathcal{Z}_{ab}$ as the quotient set (namely, the set of these equivalence classes). The composition of curves is done just as with the arrows of our pictorial description. [Question: Why is the partition into equivalence classes needed?]

(4) The tangent groupoid: Given a differentiable manifold $\mathcal{B}$, we form its tangent groupoid by considering, for each pair of points $a, b \in \mathcal{B}$ the collection of all the nonsingular linear maps between the tangent spaces $T_a\mathcal{B}$ and $T_b\mathcal{B}$.

[a] Two curves starting and ending at the same points are homotopic if, keeping these ends fixed, it is possible to transform continuously one curve into the other.

EXAMPLE 8.3. **Functionally graded materials:** A good example of this last situation is provided by the so-called *functionally graded materials*, which have smoothly varying material properties tailored to specific applications. Under certain circumstances, however, the transitivity of the material Lie groupoid of functionally graded materials can be restored by modifying the definition of material isomorphism.[8]

To summarize the results of this section we may say that the specification of the constitutive law of any material body gives rise to an associated material groupoid. Within this geometrical context, a body is (smoothly) materially uniform if, and only if, its material groupoid is a transitive Lie groupoid.

REMARK 8.5. Consider the collection $\Pi^r(\mathcal{B})$ of the r-jets $j^r\phi$ of local diffeomorphisms ϕ between all pairs of points in $\mathcal{B}$. It can be verified that $\Pi^r(\mathcal{B})$ has a natural structure of a transitive Lie groupoid called the *r-jet groupoid* of $\mathcal{B}$. For $r = 1$ we recover the tangent groupoid. Given a constitutive law for a material of grade r, we may restrict attention to those local diffeomorphisms whose r-jets are material isomorphisms. In this way, we obtain a subgroupoid of $\Pi^r(\mathcal{B})$, which we have called the material groupoid associated with the given constitutive law. Uniformity is equivalent to the fact that this groupoid is a transitive Lie groupoid. In this remark, we wanted to emphasize the fact that all material groupoids can be seen as subgroupoids of $\Pi^r(\mathcal{B})$.

8.3. The Material Principal Bundle

8.3.1. Introduction

We have already advanced the notion that, rather than comparing every pair of points, an equivalent way to describe the fact that a body is materially uniform consists of comparing the response at each point with that of an archetype. Without offering any concrete advantages, this point of view nicely corresponds to the idea that all points of a uniform body behave as some ideal piece of material, the only difference being the way in which this piece has been implanted point by point into the body to constitute the whole entity. In this section, we will see how this point of view is reflected in a precise way in the passage from the groupoid to a principal bundle. Corresponding to the fact that there is a degree of freedom in the choice of the archetype, we will find that a transitive Lie groupoid gives rise not just to one but to a whole family of equivalent principal bundles.

8.3.2. From the Groupoid to the Principal Bundle

Let $b \in \mathcal{B}$ be a fixed point in the base manifold of a transitive Lie groupoid $\mathcal{Z}$ (Figure 8.4). Consider the subset of the total set $\mathcal{Z}$ formed by the disjoint union $\tilde{\mathcal{Z}}_b$ of all the sets $\mathcal{Z}_{bx}, \forall x \in \mathcal{B}$. The elements $\tilde{z}$ of this set have the property $\alpha(\tilde{z}) = b$. The group $\mathcal{Z}_{bb}$ has a natural effective right action on $\tilde{\mathcal{Z}}_b$, as can be verified directly by composition. Moreover, two elements of $\tilde{\mathcal{Z}}_b$ that differ by the right action of an element of this group must have the same target. In other words, the equivalence classes corresponding to this action consist precisely of the sets $\mathcal{Z}_{bx}$ and, therefore, the quotient set is precisely the manifold $\mathcal{B}$. We are thus led to a principal bundle with total space $\tilde{\mathcal{Z}}_b$, structure group $\mathcal{Z}_{bb}$ and projection β (or, rather, the restriction of β to $\tilde{\mathcal{Z}}_b$).

If we were to start from a different point, c say, of $\mathcal{B}$, the previous construction would lead to a principal bundle whose structure group $\mathcal{Z}_{cc}$ is conjugate to $\mathcal{Z}_{bb}$, and it is not difficult to show that the two principal bundles are isomorphic. We see, therefore, that giving a transitive Lie groupoid is tantamount to giving an equivalence class of isomorphic principal bundles, each one conveying the same information as the groupoid. The choice of the reference point of departure is somewhat analogous to the choice of a basis in a vector space. No information is lost, but there is a certain loss of objectivity, in the sense that one is no longer working with the actual objects but rather with their representation in the chosen reference.

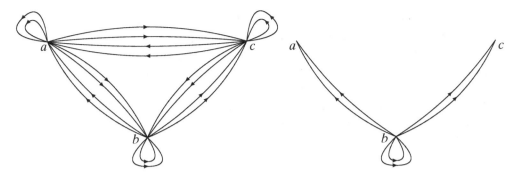

Figure 8.4. Schematic representation of a transitive groupoid $\mathcal{Z}$ (left) and the induced principal bundle $\tilde{\mathcal{Z}}_b$ (right), whose fibres are depicted as pairs of arrows projecting on the base manifold (a,b,c)

BOX 8.6. **From Principal Bundles to Groupoids**

Somewhat more surprising than the previous passage from a groupoid to any one of its representative principal bundles is the fact that, given an arbitrary principal bundle $(\mathcal{P}, \pi_P, \mathcal{B}, \mathcal{G}, \mathcal{G})$, one can construct a groupoid of which it is a representative. Indeed, define the quotient space $\mathcal{Z} = (\mathcal{P} \times \mathcal{P})/\mathcal{G}$ as the total set of the intended groupoid. The reason for this choice is clear: We want to assign an arrow between two points of the base manifold to each diffeomorphism between their fibres which is consistent with the right action of the structure group. More precisely, let $a, b \in \mathcal{B}$. The diffeomorphisms $z : \pi_P^{-1}(a) \to \pi_P^{-1}(b)$ we are referring to are those satisfying:

$$z(pg) = z(p)g, \ \forall p \in \pi_P^{-1}(a), \ g \in \mathcal{G}. \tag{8.40}$$

These are exactly the diffeomorphisms generated by pairing any two points $p \in \pi_P^{-1}(a)$, $q \in \pi_P^{-1}(b)$ (i.e., one from each fibre) and then assigning to pg the point qg. As g runs through the group the diffeomorphism is generated. In other words, we assign the same arrow to the pair $(p,q) \in \mathcal{P} \times \mathcal{P}$ as we assign to (pg, qg). Hence the quotient $\mathcal{Z} = (\mathcal{P} \times \mathcal{P})/\mathcal{G}$.

The remainder of the construction is straightforward. The inverse of the arrow with the pair (p,q) as representative is the arrow represented by the pair (q,p). The identity at $b \in \mathcal{B}$ is represented by any pair (p,p) with $\pi_P(p) = b$. The source and target maps are simply $\alpha(p,q) = \pi_P(p)$ and $\beta(p,q) = \pi_P(q)$. Finally, the composition of arrows is effected by composition of maps. More carefully, let (p,q) be an arrow between a and b, and let (r,s) be an arrow between b and c, with $a, b, c \in \mathcal{B}$. We proceed to change the representative of the equivalence class (r,s) by applying the right action of the unique $g \in \mathcal{G}$ such that $rg = q$. We obtain the new representative of the *same* arrow as (q, sg). The successive application of the arrows is the arrow from a to c whose representative is the pair (p, sg) (the common intermediate element being cancelled out, as it were).

In view of the result just obtained, we define a *material principal bundle* of a materially uniform body B as any one of the equivalent principal bundles that can be obtained from the material groupoid. We observe, however, that whereas the material groupoid always exists (whether or not the body is uniform), the material principal bundle can be defined only when the body is smoothly uniform. Then, and only then, we have a transitive Lie groupoid to work with. In conclusion, although both geometrical objects are suitable for the description of the material structure of a body, the groupoid representation is the more faithful one, since it is unique and universal.

The structure group of a material principal bundle is, according to the previous construction, nothing but the material symmetry group of the archetype. As expected, it controls the degree of freedom available in terms of implanting this archetype at the points of the body.

The material principal bundle may, or may not, admit (global) cross sections. If it does, the body is said to be *globally uniform*. This term is slightly misleading, since uniformity already implies that *all* the points of the body are materially isomorphic. Nevertheless, the term conveys the sense that the material isomorphisms can be prescribed smoothly in a single global chart of the body (which, by definition, always exists). Put in other terms, the existence of a global section implies (in a principal bundle, as we know) that the principal bundle is trivializable. A cross section of a principal bundle establishes, through the right action of the structure group, a global isomorphism between the fibres, also called a *distant parallelism*. In our context, this property will be called a *material parallelism*. If the structure group is discrete, the material parallelism is unique. Moreover, if the material symmetry group consists of just the identity, a uniform body must be globally uniform.

8.4. Flatness and Homogeneity

We have tried to provide a convincing picture of the notion of uniformity in geometrical terms, which can be summarized as follows: A smoothly uniform body, as opposed to one that is not, is automatically endowed with a strong geometric structure induced by its material response, namely, that of a transitive Lie groupoid. Our objective now is to exploit this extra structure in two directions. The first consists in exploring whether further material properties that a uniform body might enjoy are reflected in concomitant properties of the associated groupoid. The most important of these properties is that of *homogeneity*, the lack of which is related to important physical phenomena such as residual stresses, plasticity and continuous distributions of dislocations. The second, more difficult, direction is of a more technical nature, as it tries to find mathematical criteria that would allow us to ascertain whether or not these properties are fulfilled by a given groupoid. In other words, given that a body is uniform, are there clear-cut procedures to determine its homogeneity? In some cases (particularly in those that correspond to a material with a discrete symmetry group) the answer is relatively straightforward, but a general criterion is not available.

A chart in the base manifold $\mathcal{B}$ induces r-jets of diffeomorphisms between every pair of points in its domain $\mathcal{U}$ by a *local coordinate translation*. More precisely, let $a, b \in \mathcal{U}$ have coordinates X_a^I and X_b^I, respectively. We construct a diffeomorphism between a small neighbourhood of a and a small neighbourhood of b according to $X^I \mapsto X^I + X_b^I - X_a^I$, and then evaluate its r-jet. Given a transitive Lie groupoid of r-jets on $\mathcal{B}$, such as the material groupoid we have introduced, it may so happen that the r-jets induced by some coordinate chart as above belong to the groupoid. If for every point of the base $\mathcal{B}$ there exists a neighbourhood and a chart with this property, we say that the groupoid is *(locally) flat*.

We will now explore the physical meaning of the concept of *flatness*, just introduced, in terms of the material response. Since a chart of a body in Continuum Mechanics can also be seen as a configuration, we will assume that we have placed the body in a reference configuration whose restriction to $\mathcal{U}$ coincides with the given chart. We now proceed to express the constitutive law in terms of this configuration and we immediately find out that, because the principle of frame indifference excludes any dependence of the response upon translations, the expression of the constitutive law is *independent of position* in $\mathcal{U}$. In more physical terms, we have found a configuration of the body in which, at least as far as the points in $\mathcal{U}$ are concerned, all the points are not just made of the same material but also in the *same local state of deformation*. We call this property *local homogeneity* and the chart and the configuration are also called (locally) homogeneous. A uniform body that is not locally homogeneous is said to be *inhomogeneous*. If, on the other hand, there exists a homogeneous global chart, the body is *globally homogeneous*, or simply *homogeneous*.

In Chapter 9, we will explore the second direction of inquiry, to which we alluded at the beginning of this section, as it applies to simple materials.[9] Namely, we will attempt to find explicit criteria for the characterization of local homogeneity for such materials. As a preliminary step in this direction, in the next two sections we introduce the concepts of distributions and jet bundles.

8.5. Distributions and the Theorem of Frobenius

A *k-dimensional distribution* on an m-dimensional manifold $\mathcal{M}$ is a *smooth* assignment of a k-dimensional subspace D_x of the tangent space $T_x\mathcal{M}$ at each point $x \in \mathcal{M}$. The smoothness condition can be best understood as follows: For each point $x \in \mathcal{M}$ there exists a neighbourhood $\mathcal{U}$ and k smooth, linearly independent, tangent vector fields $\mathbf{X}_1, \ldots, \mathbf{X}_k$ in $\mathcal{U}$, forming at each point of $\mathcal{U}$ a basis of D_x. The distribution is then said to be spanned by these vector fields. Seen in this way, a distribution can be regarded as a generalization of the notion of vector field, which becomes the particular case $k = 1$.

[9] For second-grade materials and materials with internal structure, see Epstein M, de León M (1998), Geometrical Theory of Uniform Cosserat Media, *Journal of Geometry and Physics* **26**, 127–170. Also Epstein M, Elżanowski M (2007), op. cit.

REMARK 8.6. A 1-dimensional distribution is not quite the same as a vector field since the object prescribed at each point is not a vector but a line. A vector field spans a 1-dimensional distribution.

According to the fundamental theorem of the theory of ODEs (Theorem 5.1), every vector field gives rise to integral curves. The question, therefore, arises as to whether a similar result can be claimed for distributions of dimension k higher than 1, namely, whether there exist submanifolds of dimension k whose tangent space at each point x coincides with D_x. The answer is in general negative. The necessary and sufficient conditions under which such submanifolds exist is the content of Frobenius' theorem.

Let us assume that there exists a coordinate system y^i ($i = 1, \ldots, m$) such that (locally, on the coordinate patch) the distribution is spanned by the first k natural base vectors. If this is the case, then we the submanifolds $\mathcal{N}$ obtained in the patch by the equations

$$y^i = 0 \quad i = k+1, \ldots, m, \tag{8.41}$$

are *integral manifolds* of the distribution.[10] By this we mean that $D_x = T_x\mathcal{N}$ at each point in the coordinate chart. A distribution with this property (namely, that each point is contained in an integral manifold) is said to be *integrable*. In this case, for any $\mathbf{X}_1, \ldots, \mathbf{X}_k$ spanning the distribution, we should be able to find a nonsingular matrix function with unique entries $A_\alpha^\beta(y^1, \ldots, y^m)$ ($\alpha, \beta = 1, \ldots, k$) such that:

$$\mathbf{X}_\alpha = A_\alpha^\beta \frac{\partial}{\partial y^\beta}. \tag{8.42}$$

Let us now calculate the Lie brackets $[\mathbf{X}_\alpha, \mathbf{X}_\beta]$. Using the result of Exercise 4.10 and the fact that the Lie brackets between natural base vectors vanish, we obtain:

$$[\mathbf{X}_\alpha, \mathbf{X}_\beta] = A_\alpha^\rho \frac{\partial A_\beta^\sigma}{\partial y^\rho} \frac{\partial}{\partial y^\sigma} - A_\beta^\sigma \frac{\partial A_\alpha^\rho}{\partial y^\sigma} \frac{\partial}{\partial y^\rho}. \tag{8.43}$$

But, since the matrices $\{A_\alpha^\beta\}$ are invertible, we can express the natural base vectors as linear combinations of the (base) vectors $\mathbf{X}_1, \ldots, \mathbf{X}_k$, with the result that the Lie brackets are themselves linear combinations of $\mathbf{X}_1, \ldots, \mathbf{X}_k$, namely:

$$[\mathbf{X}_\alpha, \mathbf{X}_\beta] = C_{\alpha\beta}^\rho \mathbf{X}_\rho. \tag{8.44}$$

A distribution with this property (namely, that the Lie brackets of any two vector fields in the distribution is also in the distribution) is said to be *involutive*. We have thus proved that every integrable distribution is involutive. The converse of this result, namely, that every involutive distribution is integrable, is known as *Frobenius' theorem*.[11]

---

[10] Recall Remark 4.2 regarding the characterization of an embedded submanifold.
[11] The proof is lengthy but not very difficult. It can be found, for example, in Chern et al. (2000), op. cit., p 35.

EXERCISE 8.6. A standard example of a noninvolutive 2-distribution in $\mathbb{R}^3$ is the one spanned by the vector fields:

$$\mathbf{X}_1 = \frac{\partial}{\partial x^1} \qquad \mathbf{X}_2 = \frac{\partial}{\partial x^2} + x^1 \frac{\partial}{\partial x^3}, \qquad (8.45)$$

where x^1, x^2, x^3 are the natural coordinates of $\mathbb{R}^3$. Show that this is indeed a noninvolutive distribution. Attempt a drawing of the distribution around the origin and explain intuitively why the distribution fails to be involutive. Finally, assume that there exists a surface with equation $x^3 = \phi(x^1, x^2)$ and show that imposing the condition that its tangent plane belongs (at every point of some open domain) to the distribution leads to a contradiction.

EXERCISE 8.7. Modify the previous example by stipulating:

$$\mathbf{X}_1 = \frac{\partial}{\partial x^1} + x^2 \frac{\partial}{\partial x^3} \qquad \mathbf{X}_2 = \frac{\partial}{\partial x^2} + x^1 \frac{\partial}{\partial x^3}. \qquad (8.46)$$

Show that this distribution is involutive. Find an integral manifold through the origin in the form $x^3 = \phi(x^1, x^2)$.

8.6. Jet Bundles and Differential Equations

The notion of jets of maps between manifolds has been introduced so far as a useful device, as well as a convenient notation, to express and compose the kinematic quantities associated with local material responses of various degrees. Even from this particular application alone, one can appreciate that the theory of jets may be useful for the geometrical description of physical models that involve derivatives, a category that encompasses practically every important physical theory. The concept of *jet bundle*, which we will briefly introduce in this section, can be said to be a generalization of the geometric representation of ODEs as vector fields to the realm of PDEs. For a more complete and elegant treatment of this subject, the reader should consult the specialized literature.[12]

8.6.1. Jets of Sections

We start from a given fibre bundle $(\mathcal{C}, \pi, \mathcal{B}, \mathcal{F}, \mathcal{G})$. Recall that a cross section, $\sigma : \mathcal{B} \to \mathcal{C}$, was defined as a smooth assignment, to each point of the base manifold $\mathcal{B}$, of an element of its fibre. Cross sections as such may not exist, since there is no guarantee that such an assignment can in fact be done smoothly, or even continuously. Nevertheless, restricting the domain of the cross section to some open subset $\mathcal{U} \subset \mathcal{B}$, it is always possible to construct cross sections of the sub-bundle $\pi^{-1}(\mathcal{U})$. The reason for this statement is that, by definition, fibre bundles are locally trivializable, and

---

[12] See Saunders DJ (1989), *The Geometry of Jet Bundles*, London Mathematical Society, Lecture Note Series **142**, Cambridge University Press. Also: Kolář I, Michor PW, Slovák J (1993) *Natural Operators in Differential Geometry*, Springer.

trivial bundles always have cross sections (since they are endowed with a projection on the typical fibre).

DEFINITION 8.3. A *local cross section* of a fibre bundle $(\mathcal{C}, \pi, \mathcal{B}, \mathcal{F}, \mathcal{G})$ is a cross section of any sub-bundle of the form $\pi^{-1}(\mathcal{U})$, where $\mathcal{U}$ is an open subset of the base manifold $\mathcal{B}$.

For any point $b \in \mathcal{B}$, we denote by Γ_b (or, according to context, also $\Gamma_b(\mathcal{C})$ or $\Gamma_b(\pi)$), the set of local cross sections whose domains contain b. This set is never empty. By contrast, the set Γ of cross sections of the fibre bundle may be empty. On occasion, we will use the terminology *global cross sections* to refer to cross sections of the original fibre bundle, thus emphasizing their difference from the local counterparts. Also, the term "cross section" is often replaced by "section."

DEFINITION 8.4. An *r-jet* of a fibre bundle $\pi : \mathcal{C} \to \mathcal{B}$ at a point $b \in \mathcal{B}$ is an r-jet of local cross sections at p.

In other words, following our definition of jets (see Box 8.1), we consider equivalence classes of local sections, consisting of those that share the value of the image and of all the derivatives of orders less than or equal to r at b. An r-jet is one such equivalence class. If $\sigma \in \Gamma_b$, its corresponding r-jet is denoted by $j_b^r \sigma$.

8.6.2. Jet Bundles

The remarkable fact that allows us to work geometrically at ease with jets of sections is the statement of the following theorem.

THEOREM 8.1. *The set $J^r(\mathcal{C})$ (also denoted as $J^r(\pi)$) of all r-jets of local sections of $\pi : \mathcal{C} \to \mathcal{B}$ has a natural structure of a finite-dimensional differentiable manifold.*

PROOF. Let $\{\mathcal{U}, \psi\}$ be a local trivialisation of $(\mathcal{C}, \pi, \mathcal{B}, \mathcal{F}, \mathcal{G})$, that is, a fibre-preserving diffeomorphism from $\pi^{-1}(\mathcal{U})$ to $\mathcal{U} \times \mathcal{F}$, where $\mathcal{U}$ is an open set in $\mathcal{B}$. As we have seen in Section 7.3.1, for given systems of coordinates x^i ($i = 1, \ldots, m$) in $\mathcal{U}$ and u^α ($\alpha = 1, \ldots, n$) in $\mathcal{V} \subset \mathcal{F}$, this local trivialization induces a coordinate system in $\pi^{-1}(\mathcal{U})$ such that the coordinates of a point $p \in \pi^{-1}(\mathcal{U})$ are given by the pair (x^i, u^α). To be more precise, we should state that the domain of this adapted chart is actually not the whole subbundle $\pi^{-1}(\mathcal{U})$, but rather the open set $\psi^{-1}(\mathcal{U} \times \mathcal{V})$, where $\mathcal{V}$ is the domain of a chart in $\mathcal{F}$. In what follows, however, we do not make this distinction to avoid further encumbrance of the notation. Be that as it may, we want to show that this trivialization also induces naturally a coordinate system in $J^r(\mathcal{C})$. Consider first the case $r = 1$. Recall that an element of $J^1(\mathcal{C})$ is a 1-jet $j_b^1 \sigma$ of some local section $\sigma : \mathcal{U} \to \pi^{-1}(\mathcal{U})$ at a point $b \in \mathcal{U}$. To define a coordinate chart in $J^1(\mathcal{C})$ we must first define an open set $\bar{\mathcal{U}} \subset J^1(\mathcal{C})$. This we define as follows:

$$\bar{\mathcal{U}} = \{j_b^1 \sigma \mid \sigma(b) \in \pi^{-1}(\mathcal{U})\}. \tag{8.47}$$

For a given $j_b^1\sigma \in \bar{\mathcal{U}}$, the coordinates are defined as the triple $(x^i, u^\alpha, u_i^\alpha)$, where the coordinates u_i^α are given by:

$$u_i^\alpha(j_b^1\sigma) = \left.\frac{\partial \sigma^\alpha}{\partial x^i}\right|_b. \tag{8.48}$$

In this expression, σ^α denotes the coordinate expression of the cross section σ, namely:

$$\sigma^\alpha = u^\alpha \circ \sigma. \tag{8.49}$$

We are far from done yet. We need to show that an atlas of $J^1(\mathcal{C})$ can be constructed in this way. Moreover, we need to extend this construction for $r > 1$. The proofs are not difficult, but we shall not give the details, which can be found in the specialized literature. $\qquad\square$

With some more work, it can be shown that $J^r(\mathcal{C})$ has a natural fibred structure and that it is, in fact, a fibre bundle over $\mathcal{C}$ with projection $\pi_{r,0} : J^r(\mathcal{C}) \to \mathcal{C}$ defined by:

$$\pi_{r,0}(j_b^r\sigma) = \sigma(b). \tag{8.50}$$

DEFINITION 8.5. A *jet field* is a section $\Gamma : \mathcal{C} \to J^r(\mathcal{C})$ of the jet bundle $\pi_{r,0}$.

The same total space $J^r(\mathcal{C})$ gives rise to other natural fibre bundles with various base manifolds. In particular, we can define a fibre bundle over $\mathcal{B}$ with projection π_r given by:

$$\pi_r(j_b^r\sigma) = b. \tag{8.51}$$

8.6.3. Differential Equations

DEFINITION 8.6. A *differential equation* on the fibre bundle $\pi : \mathcal{C} \to \mathcal{B}$ is a closed embedded submanifold S of the jet manifold $J^r(\mathcal{C})$.

Notice that the usual definition of a (partial) differential equation, or a system thereof, is given by means of a functional relation between the unknown functions and their derivatives. All we have done is to reinterpret these relations as defining a closed submanifold of $J^r(\mathcal{C})$.

DEFINITION 8.7. A *solution* of the differential equation S is a local section σ such that $j_b^r\sigma \in S$ for every b in its domain.

EXERCISE 8.8. A first-order ODE: For the case of a single first-order ordinary differential equation $F(x, y(x), dy/dx) = 0$, draw a picture that illustrates the meaning of Definitions 8.6 and 8.7.

Connection, Curvature, Torsion

All the fibres of a fibre bundle are, by definition, diffeomorphic to each other. In the absence of additional structure, however, there is no canonical way to single out a particular diffeomorphism between fibres. In the case of a product bundle, for example, such a special choice is indeed available because of the existence of the second projection map onto the typical fibre. In this extreme case, we may say that we are in the presence of a *canonical distant parallelism* in the fibre bundle. An equivalent way to describe this situation is by saying that we have a canonical family of non-intersecting smooth cross sections such that each point in the fibre bundle belongs to one, and only one, of them. In a general fibre bundle we can only afford this luxury noncanonically and locally. A *connection* on a fibre bundle is, roughly speaking, an additional structure defined on the bundle that permits to establish intrinsic fibre diffeomorphisms for fibres lying along curves in the base manifold. In other words, a connection can be described as a curve-dependent parallelism. Given a connection, it may so happen that the induced fibre parallelisms turn out to be curve-independent. A quantitative measure of this property or the lack thereof is provided by the vanishing, or otherwise, of the *curvature* of the connection.

9.1. Ehresmann Connection

Although the historical development of the notion of connection witnesses an arduous ascent from the particular to the general, in hindsight it appears best, both formally and pedagogically, to proceed in the opposite direction. The notion of *Ehresmann connection*, first formulated circa 1950, is a good place to start, since it does not assume anything about the nature of the fibre bundle at hand.

9.1.1. Definition

Just as in Section 7.11.1, consider the tangent bundle TC of the total space C of an arbitrary fibre bundle (C, π, B, F, G), and denote by $\tau_C : TC \to C$ its natural projection. If the dimensions of the base manifold B and the typical fibre F are, respectively,

m and n, the dimension of $\mathcal{C}$ is $m + n$, and the typical fibre of $(T\mathcal{C}, \tau_C)$ is $\mathbb{R}^{m+n}$, with structure group $GL(m + n; \mathbb{R})$. At each point $c \in \mathcal{C}$ the tangent space $T_c\mathcal{C}$ has a canonically defined *vertical subspace* V_c, which can be identified with the tangent space $T_c\mathcal{C}_{\pi(c)}$ to the fibre of $\mathcal{C}$ at c. The dimension of V_c is n. A vector in $T_c\mathcal{C}$ belongs to the vertical subspace V_c (or is *vertical*) if, and only if, its projection by π_* is the zero vector of $T_{\pi(c)}\mathcal{B}$. If a vector in $T_c\mathcal{C}$ is not vertical, there is no canonical way to assign to it a vertical component. It is this deficiency, and only this deficiency, that the Ehresmann connection remedies. Formally, an Ehresmann connection consists of a smooth *horizontal distribution* in $\mathcal{C}$. This is a smooth assignment to each point $c \in \mathcal{C}$ of an (m-dimensional) subspace $H_c \subset T_c\mathcal{C}$ (called the *horizontal subspace at c*), such that:

$$T_c\mathcal{C} = H_c + V_c. \tag{9.1}$$

In this equation, $+$ denotes the direct sum of vector spaces. Each tangent vector $\mathbf{u} \in T_c\mathcal{C}$ is, accordingly, uniquely decomposable as the sum of a horizontal part $h(\mathbf{u})$ and a vertical part $v(\mathbf{u})$. A vector is *horizontal*, if its vertical part vanishes. The only vector that is simultaneously horizontal and vertical is the zero vector. Since H_c and $T_{\pi(c)}\mathcal{B}$ have the same dimension (m), the restriction $\pi_*|_{H_c} : H_c \to T_{\pi(c)}\mathcal{B}$, is a vector-space isomorphism. We denote its inverse by Γ_c. Thus, given a vector $\mathbf{v}$ tangent to the base manifold at a point $b \in \mathcal{B}$, there is a unique horizontal vector: $\Gamma_c\mathbf{v}$ at $c \in \pi^{-1}(\{b\})$ such that $\pi_*(\Gamma_c\mathbf{v}) = \mathbf{v}$. This unique vector is called the *horizontal lift* of $\mathbf{v}$ to c. In particular, $\Gamma_c(\pi_*(\mathbf{u})) = \Gamma_c(\pi_*(h(\mathbf{u}))) = h(\mathbf{u})$. These ideas are schematically illustrated in Figure 9.1.

EXERCISE 9.1. Show that an Ehresmann connection is the same as a *jet field*, namely, a section of the jet bundle $J^1(\mathcal{C})$.

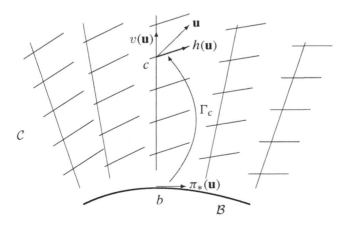

Figure 9.1. Ehresmann connection

9.1.2. Parallel Transport Along a Curve

Let:

$$\gamma : (-\epsilon, \epsilon) \longrightarrow \mathcal{B} \tag{9.2}$$

be a smooth curve in the base manifold $\mathcal{B}$ of the fibre bundle $(\mathcal{C}, \pi)$, and let $c \in \mathcal{C}_{\gamma(0)}$ be a point in the fibre at $\gamma(0)$. A *horizontal lift* of γ through c is defined as a curve:

$$\hat{\gamma} : (-\epsilon, \epsilon) \longrightarrow \mathcal{C}, \tag{9.3}$$

such that:

$$\hat{\gamma}(0) = c, \tag{9.4}$$

$$\pi\left(\hat{\gamma}(t)\right) = \gamma(t), \quad \forall t \in (-\epsilon, \epsilon), \tag{9.5}$$

and

$$\hat{\gamma}'(t) \in H_{\hat{\gamma}(t)} \quad \forall t \in (-\epsilon, \epsilon), \tag{9.6}$$

where a prime denotes the derivative with respect to the curve parameter t. A horizontal lift is thus a curve that projects onto the original curve and, moreover, has a horizontal tangent throughout.

Consider the "cylindrical" subbundle $\gamma^*\mathcal{C}$ obtained by pulling back the bundle $\mathcal{C}$ to the curve γ or, less technically, by restricting the base manifold to the curve γ. The tangent vector field of γ has a unique horizontal lift at each point of this bundle. In other words, the curve generates a (horizontal) vector field throughout this restricted bundle. By the fundamental theorem of the theory of ODEs, it follows that, at least for small enough ϵ, there is a unique horizontal lift of γ through any given point in the fibre at $\gamma(0)$, namely, the corresponding integral curve of the horizontal vector field. We conclude, therefore, that the horizontal lift of a curve through a point in a fibre bundle exists and is locally unique. As the horizontal curve issuing from c cuts the various fibres lying on γ, the point c is said to undergo a *parallel transport* relative to the given connection and the given curve. Thus, given a point $c \in \mathcal{C}$ and a curve γ through $\pi(c) \in \mathcal{B}$, we obtain a unique parallel transport of c along γ by solving a system of ODEs (so as to travel always horizontally). These concepts are illustrated schematically in Figure 9.2

It is quite remarkable that, even with just this minimum amount of structure, it is possible to define the concepts of holonomy and curvature by checking whether or not, upon parallel transport along a closed curve, one returns to the original point of departure. To avoid the use of new tools,[1] however, we now proceed to a slightly more restricted kind of connection, for which we are going to find ourselves in more familiar territory.

---

[1] Such as the Frölicher-Nijenhuis bracket.

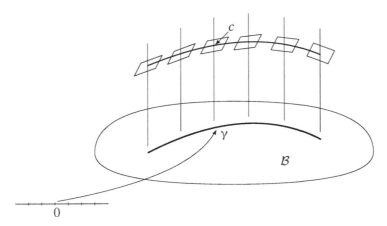

Figure 9.2. Parallel transport along a curve

9.2. Connections in Principal Bundles

9.2.1. Definition

A connection in a principal bundle $(\mathcal{P}, \pi, \mathcal{B}, \mathcal{G}, \mathcal{G})$ is an Ehresmann connection which is compatible with the right action R_g of $\mathcal{G}$ on $\mathcal{P}$, namely:

$$(R_g)_*(H_p) = H_{R_g p} \quad \forall g \in \mathcal{G} \quad p \in \mathcal{P}. \tag{9.7}$$

This condition can be stated verbally as follows: The horizontal distribution is invariant under the group action.

Recall that the group $\mathcal{G}$ acts freely (to the right) on $\mathcal{P}$. Consequently, the fundamental vector field $\mathbf{v_g}$ associated with any nonzero vector $\mathbf{g}$ in the Lie algebra $\mathfrak{g}$ of $\mathcal{G}$ does not vanish anywhere on $\mathcal{P}$. Moreover, since the action of $\mathcal{G}$ maps fibres into themselves, the fundamental vector fields are all vertical. The correspondence between vectors in the Lie algebra and tangent vectors to the fibre at any point is clearly linear and one-to-one. Since the dimension of $\mathcal{G}$ is equal to the dimension of each fibre, we conclude that the map $\mathfrak{g} \to V_p$ given by $\mathbf{g} \mapsto \mathbf{v_g}(p)$ is a linear isomorphism between the Lie algebra and each of the vertical subspaces of the principal bundle.

Let $\mathbf{v} \in T\mathcal{P}$ be any tangent vector to the fibre bundle. A connection Γ assigns to it a unique vertical part and, as we have just seen, the action of the group assigns to this vertical part an element of the Lie algebra $\mathfrak{g}$. This means that we have a well-defined linear map:

$$\omega : T\mathcal{P} \longrightarrow \mathfrak{g}, \tag{9.8}$$

associated with a given connection in a principal bundle. This map can be regarded as a *Lie-algebra valued 1-form*. It is called the *connection form* associated with Γ. On the other hand, let a Lie-algebra valued 1-form ω be defined on a principal bundle

$\mathcal{P}$. In order to qualify as a connection form, it will have to satisfy the following conditions. First, when acting on a vertical vector $\mathbf{v_g}$ it assigns to it the vector $\mathbf{g}$ in the Lie algebra of which $\mathbf{v_g}$ is the local value of the fundamental vector field. Second, in order to be consistent with the invariance of horizontal spaces, we must have:

$$(R_h)^* \omega = Ad(h^{-1})\omega. \tag{9.9}$$

This condition, which arises from Lemma 5.14, can also be expressed as:

$$\omega((R_h)_*\mathbf{X}) = Ad(h^{-1})[\omega(\mathbf{X})]. \tag{9.10}$$

9.2.2. Existence

A connection in a principal fibre bundle, whether seen as a horizontal distribution or as a Lie-algebra valued 1-form, is not as flexible as it might appear. This is due to the fact that it has to be compatible with the right action of the structure group. In other words, once a horizontal subspace is chosen at one point of a fibre, all the horizontal subspaces on this fibre are uniquely determined. This restriction raises the question as to whether or not connections exist, a legitimate question to ask for any entity that one may care to define. The proof of existence of connections in principal bundles is surprisingly simple and, in fact, makes use of the very restriction mentioned above.

THEOREM 9.1. *A principal fibre bundle admits a connection.*

PROOF. The only condition needed to prove the theorem is that the base manifold be paracompact so that it admits a partition of unity, a condition that we have already incorporated in our very definition of a manifold. Let $(\mathcal{U}, \phi)$ be a local trivialization of the bundle. In other words, $\mathcal{U}$ is an open set in the base manifold $\mathcal{B}$ and ϕ is a fibre-preserving diffeomorphism:

$$\phi : \pi^{-1}(\mathcal{U}) \longrightarrow \mathcal{U} \times \mathcal{G}, \tag{9.11}$$

where the restriction of ϕ to each fibre preserves the right group action. Since the product bundle has a standard constant cross section at the group unit e, we have a (trivialization-dependent) cross section σ of $\pi^{-1}(\mathcal{U})$ which is mapped therein. We now choose, as a point of departure, *any* 1-form α on $\mathcal{B}$ with values in the Lie algebra $\mathfrak{g}$. We want to define, on the basis of this form, a $\mathfrak{g}$-valued 1-form ω on $\pi^{-1}(\mathcal{U})$. Let $\mathbf{v}$ be a tangent vector at a point $p \in \sigma(\mathcal{U})$. The tangent map ϕ_* takes this vector to the vector $\phi_*(\mathbf{v})$ tangent to the product manifold $\mathcal{U} \times \mathcal{G}$, where a standard horizontal distribution already exists. The vertical component, which we denote by $\mathbf{w}(\mathbf{v})$, being an element of $T_e\mathcal{G}$ belongs to the Lie algebra. Define now at every point of σ:

$$\omega(\mathbf{v}) := \alpha(\pi_*\mathbf{v}) + \mathbf{w}(\mathbf{v}) \quad \forall \, \mathbf{v} \in T_\sigma\pi^{-1}(\mathcal{U}). \tag{9.12}$$

So far, we have defined a connection form on only one point of each fibre. Recall that the right action of the structure group on a principal bundle is transitive. This implies that we can extend the definition to all other points of the fibre by the group consistency condition (9.9). Thus, we have obtained a connection 1-form over $\pi^{-1}(\mathcal{U})$. If $(\mathcal{U}_i, \phi_i)$ is a trivializing covering, and if ω_i denotes a connection form defined on each $\pi^{-1}(\mathcal{U}_i)$ by the procedure described above, the form defined at each point p by:

$$\omega(p) = \sum_i (\theta_i(\pi(p)))\, \omega_i(p), \tag{9.13}$$

where θ_i is a partition of unity subordinate to the covering, is a connection 1-form on the principal bundle. $\qquad\square$

9.2.3. Curvature

Suppose that we draw through a point b of the base manifold $\mathcal{B}$ a small closed curve γ. If we now choose a point p in the fibre on b, we have learned that there exists a unique horizontal lift $\tilde{\gamma}$, namely, a horizontal curve containing p and projecting on γ. Is this curve closed? To clarify the meaning of this question and its possible answer, recall that a connection on a principal fibre bundle is a special case of a distribution, which we have called horizontal (the dimension of the horizontal distribution equals the dimension of the base manifold and is thus strictly smaller than the dimension of the fibre bundle, assuming that the typical fibre is of dimension greater than zero). Clearly, if the horizontal distribution is involutive, any horizontal lift of a small curve in the base manifold will lie entirely on an integral submanifold and, therefore, will be closed. This observation suggests that a measure of the lack of closure of the horizontal lift of closed curves is the fact that the Lie bracket between horizontal vector fields has a vertical component. We want to see now how to extract this information from the connection itself. More particularly, since a connection is specified by its connection form ω, we want to extract this information from ω alone.

Consider two horizontal vector fields $\mathbf{u}$ and $\mathbf{v}$. Let us evaluate the 2-form[2] $d\omega$ on this pair. Using Proposition 4.3, we obtain:

$$\langle d\omega \mid \mathbf{u} \wedge \mathbf{v} \rangle = \mathbf{u}\,(\langle \omega \mid \mathbf{v} \rangle) - \mathbf{v}\,(\langle \omega \mid \mathbf{u} \rangle) - \langle \omega \mid [\mathbf{u}, \mathbf{v}] \rangle, \tag{9.14}$$

which, in view of the fact that $\mathbf{u}$ and $\mathbf{v}$ are assumed to be horizontal, yields:

$$\langle d\omega \mid \mathbf{u} \wedge \mathbf{v} \rangle = -\langle \omega \mid [\mathbf{u}, \mathbf{v}] \rangle. \tag{9.15}$$

The right-hand side of this equation will vanish if, and only if, the Lie bracket is horizontal. This means that we have found a way to extract the right information from ω by just taking its exterior derivative and applying it to two horizontal vector fields. Notice, however, that $d\omega$ can be applied to arbitrary pairs of vector fields, not necessarily horizontal. To formalize this point, given a connection, we define the *exterior covariant derivative* $D\alpha$ of an r-form α as the $(r+1)$-form given by:

$$\langle D\alpha \mid \mathbf{U}_1 \wedge \cdots \wedge \mathbf{U}_{r+1} \rangle = \langle d\alpha \mid h(\mathbf{U}_1) \wedge \cdots \wedge h(\mathbf{U}_{r+1}) \rangle, \tag{9.16}$$

---

[2] Notice that this is a Lie-algebra valued differential form.

where $h(.)$ denotes the horizontal component of a vector. Accordingly, we define the *curvature 2-form* Ω of a connection ω on a principal fibre bundle as:

$$\Omega := D\omega. \tag{9.17}$$

9.2.4. Cartan's Structural Equation

Our definition of curvature, by using both the connection 1-form and the horizontal projection map, is a hybrid that mixes both (equivalent) definitions of a connection. It is possible, on the other hand, to obtain an elegant formula that involves just the connection 1-form. This formula, known as *Cartan's structural equation*, reads:

$$\Omega = d\omega + \frac{1}{2}\,[\omega,\omega], \tag{9.18}$$

or, more precisely, for any two vectors $\mathbf{U}$ and $\mathbf{V}$ at a point[3] of the frame bundle:

$$\langle \Omega \mid \mathbf{U} \wedge \mathbf{V} \rangle = \langle d\omega \mid \mathbf{U} \wedge \mathbf{V} \rangle + \frac{1}{2}\,[\omega(\mathbf{U}),\omega(\mathbf{V})]. \tag{9.19}$$

The proof of this formula, whose details we omit, is based on a careful examination of three cases: (i) $\mathbf{U}$ and $\mathbf{V}$ are horizontal, whereby the formula is obvious; (ii) $\mathbf{U}$ is horizontal and $\mathbf{V}$ is vertical, in which case one can extend them, respectively, to a horizontal and a fundamental (vertical) vector field; (iii) $\mathbf{U}$ and $\mathbf{V}$ are both vertical, in which case they can both be extended to fundamental fields.

To express Cartan's structural equation in components, let $\mathbf{e}_i$ $(i = 1,\ldots,n)$ be a basis of the Lie algebra $\mathfrak{g}$ and let c^i_{jk} be the structure constants of $\mathfrak{g}$ in this basis (see Section 5.12.1). Moreover, since ω and Ω take values in the Lie algebra, there exist smooth (ordinary, real valued) forms ω^i and Ω^i such that:

$$\omega = \omega^i \mathbf{e}_i, \qquad \Omega = \Omega^i \mathbf{e}_i. \tag{9.20}$$

The structural Equation (9.18) can be written as:

$$\Omega^i = d\omega^i + \frac{1}{2}\,c^i_{jk}\omega^j \wedge \omega^k. \tag{9.21}$$

9.2.5. Bianchi Identities

Unlike the ordinary exterior derivative d, the operator D (of exterior covariant differentiation) is not necessarily nilpotent, namely, in general $D^2 \neq 0$. Therefore, there is no reason to expect that $D\Omega$, which is equal to $D(D\omega)$, will vanish identically. But in fact it does. To see that this is the case, notice that, by definition of D, we need only verify the vanishing of $D\Omega$ on an arbitrary triple of *horizontal* vectors. Working, for clarity, in components, we obtain from (9.21):

$$d\Omega^i = \frac{1}{2}\,c^i_{jk}d\omega^j \wedge \omega^k + \frac{1}{2}\,c^i_{jk}\omega^j \wedge d\omega^k. \tag{9.22}$$

---

[3] Notice that this formula is valid pointwise, since the Lie bracket on the right-hand side is evaluated in the Lie algebra, not in the manifold.

Since each term has a factor ω, the right-hand side will vanish when applied to any horizontal vector, thus establishing the result:

$$D\Omega = 0. \tag{9.23}$$

In terms of components, we obtain differential identities to be satisfied by any curvature form. They are known as the *Bianchi identities*.

EXAMPLE 9.1. In the theory of General Relativity, where the curvature of the space-time manifold is related to the presence of matter, the Bianchi identities play the role of an automatic conservation law. In Continuum Mechanics, the so-called compatibility conditions of the strain (simply establishing that the Riemannian curvature of space is zero) are subject to the Bianchi identities, thus reducing the number of independent compatibility conditions from six to three.

9.3. Linear Connections

A connection on the bundle of linear frames $F\mathcal{B}$ is called a *linear connection* on $\mathcal{B}$. Among principal bundles, the bundle of linear frames occupies a special position for various reasons. In the first place, the bundle of linear frames is canonically defined for any given base manifold $\mathcal{B}$. Moreover, the associated bundles include all the tensor bundles, thus allowing for a unified treatment of all such entities. Another way to express this peculiar feature of the bundle of linear frames is that, whereas the quantities parallel-transported along curves in a general principal bundle are of a nature not necessarily related to the base manifold, in the case of the bundle of linear frames the quantities transported are precisely the very frames used to express the components of vectors and forms defined on the base manifold. An elegant manifestation of this property is the existence of a canonical 1-form that ties everything together. A direct consequence of the existence of this 1-form is the emergence of the idea of the *torsion* of a connection. We start the treatment of linear connections by lingering for a while on the definition of the canonical 1-form.

9.3.1. The Canonical 1-Form

Given a tangent vector $\mathbf{v} \in T_x\mathcal{B}$ at a point x in the base manifold, and a point $p \in F_x\mathcal{B}$ in the fibre over x, and recalling that p consists of a frame (or basis) $\{\mathbf{e}_1,\ldots,\mathbf{e}_m\}$ of $T_x\mathcal{B}$, we can determine uniquely the m components of $\mathbf{v}$ in this frame, namely:

$$\mathbf{v} = v^a \mathbf{e}_a. \tag{9.24}$$

In other words, at each point $p \in F\mathcal{B}$, we have a well-defined nonsingular linear map[4]:

$$u(p) : T_{\pi(p)}\mathcal{B} \longrightarrow \mathbb{R}^m. \tag{9.25}$$

---

[4] This map is, in fact, an alternative definition of a linear frame at a point of a manifold $\mathcal{B}$.

The *canonical 1-form* θ on $F\mathcal{B}$ is defined as:

$$\theta(\mathbf{V}) := u(p) \circ \pi_*(\mathbf{V}) \quad \forall \, \mathbf{V} \in T_p(F\mathcal{B}). \tag{9.26}$$

Note that this is an $\mathbb{R}^m$-valued form. The canonical 1-form of the frame bundle is a particular case of a more general construct known as a *soldering form*.

It may prove instructive to exhibit the canonical form in components. Let $x^1, \ldots, x^m$ be a local coordinate system on $\mathcal{U} \subset \mathcal{B}$. Every frame $\{\mathbf{e}_1, \ldots, \mathbf{e}_m\}$ at $x \in \mathcal{U}$ can be expressed uniquely by means of a nonsingular matrix with entries x_j^i as:

$$\mathbf{e}_a = x_a^i \frac{\partial}{\partial x^i}. \tag{9.27}$$

This means that the $m + m^2$ functions $\{x^i, x_a^i\}$ constitute a coordinate system for the linear frame bundle $\pi^{-1}(\mathcal{U})$. We call it the coordinate system *induced* by x^i. The projection map $\pi : F\mathcal{B} \to \mathcal{B}$ has the coordinate representation:

$$(x^i, x_a^i) \mapsto \pi(x^i, x_a^i) = (x^i), \tag{9.28}$$

with some notational abuse.

Consider now the tangent bundle $TF(\mathcal{B})$ with projection $\tau : TF(\mathcal{B}) \to F(\mathcal{B})$. The coordinate system $\{x^i, x_a^i\}$ induces naturally a coordinate system in $TF(\mathcal{B})$. A vector $\mathbf{X} \in TF(\mathcal{B})$ is expressed in these coordinates as follows:

$$\mathbf{X} \mapsto \left(x^i, x_a^i, X^i \frac{\partial}{\partial x^i} + X_a^i \frac{\partial}{\partial x_a^i} \right) = \left(x^i, x_a^i, X^i, X_a^i \right). \tag{9.29}$$

The derivative of the projection π is a map: $\pi_* TF(\mathcal{B}) \to T\mathcal{B}$. Its coordinate representation is:

$$\left(x^i, x_a^i, X^i, X_a^i \right) \mapsto \left(x^i, X^i \right). \tag{9.30}$$

The map u defined in Equation (9.25) is given in coordinates by:

$$[u(x^i, x_a^i)](x^j, w^j) = x_i^{-a} w^i \quad (a = 1, \ldots, m), \tag{9.31}$$

where we have denoted by x_i^{-a} the entries of the inverse of the matrix with entries x_a^i. Combining (9.30) and (9.31), we obtain from (9.26) the following coordinate representation of the ($\mathbb{R}^m$)-valued canonical form θ:

$$\theta^a = x_i^{-a} \, dx^i \quad (a = 1, \ldots, m). \tag{9.32}$$

9.3.2. The Christoffel Symbols

The canonical form θ exists independently of any connection. Let us now introduce a connection on $F(\mathcal{B})$, that is, a linear connection on $\mathcal{B}$. If we regard a connection as a horizontal distribution, there must exist non-singular linear maps $\Gamma(x,p)$ from each $T_x\mathcal{B}$ to each of the tangent spaces $T_pF(\mathcal{B})$ (with $\pi(p) = x$) defining the distribution. Noticing that the same distribution may correspond to an infinite number of such maps, we pin down a particular one by imposing the extra condition that they must be also horizontal lifts. In other words, we demand that:

$$\pi_* \circ \Gamma(x,p) = id_{T_x \mathcal{B}}. \tag{9.33}$$

The implication of this condition is that, when written in components, we must have:

$$\Gamma(x,p)\left(v^i \frac{\partial}{\partial x^i}\right) = v^i \frac{\partial}{\partial x^i} - \hat{\Gamma}^j_{ia}(x,p)\, v^i \frac{\partial}{\partial x^j_a}, \tag{9.34}$$

where $\hat{\Gamma}^j_{ia}(x,p)$ are smooth functions of x and p. The minus sign is introduced for convenience.

These functions, however, cannot be arbitrary, since they must also satisfy the compatibility condition (9.7). It is not difficult to verify that this is the case if, and only if, the functions Γ^j_{ik} defined by:

$$\Gamma^j_{ik} := \hat{\Gamma}^j_{ia}(x,p)\, x_k^{-a}(p) \tag{9.35}$$

are independent of p along each fibre.

EXERCISE 9.2. Prove the above assertion.

We conclude that a linear connection is completely defined (on a given coordinate patch) by means of m^3 smooth functions. These functions are known as the *Christoffel symbols* of the connection.

9.3.3. Parallel Transport and the Covariant Derivative

Now that we are in possession of explicit coordinate expressions for the horizontal distribution, we can write explicitly the system of ODEs that effects the horizontal lift of a curve in $\mathcal{B}$. A solution of this system is a one-parameter family of frames being parallel-transported along the curve. Let the curve γ in the base manifold be given by:

$$x^i = x^i(t). \tag{9.36}$$

On this curve, the connection symbols are available as functions of t, by composition. The nontrivial part of the system of equations is given by:

$$\frac{dx^i_a(t)}{dt} = -\Gamma^i_{jk}(t)\, x^k_a(t)\, \frac{dx^j(t)}{dt}. \tag{9.37}$$

The local solution of this system with given initial condition (say, $x^i_a(0) = \bar{x}^i_a$) is the desired curve in $F(\mathcal{B})$, representing the parallel transport of the initial frame along the given curve.

Let now $\bar{v}$ be a vector in $T_{x^i(0)}\mathcal{B}$, that is, a vector at the initial "time" $t = 0$. We say that the curve $v = v(t)$ in $T\mathcal{B}$ is the parallel transport of $\bar{v}$ if it projects on γ, with $v(0) = \bar{v}$, and if the components of $v(t)$ in a parallel-transported frame along γ are constant.[5] For this definition to make sense, we must make sure that the constancy of the components is independent of the particular initial frame chosen. This, however, is a direct consequence of the fact that our linear connection is, by definition, consistent with the right action of the group.

EXERCISE 9.3. Prove the assertion just made.

To obtain the system of ODEs corresponding to the parallel transport of $\bar{v}$ along γ, we enforce the constancy conditions:

$$v^i(t)\, x_i^{-a}(t) = C^a, \tag{9.38}$$

where each C^a $(a = 1,\ldots,m)$ is a constant and where v^i denotes components in the coordinate basis. Differentiating this equation with respect to t and invoking Equation (9.37), we obtain:

$$\frac{dv^i}{dt} + \Gamma^i_{jk} \frac{dx^j}{dt}\, v^k = 0. \tag{9.39}$$

A vector field along γ satisfying this equation is said to be *covariantly constant*. For a given vector field $\mathbf{w}$ on $\mathcal{B}$, the expression on the left-hand side makes sense in a pointwise manner whenever a vector $\mathbf{u}$ is defined at a point (whereby the curve γ can be seen as a representative at $t = 0$). The expression:

$$\nabla_{\mathbf{u}}\mathbf{w} := \left(\frac{dw^i}{dt} + \Gamma^i_{jk}\, u^j\, w^k \right) \frac{\partial}{\partial x^i} \tag{9.40}$$

is called the *covariant derivative* of $\mathbf{v}$ in the direction of $\mathbf{u}$. From the treatment above, it can be seen that the covariant derivative is precisely the limit:

$$\nabla_{\mathbf{u}}\mathbf{w} = \lim_{t \to 0} \frac{\rho_{t,0}\mathbf{w} - \mathbf{w}(0)}{t}, \tag{9.41}$$

where $\rho(a,b)$ denotes the parallel transport along γ from $t = b$ to $t = a$.

EXERCISE 9.4. Taking the covariant derivatives of a vector field $\mathbf{w}$ at a point in the direction of each of the coordinate lines, we obtain a matrix with entries $\nabla_i w^j$, also denoted as $w^i_{;j}$. Show that these quantities behave as the components of a tensor of type $(1,1)$.

EXERCISE 9.5. Derive the formula for the parallel transport and the covariant derivative of a 1-form. Induce the formula for the covariant derivative of a tensor of any type.

---

[5] The same criterion for parallel transport that we are using for the tangent bundle can be used for any associated bundle.

9.3.4. Curvature and Torsion

To obtain an explicit equation for the curvature form Ω, we should start by elucidating the connection form ω on the basis of the connection symbols Γ. Given a vector $\mathbf{X} \in T_p F\mathcal{B}$, we know that its horizontal component $h(\mathbf{X})$ is given by:

$$h(\mathbf{X}) = \Gamma(\pi(p),p) \circ \pi_* (\mathbf{X}). \tag{9.42}$$

Its vertical component must, therefore, be given by:

$$v(\mathbf{X}) = \mathbf{X} - h(\mathbf{X}) = \mathbf{X} - \Gamma(\pi(p),p) \circ \pi_* (\mathbf{X}). \tag{9.43}$$

EXERCISE 9.6. Prove that the previous expression indeed delivers a vertical vector. [Hint: use Equation (9.33).]

Recall that the connection form ω assigns to $\mathbf{X}$ the vector in $\mathfrak{g}$ such that $v(\mathbf{X})$ belongs to its fundamental vector field. Let the coordinates of p be (x^i, x_a^i). The right action of $GL(m;\mathbb{R})$ is given by:

$$\left(R_g(p)\right)_a^i = x_b^i \, g_a^b, \tag{9.44}$$

where we have shown only the action on the fibre component and where g_a^b is the matrix corresponding to $g \in GL(m;\mathbb{R})$. Consequently, if $g(t)$ is a one-parameter subgroup represented by the vector:

$$\hat{g}_b^a = \left. \frac{dg_b^a(t)}{dt} \right|_{t=0}, \tag{9.45}$$

the value of the corresponding fundamental vector field at p is:

$$\tilde{g}_a^i = x_b^i \, \hat{g}_a^b. \tag{9.46}$$

The coordinate expression of Equation (9.43) is:

$$(v(\mathbf{X}))_a^i = X_a^i - h(\mathbf{X}) = \mathbf{X} - \Gamma(\pi(p),p) \circ \pi_* (\mathbf{X}). \tag{9.47}$$

Let the main part of the vector $\mathbf{X}$ be given by:

$$\mathbf{X} = v^i \frac{\partial}{\partial x^i} + X_a^i \frac{\partial}{\partial x_a^i}. \tag{9.48}$$

Then, Equation (9.47) delivers:

$$(v(\mathbf{X}))_a^i = X_a^i + \Gamma_{ik}^j \, v^i \, x_a^k. \tag{9.49}$$

According to Equation (9.46), the corresponding element of the Lie algebra is:

$$\hat{g}_a^b = \left(X_a^j + \Gamma_{ik}^j \, v^i \, x_a^k \right) x_j^{-b}. \tag{9.50}$$

Accordingly, the Lie-algebra valued connection form ω is given by:

$$\omega_a^b = \Gamma_{ik}^j \, x_a^k \, x_j^{-b} \, dx^i + x_j^{-b} \, dx_a^j. \tag{9.51}$$

The exterior derivative is given by:

$$d\omega_a^b = \frac{\partial \Gamma_{ik}^j}{\partial x^m} x_a^k x_j^{-b} \, dx^m \wedge dx^i$$

(9.52)

$$+ \Gamma_{ik}^j x_j^{-b} \, dx_a^k \wedge dx^i - \Gamma_{ik}^j x_a^k x_s^{-b} x_j^{-c} dx_c^s \wedge dx^i - x_j^{-c} x_s^{-b} \, dx_c^s \wedge dx_a^j.$$

A vector such as (9.48) has the following horizontal component:

$$h(\mathbf{X}) = v^i \frac{\partial}{\partial x^i} - \Gamma_{ik}^j x_a^k v^i \frac{\partial}{\partial x_a^j}.$$

(9.53)

With a similar notation, the horizontal component of another vector $\mathbf{Y}$ is given by:

$$h(\mathbf{Y}) = w^i \frac{\partial}{\partial x^i} - \Gamma_{ik}^j x_a^k w^i \frac{\partial}{\partial x_a^j}.$$

(9.54)

Consider now the following evaluations:

$$\langle dx^j \wedge dx^i \mid h(\mathbf{X}) \wedge h(\mathbf{Y}) \rangle = v^j w^i - v^i w^j,$$

(9.55)

$$\langle dx_a^k \wedge dx^i \mid h(\mathbf{X}) \wedge h(\mathbf{Y}) \rangle = -\Gamma_{rs}^k x_a^s \left(v^r w^i - v^i w^r \right),$$

(9.56)

and

$$\langle dx_c^j \wedge dx_a^s \mid h(\mathbf{X}) \wedge h(\mathbf{Y}) \rangle = -\Gamma_{rn}^j x_c^n \, \Gamma_{ik}^s x_a^k \left(v^r w^i - v^i w^r \right).$$

(9.57)

Putting all these results together, we obtain:

$$\langle \Omega \mid \mathbf{X} \wedge \mathbf{Y} \rangle = \langle \omega \mid h(\mathbf{X}) \wedge h(\mathbf{Y}) \rangle = x_a^k x_j^{-b} R_{kri}^j v^r w^i,$$

(9.58)

where

$$R_{kri}^j = \frac{\Gamma_{ik}^j}{\partial x^r} - \frac{\Gamma_{rk}^j}{\partial x^i} + \Gamma_{rh}^j \Gamma_{ik}^h - \Gamma_{ih}^j \Gamma_{rk}^h$$

(9.59)

is called the *curvature tensor* of the linear connection.

EXERCISE 9.7. Complete the missing steps in the derivation.

In analogy with the concept of curvature form, we define the *torsion form* of a connection as:

$$\Theta := D\theta.$$

(9.60)

Notice that the coupling with the connection is in the fact that the operator D is the exterior *covariant* derivative, which involves the horizontal projection. To understand the meaning of the torsion, consider a case in which the curvature vanishes. This means that there exists a *distant* (or curve independent) parallelism in the manifold $\mathcal{B}$. Thus, fixing a basis of the tangent space at any one point $x_0 \in \mathcal{B}$, a field of bases is uniquely determined at all other points. The question that the torsion tensor addresses is the following: Does a coordinate system exist such that these bases coincide at each point with its natural basis? [Question: Does the answer depend on the initial basis chosen?] An interesting example can be constructed in $\mathbb{R}^3$ as follows. Starting from the standard coordinate system, move up the x^3 axis and, while so doing, apply a linearly increasing rotation to the horizontal planes, imitating the action of a corkscrew. Thus, we obtain a system of (orthonormal) bases which are perfectly Cartesian plane by horizontal plane, but twisted with respect to each other as we ascend. These frames can be used to define a distant parallelism (two vectors are parallel if they have the same components in the local frame). It is not difficult to show (or to see intuitively) that there is no coordinate system that has these as natural bases (use, for example, Frobenius' theorem). This example explains the terminology of "torsion."

EXERCISE 9.8. Discuss how the above interpretation is supported by Proposition 4.3.

To obtain the coordinate expression of the torsion form, we start by calculating the exterior derivative of Equation (9.32) as:

$$d\theta^a = dx_i^{-a} \wedge dx^i = -x_j^{-a} x_i^{-b} dx_b^j \wedge dx^i. \tag{9.61}$$

Using Equation (9.56), we obtain:

$$\langle D\theta \mid \mathbf{X} \wedge \mathbf{Y} \rangle = \langle d\theta \mid h(\mathbf{X}) \wedge h(\mathbf{Y}) \rangle = x_j^{-a} \, T_{ri}^j \, v^r w^i, \tag{9.62}$$

where

$$T_{ri}^j := \Gamma_{ri}^j - \Gamma_{ir}^j \tag{9.63}$$

are the components of the *torsion tensor* of the connection.

EXERCISE 9.9. Suppose that a linear connection with vanishing curvature has been specified on the manifold $\mathcal{B}$, and let $\mathbf{e}_1, \ldots, \mathbf{e}_n$ be a field of parallel frames on the manifold. Show that the Christoffel symbols of the connection are given by the formula:

$$\Gamma_{kj}^i = -e_k^{-a} \frac{\partial e_a^i}{\partial x^j}, \tag{9.64}$$

where e_a^i are the components of the frame in the natural basis of a coordinate system $x^1, \ldots, x^m$. Verify that, in this case, the components of the torsion tensor are proportional to the components of the Lie brackets of corresponding pairs of vectors of the frames. Invoke the theorem of Frobenius to conclude that, unless the torsion vanishes identically, there is no coordinate system adapted to the moving parallel frame defining the curvature-free connection.

9.4. *G*-Connections

When describing the general geometrical setting for the theory of material inhomogeneities (in Chapter 8), we encountered the idea of a principal bundle whose structure group is the material symmetry group of the archetypal material point. Since, ultimately, all material symmetry groups are subgroups of $GL(3; \mathbb{R})$, it is appropriate to say that all our material principal bundles are *reductions* of the linear frame bundle to a subgroup. These reductions are known as *G-structures*.[6] A connection on a *G*-structure is called a *G-connection*.

9.4.1. Reduction of Principal Bundles

We have already introduced the notion of *principal-bundle morphism* in Chapter 7 as (i) a smooth fibre-preserving[7] map $f : C \rightarrow C'$ between the principal bundles (C, π_C, B, G, G) and (C', π'_C, B', G', G'); and (ii) a Lie-group homomorphism $\hat{f} : G \rightarrow G'$ such that (according to Equation (7.38)):

$$f(R_g p) = R'_{g'} f(p), \quad \forall p \in P, g \in G, \tag{9.65}$$

where $g' = \hat{f}(g)$, or, more elegantly:

$$f(pg) = f(p)\hat{f}(g). \tag{9.66}$$

If the map f is a diffeomorphism, the morphism is called a *principal-bundle isomorphism*.

Consider now the special case in which $B = B'$, namely, the two principal bundles have the same base manifold. Moreover, let the group G be a subgroup of G' and let the group homomorphism $\hat{f} : G \rightarrow G'$ be the inclusion map (assigning to each element in G the same element in G'). In this case, the principal-bundle morphism is called a *reduction* of the bundle C' to the bundle C. Accordingly, a principal bundle is said to be *reducible* to the subgroup G if there exists such a reduction.

EXERCISE 9.10. Show that a principal bundle is reducible to the trivial subgroup (namely, the group consisting of just the identity) if, and only if, it admits a global cross section. Equivalently, the principal bundle is trivializable (i.e., isomorphic to the product of the base manifold with the typical fibre).

Notice that each fibre of the reduced bundle can be seen as a submanifold of the corresponding fibre in the original bundle. The following proposition is important in understanding the notion of material principal bundles.

PROPOSITION 9.2. *A principal bundle (C', π', G') is reducible to a subgroup $G \subset G'$ if, and only if, it admits a trivializing covering (U_α, ψ_α) whose transition functions take values in G.*

PROOF. Although the proof involves some technicalities,[8] the basic idea can be explained intuitively. Assume first that the bundle (C', π', G') is reducible to G. The trick consists of showing that every trivializing covering of the reduced bundle (C, π, G) (whose transition functions, by definition, take values in G) can be extended

---

[6] An excellent reference for an introduction to this topic is Sternberg S (1983), *Lectures on Differential Geometry*, Chelsea. A detailed treatment can also be found in Fujimoto A (1972), *Theory of G-structures*, Publications of the Study Group of Geometry, Vol. I. Tokyo.

[7] Fibre preservation is a consequence of part ii of the definition. See Exercise 7.8.

[8] See Sternberg (1983), op. cit., p 296.

to a trivializing covering of the larger bundle. This follows from the fact that the right action is transitive. To see how this works, consider in the reduced bundle the trivializing pair $(\mathcal{U}_\alpha, \psi_\alpha)$ for a particular α. We want to define an extension ψ'_α of the function ψ_α to $\pi'^{-1}(\mathcal{U}_\alpha)$. Let $p' \in \pi'^{-1}(\mathcal{U}_\alpha)$. By transitivity of the right action, for any point $p \in \pi^{-1}(\{\pi'(p')\})$ there exists a (unique) $a \in \mathcal{G}'$ such that $p' = pa$. Let us define, accordingly, $\psi'(p') = \psi(p)a$. The problem with this definition is that it may be dependent on the particular choice of point p in the fibre. But it turns out not to be so. Indeed, let q be a different point on the same fibre and let b be the element of $\mathcal{G}'$ such that $p' = qb$. It follows that $p = qba^{-1}$. Recalling that the functions ψ_α themselves are consistent with the group action, we can write:

$$\psi'(p') = \psi(p)a = \psi(qba^{-1})a = \psi(q)b, \tag{9.67}$$

which shows that the definition of ψ' is indeed independent of the point p chosen in the fibre of the reduced bundle, and the map ψ' is a well-defined local trivialization of $\pi'^{-1}(\mathcal{U}_\alpha)$. We still need to check the transition functions $\phi'_{\alpha,\beta}$. In fact, because of the definition of ψ', these functions are identical with the corresponding transition functions $\psi_{\alpha,\beta}$ in the reduced bundle, which take values in $\mathcal{G}$. This completes the proof of necessity.

To prove sufficiency, assume that there exists a trivializing covering $(\mathcal{U}_\alpha, \psi'_\alpha)$ of $\mathcal{C}'$ whose transition functions happen to take values in the subgroup $\mathcal{G}$. In this case, we can use their restrictions as a trivializing covering of $\mathcal{C}$, since all the conditions for such a covering are satisfied.   □

A useful construction allows us to use a principal-bundle morphism $f : \mathcal{C} \to \mathcal{C}'$ to induce a connection in $\mathcal{C}'$ from a given connection in $\mathcal{C}$. Indeed, let H be a horizontal distribution in $\mathcal{C}$ compatible with the right action of $\mathcal{G}$. Let H_c denote the horizontal space at $c \in \mathcal{C}$. We define the (pushed-forward) horizontal space $H'_{c'}$ at $c' = f(c) \in \mathcal{C}'$ by:

$$H'_{c'} = f_* H_c. \tag{9.68}$$

For this formula to provide a viable connection in $\mathcal{C}'$ we need to verify three conditions, as follows: (i) Since the function f is not necessarily injective (one-to-one), we need to check that, if $f(c_1) = f(c_2)$, Equation (9.68) still delivers the same horizontal hyperplane. This can be verified by a direct calculation. (ii) The second condition is that the assignment (9.68) must be consistent with the right action of the group $\mathcal{G}'$. We know that, for the original connection in $\mathcal{C}$, the group consistency applies, viz.:

$$R_{a*} H_c = H_{ca} \quad \forall c \in \mathcal{C}, \ a \in \mathcal{G}. \tag{9.69}$$

Moreover, the principal-bundle morphism f entails a group homomorphism $\hat{f} : \mathcal{G} \to \mathcal{G}'$ satisfying Equation (9.65). This implies that:

$$f_*(H_{ca}) = f_*(R_{a*} H_c) = (f R_a)_* H_c = (R_{\hat{f}(a)} f)_* H_c = R_{\hat{f}_{a*}}[f_*(H_c)], \tag{9.70}$$

thus showing that the induced horizontal distribution satisfies the consistency condition wherever defined. (iii) The third condition to be satisfied is that the distribution must be defined over the whole of $\mathcal{C}'$, whereas we have so far defined it only over the image $f(\mathcal{C})$. But this extension can be easily achieved by using the right action of the group $\mathcal{G}'$.

The previous construction can be used, in particular, to induce a connection on $\mathcal{C}'$ from a connection over one of its reductions to a subgroup. In this spirit, we can say that a connection H' in $\mathcal{C}'$ is reducible to a subgroup $\mathcal{G}$ if there exists a connection H in the reduced bundle such that H' is induced by the reduction morphism.

9.4.2. G-structures

DEFINITION 9.1. A *G-structure* $\mathcal{B}_G$ over an m-dimensional manifold $\mathcal{B}$ is a reduction of the linear frame bundle $F(\mathcal{B})$ to a subgroup of $GL(m;\mathbb{R})$.

As a point of notation, it is sometimes convenient to include the symbol for the subgroup in the name of the G-structure. Thus, we may speak of an $\mathcal{H}$-structure if $\mathcal{H} \subset GL(m;\mathbb{R})$ is the subgroup to which $F(\mathcal{B})$ has been reduced. This policy is not always the most convenient, and we may use the term G-structure generically. A convenient, equivalent, way to look at a $\mathcal{G}$-structure over $\mathcal{B}$ is as a submanifold $\mathcal{B}_G$ of $F(\mathcal{B})$ such that, for all $p \in \mathcal{B}$ and any $g \in GL(m;\mathbb{R})$, we have that $pg \in \mathcal{B}_G$ if, and only if, $g \in \mathcal{G}$.

The notion of a G-structure, as so many concepts of modern mathematics, serves to unify the treatment of many entities which would otherwise appear completely unrelated, as will be demonstrated with a few examples below. It is important to remark, however, that, given a manifold and a subgroup of the corresponding general linear group, it is not necessarily true that a corresponding G-structure exists.

EXAMPLE 9.2. **Material G-structures:** The theory of material uniformity is one of the best engineering examples of the application of the theory of G-structures. Material uniformity is tantamount to the fact that the linear frame bundle is reducible to the material symmetry group of the archetype, as we have discussed in Chapter 8. Since we are considering only material isomorphisms that preserve orientation, we need only discuss subgroups of $GL^+(3;\mathbb{R})$ and, more particularly, of $SL^+(3;\mathbb{R})$. We distinguish the case in which the symmetry group of the archetype (to which we reduce the bundle of linear frames) is discrete from the case in which it is continuous. In the discrete case, let us assume that the symmetry group consists just of the group unit. In this case, we have a smooth global section or, in other words, a global distant parallelism. The associated connection, called a *material connection*, is uniquely determined and curvature free. For other discrete subgroups the existence of a global smooth cross section cannot be guaranteed and, in fact, counterexamples can be constructed. Nevertheless, the associated material connection is unique (and curvature free) as can be seen by considering that, given a curve in the base manifold and an initial point in a fibre, there can be only one horizontal lift (recall that the fibre consists of a finite number of points in this case). For continuous symmetry groups, the material connection is no longer unique. Consider the case of an isotropic solid, whereby the symmetry group is the proper orthogonal group. Each fibre of the G-structure consists then of a collection of frames differing from each other by a proper orthogonal transformation. Thus, we have a Riemannian metric on the base manifold. Conversely, if a Riemannian metric is given in the base manifold, we have a means to determine which are the orthogonal frames. That this can be done smoothly means that there exists a trivializing covering of the linear frame bundle with transition functions that take values in the orthogonal group. This reasoning shows that specifying a Riemannian metric and giving a reduction to a G-structure governed by the orthogonal group are one and the same thing. Notice that if the linear frame bundle is reducible to a subgroup of the orthogonal group, it is automatically reducible to the orthogonal group. This means that every solid body has a material Riemannian metric. But a given Riemannian metric would correspond to many different G-structures if the structure group is strictly smaller than the proper orthogonal group. Consider now the case of a fluid, whereby the symmetry group is the unimodular group $SL^+(3; \mathbb{R})$. By a reasoning similar to the preceding one, we conclude that the reduction in this case is tantamount to the specification of a (nowhere vanishing) volume form on the base manifold.

We emphasize the fact that the G-structure construct conveys geometrically all the information contained in the constitutive equation, at least as far as the material uniformity of the body is concerned. On the other hand, the material G-structure is not unique, in the sense that it depends on the choice of an archetype (namely, a frame at a body point). Each material G-structure, however, is conjugate to all others (see Box 9.1), as follows from our considerations in Section 8.3. It is remarkable that there exists another, more general, differential geometric object that encompasses all these conjugate G-structures under a single umbrella. This object, as we have learned in Chapter 8, is the *material groupoid*.

9.4.3. Local Flatness

A particularly simple example of a principal bundle is given by $F(\mathbb{R}^m)$, the bundle of frames of $\mathbb{R}^m$. Since $\mathbb{R}^m$ is endowed at each point with a standard natural basis, there exists a natural global parallelism on $\mathbb{R}^m$. In other words, $F(\mathbb{R}^m)$ admits smooth global constant sections. We will, accordingly, call $F(\mathbb{R}^m)$ the *standard frame bundle* (in m dimensions).

We have already pointed out that, given the frame bundle of an m-dimensional manifold and a subgroup $\mathcal{G}$ of the general linear group $GL(m; \mathbb{R})$, it is not necessarily true that the frame bundle is reducible to this subgroup. The standard frame bundle $F(\mathbb{R}^m)$, however, is reducible to *any* subgroup $\mathcal{G}$ of $GL(m; \mathbb{R})$. To see why this is the case, we need only consider at each point of $\mathbb{R}^m$ the standard natural basis and construct the set of all bases obtained from it by the right action of all elements of $\mathcal{G}$. The collection of bases thus obtained can be regarded as a reduction of $F(\mathbb{R}^m)$ to the subgroup $\mathcal{G}$, namely, as a $\mathcal{G}$-structure, since (by construction) it is stable under the action of this subgroup. Note that the availability of a distinguished global section is

BOX 9.1. **Conjugate G-Structures**

Two G-structures, $\mathcal{B}_{\mathcal{G}_1}$ and $\mathcal{B}_{\mathcal{G}_2}$, with structure groups $\mathcal{G}_1$ and $\mathcal{G}_2$, respectively, over the same m-dimensional base manifold $\mathcal{B}$, are said to be *conjugate* if there exists a fixed element $g \in GL(m; \mathbb{R})$ such that:

$$\mathcal{B}_{\mathcal{G}_2} = R_g(\mathcal{B}_{\mathcal{G}_1}), \tag{9.71}$$

where R_g is the right action of g on the frame bundle $F(\mathcal{B})$.

The reason for this terminology is the following: Let two elements, p_1 and p_2, lying on the same fibre of $\mathcal{B}_{\mathcal{G}_1}$ be related by:

$$p_2 = p_1\, a, \tag{9.72}$$

where $a \in \mathcal{G}_1$ and where we have indicated the right action by apposition (i.e., $p_1\, a = R_a(p_1)$). The images of these two points, lying on the same fibre of $\mathcal{B}_{\mathcal{G}_2}$, are, respectively,

$$p_1' = p_1\, g, \tag{9.73}$$

and

$$p_2' = p_2\, g. \tag{9.74}$$

It follows, then, that:

$$p_2' = p_1'\, g^{-1}\, a\, g. \tag{9.75}$$

Refining this argument slightly, it is not difficult to prove that in fact the two structure groups *must be conjugate of each other*. More precisely:

$$\mathcal{G}_2 = g^{-1}\, \mathcal{G}_1\, g. \tag{9.76}$$

The converse, however, is not true: Two G-structures (over the same manifold) having conjugate structure groups are not necessarily conjugate. For example, two different global sections of the frame bundle of a trivial manifold can be regarded as two different G-structures with the same (trivial) structure group. Nevertheless, unless these two sections happen to differ by the right action of a *constant* element of the general linear group, the two G-structures are clearly not conjugate.

crucial for this procedure to work. We will call the $\mathcal{G}$-structure obtained in this way the *standard (flat) $\mathcal{G}$-structure* (in m dimensions).

Having thus established the existence of these standard $\mathcal{G}$-structures, we are in a position of inquiring whether or not a given nonstandard $\mathcal{G}$-structure "looks" (at least locally) like its standard counterpart. This question is a particular case of the more general question of *local equivalence* between G-structures, which we now tackle. We commence by remarking that, given two m-dimensional manifolds, $\mathcal{B}$ and $\mathcal{B}'$, a diffeomorphism:

$$\phi : \mathcal{B} \longrightarrow \mathcal{B}', \tag{9.77}$$

maps, through its tangent map ϕ_*, every frame $\{e_i\}$ $(i = 1, \ldots, m)$ at each point $b \in \mathcal{B}$ to a frame $\{e_i'\} = \{\phi_*(e_i)\}$ $(i = 1, \ldots, m)$ at the image point $b' = \phi(b)$. Now let $\mathcal{C}$ be a $\mathcal{G}$-structure over $\mathcal{B}$. Clearly, the procedure just described generates at each point $b' \in \mathcal{B}'$ a collection of frames. The union $\mathcal{C}'$ of all these image bases can be considered as a $\mathcal{G}$-structure over $\mathcal{B}'$ having *the same structure group* $\mathcal{G}$ as the original $\mathcal{G}$-structure. The $\mathcal{G}$-structure thus constructed is said to be *induced* or *dragged* by the diffeomorphism ϕ between the base manifolds. The induced $\mathcal{G}$-structure is denoted by $\mathcal{C}' = \phi_*(\mathcal{C})$. Clearly, this is a particular example of a principal-bundle morphism.

EXERCISE 9.11. Verify that the set of bases $\phi_*(\mathcal{C})$ is indeed a $\mathcal{G}$-structure and that its structure group is indeed the original $\mathcal{G}$. In other words, verify that the set is stable under the right action of this group.

Given two $\mathcal{G}$-structures, $\mathcal{C}$ and $\mathcal{C}'$, defined, respectively, over the base manifolds $\mathcal{B}$ and $\mathcal{B}'$, we say that they are (globally) *equivalent* if there exists a diffeomorphism $\phi : \mathcal{B} \to \mathcal{B}'$ such that $\mathcal{C}' = \phi_*(\mathcal{C})$. Two $\mathcal{G}$-structures are said to be *locally equivalent* at the points $b \in \mathcal{B}$ and $b' \in \mathcal{B}'$ if there exists an open neighbourhood $\mathcal{U}$ of b and a diffeomorphism $\phi : \mathcal{B} \to \mathcal{B}'$ with $\phi(b) = b'$ such that:

$$\mathcal{C}'|_{\phi(\mathcal{U})} = \phi_*(\mathcal{C}|_{\mathcal{U}}), \tag{9.78}$$

where $\mathcal{C}|_{\mathcal{U}}$ denotes the subbundle of $\mathcal{C}$ obtained by restricting the base manifold to $\mathcal{U}$.

DEFINITION 9.2. A $\mathcal{G}$-structure is called *locally flat* if it is locally equivalent to the standard (flat) $\mathcal{G}$-structure.

9.5. Riemannian Connections

A *Riemannian connection* is a linear connection on a Riemannian manifold. The most important basic result for Riemannian connections is contained in the following theorem:

THEOREM 9.3. *On a Riemannian manifold there exists a unique linear connection with vanishing torsion such that the covariant derivative of the metric vanishes identically.*

We omit the proof. The curvature tensor associated with this special connection is called the *Riemann-Christoffel* curvature tensor. A Riemannian manifold is said to be *locally flat* if, for each point, a coordinate chart can be found whereby the metric tensor components everywhere in the chart reduce to the identity matrix. It can be shown that local flatness is equivalent to the identical vanishing of the Riemann-Christoffel curvature tensor.

9.6. Material Homogeneity

9.6.1. Uniformity and Homogeneity

We have interpreted the notion of material uniformity as the formalization of the statement: "All points of the body are made of the same material." Isn't that homogeneity? It is certainly a precondition of homogeneity, but when engineers tell you that a body is homogeneous they certainly mean more than just that. They mean, tacitly perhaps, that the body can be put in a configuration such that a mere translation of the neighbourhood of any point to that of any other point will do as a material isomorphism. Thus, in that particular configuration, all the points are indistinguishable from each other as far as the constitutive equation is concerned. If such a configuration exists and if it were to be used as a reference configuration, the constitutive law in a Cartesian coordinate system would become independent of position. We call such a configuration a (globally) *homogeneous configuration*.

The formalization of this notion of homogeneity is not very difficult. We will motivate the derivation by working in coordinates and concentrating on a first-grade material. Notice that, using the notation of Equation (8.30), the uniformity condition in terms of tensor components reads:

$$\psi(F^i_I, X^J) = J_P^{-1} \bar{\psi}(F^i_I P^I_\alpha(X^J)). \tag{9.79}$$

The meaning of the entries P^I_α is the following: If we choose a fixed basis $\{\mathbf{E}_\alpha\}$ at the archetype, the implant $\mathbf{P}(X)$ induces a moving frame $\{\mathbf{e}_\alpha\}(X)$ in the body. Such a field of bases is called a *uniformity field*. The quantities P^I_α represent the components of the base vectors $\mathbf{e}_\alpha$ of this uniformity field in the coordinate system X^J. Suppose now that there exists a uniformity field and a coordinate chart such that the natural basis of the coordinate system happens to coincide, at each point, with the uniformity basis at that point. We say that the uniformity basis in this case is *integrable*.[9] Then, clearly, the components P^I_α in this coordinate system become:

$$P^I_\alpha = \delta^I_\alpha, \tag{9.80}$$

whence it follows that the constitutive Equation (9.79) in this coordinate system is independent of X. Since we can regard a coordinate system as a reference configuration, it follows that the engineering notion of homogeneity is satisfied in this case. The reverse is clearly true. Recall, however, that both a coordinate system and a smooth uniformity field are, in general, available in open sets that may not be as large as the whole body. If for every point of the body there exists a neighbourhood within which an integrable uniformity basis field can be found, we say that the uniform body is *locally homogeneous*. A uniform body is *globally homogeneous*, or simply *homogeneous*, if the previous condition can be satisfied on a neighbourhood consisting of the whole body.

---

[9] The terminology *holonomic* can also be used, but we reserve this term for a different concept to be applied to materials of higher grade.

REMARK 9.1. To illustrate the difference between local and global homogeneity, consider a long cylindrical uniform solid body in a natural (stress-free) configuration. Assume that the material is fully isotropic. This body is clearly homogeneous and the given cylindrical configuration is a homogeneous configuration. If this body is bent into a torus (by welding the two ends together) we obtain a new body $\mathcal{B}$ which, clearly, is not stress-free any longer. Any simply connected chunk of this body can be "straightened" and thus brought back locally into a homogeneous configuration, but not so the whole body. In other words, $\mathcal{B}$ is only locally homogeneous.

A uniform body which is not locally homogeneous is said to be *inhomogeneous* or to contain a distribution of inhomogeneities. Historically, the theory of continuous distributions of inhomogeneities in material bodies stemmed from various attempts at generalizing the geometric ideas of the theory of isolated dislocations arising in crystalline solids. As should be clear from the previous presentation, however, the theory of uniform and inhomogeneous bodies has a life of its own, whose mathematical and physical apparatus is completely and exclusively based upon the constitutive law of a continuous medium.

REMARK 9.2. A simple example of an inhomogeneous elastic body is provided by a uniform thick spherical cap made of a transversely isotropic solid material whose axis of transverse isotropy $\mathbf{e}_1(X)$ is aligned at each point X with the local radial direction of the sphere. If we assume that in this configuration the material is stress free, the material isomorphisms between any two points consist of rotations that bring the corresponding axes of transverse isotropy into coincidence. In a (locally) homogeneous configuration, we should have that the field $\mathbf{e}_1(X)$ becomes a Euclidean-parallel field over some open subbody. This can be done, but only at the expense of stretching the points within each spherical layer by variable amounts (recall that no portion of a spherical surface can be mapped isometrically into a plane). Thus, any configuration in which the axes of transverse isotropy become parallel in some open neighbourhood, will necessarily contain points in that neighbourhood that are differently stretched. In other words, no homogeneous configuration can exist, even locally.

9.6.2. Homogeneity in Terms of a Material Connection

Given a uniformity field on an open set $\mathcal{U}$, how can we ascertain whether or not it is integrable? In other words, given a uniformity field expressed in terms of some coordinate system X^I, how can we know whether or not there exists a local coordinate system such that the given uniformity bases are adapted to it? To obtain a necessary condition for this question of *integrability*, let us assume that such a coordinate system, say Y^M, does exist and let us write the change of coordinates to our system X^I by means of three smooth and smoothly invertible functions:

$$X^I = X^I(Y^M) \quad (I, M = 1, 2, 3). \tag{9.81}$$

Any given vector field $\mathbf{w}$ can be expressed in terms of local components in either coordinate basis, say w^I and $\hat{w}^M$, respectively, for the coordinate systems X^I and Y^M. These components are related by the formula:

$$\hat{w}^M = w^I \frac{\partial Y^M}{\partial X^I}. \tag{9.82}$$

When it comes to each of the three vectors $\mathbf{p}_\alpha$ ($\alpha = 1,2,3$) of the given uniformity field, with components P^I_α in the X^I-coordinate system, their components in the Y^M-coordinate system are, according to our starting assumption, given by δ^M_α (since they coincide with the natural basis of this system). By Equation (9.82), therefore, we must have:

$$\delta^M_\alpha = P^I_\alpha \frac{\partial Y^M}{\partial X^I}. \tag{9.83}$$

Taking the partial derivative of this equation with respect to X^J, we obtain:

$$0 = P^I_\alpha \frac{\partial^2 Y^M}{\partial X^J \, \partial X^I} + \frac{\partial P^I_\alpha}{\partial X^J} \frac{\partial Y^M}{\partial X^I}, \tag{9.84}$$

or, multiplying through by $P^{-\alpha}_K$:

$$0 = \frac{\partial^2 Y^M}{\partial X^J \, \partial X^K} + P^{-\alpha}_K \frac{\partial P^I_\alpha}{\partial X^J} \frac{\partial Y^M}{\partial X^I}. \tag{9.85}$$

Using the expression (9.64) for the Christoffel symbols of the given material parallelism, we obtain:

$$\frac{\partial^2 Y^M}{\partial X^J \, \partial X^K} = \Gamma^I_{KJ} \frac{\partial Y^M}{\partial X^I}. \tag{9.86}$$

But, by the equality of mixed partial derivatives, the left-hand side is symmetric, implying that so must the Christoffel symbols be, namely:

$$\Gamma^I_{KJ} = \Gamma^I_{JK}, \tag{9.87}$$

throughout $\mathcal{U}$. We have only shown that this is a necessary condition for integrability, but it can be proven that under certain restrictions (on the connectedness and shape of the domain $\mathcal{U}$) the symmetry of the Christoffel symbols is also a sufficient condition for the existence of an adapted coordinate system. We conclude that the body is locally homogeneous if, and only if, there exists for each point a neighbourhood and a (local) material parallelism whose Christoffel symbols are symmetric.

A more geometric way to view the local homogeneity condition just discovered is to recall that the *torsion* of a linear connection is measured by the tensor with coordinate components given by:

$$T^I_{JK} = \Gamma^I_{JK} - \Gamma^I_{KJ}, \tag{9.88}$$

as we have obtained before in Equation (9.63). It can be shown, indeed, that although the Christoffel symbols are not the components of a tensor, their skew-symmetrized components (9.88) are. For a body to be locally homogeneous, therefore, it will be required that for each point there exists a neighbourhood in which a (local) material parallelism can be defined with vanishing torsion.

We have already seen that, if the symmetry group of the uniform body is discrete, the material connection is unique. This means that for such materials the question of homogeneity is settled once and for all by checking the torsion of the unique material connection. Equivalently, we may say that these geometric quantities are a true measure of the presence of inhomogeneities. For materials with a continuous symmetry group, on the other hand, the fact that a given material connection has a nonvanishing torsion is not necessarily an indication of inhomogeneity. Indeed, the nonvanishing of the torsion may be due to an unhappy choice of material connection. For materials with continuous symmetry groups, therefore, the criteria for homogeneity are mathematically more sophisticated, as we shall see.

REMARK 9.3. An important remark needs to be made regarding the choice of an archetype, that is, an archetypal material point and a basis therein. Could it happen that perhaps the frame fields induced by a certain archetype satisfy the homogeneity criteria, whereas the frame fields generated by a different archetype and frame do not? The answer in the case of simple materials is negative. Indeed, if a coordinate system exists such that its natural basis is a uniformity field of bases for some archetype, then any linear change of coordinates immediately provides a new coordinate system whose natural basis corresponds to a linear transformation of the archetype. The situation is different in the case of higher-grade materials.

9.6.3. Homogeneity in Terms of a Material G-structure

Since we have repeatedly said that a genuine representation of the inhomogeneity of a material body is not given by the (possibly nonunique) local material connections but rather by any one of its conjugate material G-structures, it is incumbent upon us now to explicitly show what feature of a material G-structure corresponds exactly to the notion of local homogeneity. To this effect, we recall Definition 9.2, according to which a G-structure is flat if it is equivalent to the corresponding standard G-structure. The equivalence is ascertained by means of a diffeomorphism ϕ from the base manifold $\mathcal{B}$ onto a subset of $\mathbb{R}^3$, as shown in Equation (9.78). From the material point of view, the diffeomorphisms $\phi : \mathcal{B} \to \mathbb{R}^3$ can be seen as changes of reference configuration of the body $\mathcal{B}$. We conclude, therefore, that local flatness of a material $\mathcal{G}$-structure means that a mere change of reference configuration renders the material isomorphisms (within an open neighbourhood) Euclidean parallelisms, which is precisely what local homogeneity is intended to convey! So, local flatness of a material $\mathcal{G}$-structure is the precise mathematical expression of local homogeneity.

REMARK 9.4. Although the material $\mathcal{G}$-structure is not unique (depending as it does on the choice of archetype), local flatness is independent of the particular $\mathcal{G}$-structure chosen. To prove this assertion, we recall that all the possible material $\mathcal{G}$-structures are conjugate of each other. Suppose that one of the material

$\mathcal{G}$-structures is locally flat. Any other material $\mathcal{G}$-structure (see Box 9.1) can, therefore, be obtained from the given one through the right action R_g of a fixed element g of $GL(3;\mathbb{R})$. This element can be represented as a 3×3 matrix, which we denote by $[g]$. We now construct the following affine transformation of $\mathbb{R}^3$:

$$\psi : \mathbb{R}^3 \longrightarrow \mathbb{R}^3$$

$$\{x\} \mapsto \{a\} + [g]\{x\}, \tag{9.89}$$

where $\{x\} = \langle x^1, x^2, x^3 \rangle^T$ is the generic position vector in $\mathbb{R}3$, and $\{a\} = \langle a^1, a^2, a^3 \rangle^T$ is a fixed vector. Let ϕ denote a local diffeomorphism (such as in Equation (9.78)) used in establishing the local flatness of the original $\mathcal{G}$-structure. Then, the map $\psi^{-1} \circ \phi$ can be used in establishing the local flatness of the new $\mathcal{G}$-structure, as it can be verified by a direct calculation. A quick symbolic way to carry out this calculation is the following: Let $\{\mathbf{f}\}$ be a frame belonging to the original $\mathcal{G}$-structure. Its counterpart in the new (conjugate) $\mathcal{G}$-structure is then $R_g\{\mathbf{f}\}$. This frame is dragged by $\psi^{-1} \circ \phi$ according to:

$$(\psi^{-1} \circ \phi)_*(R_g\{\mathbf{f}\}) = (\psi^{-1})_* \circ \phi_* \circ R_g(\{\mathbf{f}\}) = L_{g^{-1}} \circ R_g(\phi(\{\mathbf{f}\})). \tag{9.90}$$

This equation shows that the previous standard $\mathcal{G}$-structure moves (always within the frame bundle of $\mathbb{R}^3$) to the standard $\mathcal{G}$-structure corresponding to the new (conjugate) group, as desired.[10]

To connect this characterization of local homogeneity with the coordinate-based one presented in the previous section, we need only note that a diffeomorphism ϕ of an open neighbourhood in $\mathcal{B}$ with an open neighbourhood in $\mathbb{R}^3$ can be regarded as a coordinate chart $x^i = \phi^i(b)$. Moreover, the inverse image by ϕ of the standard basis in $\mathbb{R}^3$ provides, at each point $b \in \mathcal{B}$, the natural basis $\{\frac{\partial}{\partial x^i}\}$ associated with this coordinate chart. In other words, a $\mathcal{G}$-structure is locally flat if, and only if, there exists an atlas $\{\mathcal{U}_\alpha, \phi_\alpha\}$ such that its natural frames belong to the $\mathcal{G}$-structure.[11]

9.6.4. Homogeneity in Terms of the Material Groupoid

As we know, all the conjugate material $\mathcal{G}$-structures of a material body $\mathcal{B}$ are implicitly contained in the material groupoid $\mathcal{Z}$. Now, this groupoid is necessarily a subgroupoid[12] of the tangent groupoid $\mathcal{T}(\mathcal{B})$. We recall that the tangent groupoid of a differentiable manifold $\mathcal{B}$ is obtained by considering, for each pair of points $a, b \in \mathcal{B}$, the collection of all the nonsingular linear maps between the tangent spaces $T_a\mathcal{B}$ and $T_b\mathcal{B}$. The material groupoid $\mathcal{Z}$ is obtained by retaining only those linear maps that represent material isomorphisms.

---

[10] Notice that the corresponding standard $\mathcal{G}$-structures are not conjugate of each other, since, by definition, they all share the standard (unit) section.

[11] For this to be the case, according to Frobenius' theorem, we need the Lie brackets of pairs of base vectors to vanish identically, which is precisely the same as the vanishing of the torsion of the local material connections induced by these bases in their respective neighbourhoods.

[12] A subset of a groupoid is called a subgroupoid if it is a groupoid with respect to the law of composition of the original groupoid.

Let $\mathcal{Z}$ be a transitive Lie groupoid with base manifold $\mathcal{B}$. Given an atlas of $\mathcal{B}$, every chart (via the natural basis of the coordinate system) induces a local parallelism. In other words, for every pair of points in the domain of the chart there exists a distinguished (chart-related) element of the tangent groupoid $\mathcal{T}(\mathcal{B})$. We say that the groupoid $\mathcal{Z}$ is *locally flat* if there exists an atlas such that, for every pair of points in the domain of every chart of the atlas, this distinguished chart-induced element of $\mathcal{T}(\mathcal{B})$ belongs to $\mathcal{Z}$. We conclude, therefore, that a material body $\mathcal{B}$ is locally homogeneous if, and only if, its material groupoid is locally flat.

9.7. Homogeneity Criteria

9.7.1. Solids

We recall that, by definition, an elastic material point is a *solid point* if its material symmetry group is a conjugate of a subgroup of the orthogonal group $\mathcal{O}$. Solid points, moreover, usually possess *natural states*, that is, configurations in which they are free of stress. Given a natural state of a solid point, all other configurations obtained by applying to the given natural state an arbitrary orthogonal transformation are also natural states. The symmetry group of a solid point in any natural state is a subgroup of the orthogonal group. An elastic body is said to be a *solid body* if all its points are solid points. In a uniform solid body it is always possible and often convenient to adopt an archetype that is in a natural state. For simplicity, we will assume that the natural state is unique to within an orthogonal transformation.

Consider now a uniform solid body $\mathcal{B}$ in which a natural-state archetype has been chosen with symmetry group $\bar{\mathcal{G}} \subset \mathcal{O}$. Let $\mathbf{P}(X)$ be a uniformity field (i.e., a field of implants of the archetype over the body). We claim that the standard (Cartesian) inner product at the archetype induces a unique inner product in each of the tangent spaces $T_X\mathcal{B}$. In other words, we claim that a uniform solid body is automatically endowed with a specific materially induced *intrinsic Riemannian metric structure* $\mathbf{g}$. Let $\mathbf{u}, \mathbf{v} \in T_X\mathcal{B}$ be two vectors at a point $X \in \mathcal{B}$. We define their inner product $*$ by the following formula:

$$\mathbf{u} * \mathbf{v} \equiv (\mathbf{P}^{-1}\mathbf{u}) \cdot (\mathbf{P}^{-1}\mathbf{v}), \qquad (9.91)$$

where $\cdot$ denotes the ordinary Cartesian inner product at the archetype. We need to show that this new inner product is independent both of the particular uniformity field chosen and of the particular natural state of the archetype. In each case, the change in the uniformity field consists of a multiplication of $\mathbf{P}$ to the right by a member of the archetypal symmetry group (for the degree of freedom permitted by the material symmetry) or a member of the general linear group (for the degree of freedom permitted by an arbitrary change of archetype). Since we are only considering natural-state archetypes, the latter degree of freedom reduces to a multiplication by an arbitrary member of the orthogonal group. Moreover, since the body is an elastic solid, the former degree of freedom is a subgroup of the orthogonal group. But the right-hand side of Equation (9.91) is invariant under multiplication of $\mathbf{P}$ to the right by any member of the orthogonal group, which proves our claim.

REMARK 9.5. The preceding argument is, in fact, equivalent to the fact that an $\mathcal{O}$-structure is tantamount to a Riemannian metric, as demonstrated already in Example 9.2.

The material Riemannian metric **g** will, in general, have a nonvanishing Riemann-Christoffel curvature tensor R_g. The special case in which $R_g \equiv 0$ deserves further consideration. In this case, every body point has an open neighbourhood for which an appropriate change of configuration renders the metric **g** equal to the Cartesian identity. In other words, in this configuration every material frame (as induced by all possible transplants of the archetypal frame) is an orthonormal frame in the Cartesian sense. From the physical point of view, we have that the neighbourhood in question can be brought to a configuration in which every point in the neighbourhood is in a natural state. This special case of inhomogeneity is called (local) *contorted aelotropy*.[13] Since, clearly, if the body enjoys local contorted aelotropy the curvature tensor of the intrinsic material metric vanishes, we have that:

$$R_g = 0 \Leftrightarrow \text{Contorted aelotropy.} \tag{9.92}$$

Assume now that the body is isotropic, namely, $\bar{\mathcal{G}} = \mathcal{O}$. Then, in any of the special configurations of contorted aelotropy, we can make use of the material symmetry to give a further rotation to the material frames so as to render them all parallel to each other (in the Euclidean sense). We have just proved that for an isotropic solid:

$$R_g = 0 \Leftrightarrow \text{Local homogeneity.} \tag{9.93}$$

REMARK 9.6. The case of the intrinsic metric just presented is a particular case of a *characteristic object*.[14]

At this point it is worth recalling that for solid bodies with a discrete symmetry group $\bar{\mathcal{G}}$, since the local material connections are uniquely defined, the criterion of local homogeneity simply boils down to the vanishing of the torsion thereof. Besides the case of full isotropy already discussed, the only other possible continuous symmetry group of a solid point consists of *transverse isotropy*, whereby the symmetry group is (a conjugate of) the group of rotations about a fixed unit vector **e**. Choosing a natural-state archetype, it is clear that the condition $R_g = 0$ is necessary, but no longer sufficient, to ensure local homogeneity. The unit vector **e** at the archetype gives rise, via the implants $\mathbf{P}(X)$, to a smooth vector field in the neighbourhood

---

[13] This terminology is due to Noll W (1967), Materially Uniform Bodies with Inhomogenenities, *Archive for Rational Mechanics and Analysis* **27**, 1-32. A somewhat more general case, in which the archetype is not necessarily in a natural state, has been referred to as a body endowed with states of *constant strain* (see Epstein M (1987), A Question of Constant Strain, *Journal of Elasticity* **17**, 23–34). The common feature is that the body can be brought locally into configurations in which all the material isomorphisms are orthogonal transformations.

[14] For a detailed treatment of this concept and its relevance to the problem of finding integrability conditions of G-structures, see Elżanowski M, Epstein M, Śniatycki J (1990), G-structures and Material Homogeneity, *Journal of Elasticity* **23**, 167–180. See also Chapter 5 of Wang C-C, Truesdell C (1973), *Introduction to Rational Elasticity*, Noordhoff. For the related concept of a G-structure generated by a tensor, see Fujimoto (1972), op. cit.

of each point X. If the covariant derivative of this vector field (with respect to the symmetric metric connection generated by $\mathbf{g}$) vanishes identically over this neighbourhood, at a configuration of contorted aelotropy, we will have that these vectors become actually parallel (in the Euclidean sense). Using the material symmetry at hand so as to give a further rotation about this axis at each point, we can obtain a field of parallel material frames, so that the body is locally homogeneous. The converse is clearly true. We have shown that for a transversely isotropic body:

$$R_g = 0, \nabla_g \mathbf{e} = 0 \Leftrightarrow \text{Local homogeneity.} \tag{9.94}$$

9.7.2. Fluids

An *elastic fluid point* is an elastic material point whose symmetry group is the unimodular group, namely, the group of all matrices with unit determinant. This group is the same in all configurations, since a conjugation of the unimodular group by any nonsingular matrix delivers again the unimodular group. For physical reasons (e.g., cavitation), elastic fluid points do not have natural (i.e., stress-free) states. An elastic body is said to be an *elastic fluid* if all its points are elastic fluid points. It is not difficult to prove that the constitutive law of a fluid point entails only a dependence on the determinant of the deformation gradient. Let κ_0 be a reference configuration of a uniform elastic fluid body $\mathcal{B}$ and let $\mathbf{P}(X)$ denote a field of implants from some archetype. The constitutive law of the body relative to this reference configuration must be of the form:

$$\psi_0(\mathbf{F}, X) = \bar{\psi}(\det(\mathbf{F}\mathbf{P}(X))), \tag{9.95}$$

where, as usual, we have denoted by $\bar{\psi}$ the constitutive law of the archetype. Let κ_1 denote another reference configuration and let:

$$\lambda = \kappa_1 \circ \kappa_0^{-1} : \kappa_0(\mathcal{B}) \longrightarrow \kappa_1(\mathcal{B}) \tag{9.96}$$

denote the corresponding change of reference configuration. This map λ is necessarily a diffeomorphism and the corresponding Jacobian determinant is a smooth function of position. The constitutive law of the body relative to the reference configuration κ_1 is given by:

$$\psi_1(\mathbf{F}, \lambda(X)) = \psi_0(\mathbf{F}\lambda_*, X) = \bar{\psi}(\det(\mathbf{F}\lambda_* \mathbf{P}(X))), \tag{9.97}$$

where λ_* denotes the derivative map of λ at $\mathbf{X}$. Obviously, the body is homogeneous if, and only if, a change of reference configuration, λ, can be found such that:

$$\det(\lambda_* \mathbf{P}(X)) = \text{constant.} \tag{9.98}$$

Moser's Lemma[15] guarantees that this can always be done, at least locally. We conclude, therefore, that for an elastic fluid:

$$\text{Uniformity} \Rightarrow \text{Local homogeneity.} \tag{9.99}$$

---

[15] Moser J (1965), On the Volume Elements of a Manifold, *Transactions of the American Mathematical Society* **120**, 286–294.

9.7.3. Fluid Crystals

An elastic material point that is not a solid is, in Wang's terminology,[16] a *fluid crystal point*. A body all of whose points are fluid crystal points is called a *fluid crystal* body. As a consequence of a theorem in group theory,[17] a fluid crystal cannot be isotropic, unless it is a fluid point. In other words, the symmetry group of a proper fluid crystal point (one that is neither a solid nor a fluid) cannot contain the orthogonal group (or a conjugate thereof).

As an example of a fluid crystal point, consider a material with a preferred direction. Its symmetry group consists of all unimodular transformations that leave that direction invariant (i.e., all unimodular matrices with that direction as an eigenvector, with an arbitrary eigenvalue). Given a uniform fluid crystal body of this type in some reference configuration, we can always find a change of reference configuration that renders the field of preferred directions parallel in a neighbourhood of a given point. After this has been achieved, a further unimodular transformation keeping this field of directions intact can be used to bring all points in the neighbourhood to the same state. We conclude, therefore, that for this type of fluid crystal the implication (9.99) holds. In other words, every uniform fluid crystal of this type is automatically locally homogeneous.

A different type of fluid crystals of interest consists of material points with a preferred plane. The symmetry group in this case consists of all unimodular transformations having that plane as an eigenspace. Attaching to each point the preferred plane, we obtain a 2-dimensional distribution over the body manifold. It can be shown[18] that a uniform body of this type is locally homogeneous if, and only if, this distribution is involutive.[19]

---

[16] Wang C-C (1967), On the Geometric Structure of Simple Bodies, a Mathematical Foundation for the Theory of Continuous Distributions of Dislocations, *Archive for Rational Mechanics and Analysis* **27**, 33–94.

[17] See Noll W (1965), Proof of the Maximality of the Orthogonal Group in the Unimodular Group, *Archive for Rational Mechanics and Analysis* **18**, 100–102.

[18] See Elżanowski, Epstein Śniatycki (1990), op. cit.

[19] A detailed study of the local homogeneity conditions for all possible types of fluid crystals can be found in Marín D, de León M (2004), Classification of Material G-structures, *Mediterranean Journal of Mathematics* **1**, 375–416.

A Primer in Continuum Mechanics

The principal aim of this book is to emphasize the geometric structure of Continuum Mechanics, but a reader not familiar with the by now standard presentation of this discipline is likely to miss the punch line. In part to avoid this unintended situation and in part to have a basic conceptual and terminological framework for the rest of the book, in this Appendix we provide a concise presentation of the subject as it can be found in more or less standard elementary textbooks. The level of mathematical sophistication is kept as low as possible: vectors are arrows, Pythagoras reigns supreme, everything is nice and smooth.

A.1. Bodies and Configurations

The passage from the classical mechanics of finite systems of particles and rigid bodies to the mechanics of deformable continua is not a trivial one. Already at the beginning of the subject we find ourselves confronted with the problem of defining the main concept: the body, or material continuum. Is it merely an infinite collection of particles? Because we must, at the very least, be able to define *fields* (temperature, velocity, and so on) over this entity, it is clear that we need a rigorous definition. This need is all the more pressing since, from our experience with centuries-old particular theories (hydrodynamics, linear elasticity, and so on), we know that soon enough temporal and spatial derivatives of these fields will enter the scene. Leaving aside the definition of the technical concept (namely, a *(trivial) differentiable manifold*), we will content ourselves for now by saying that a material body B is something that looks very much like a 3-dimensional connected open set. The shape itself is not part of the bargain, only the connectivity and continuity. The fact that the set is open (like a peeled apple) is part of the bargain, at least in the elementary treatment.

The material body, whose points are called *material particles*, does not live permanently in the 3-dimensional space of our physical experience, but dwells in some Platonic world of similar entities, devoid of any particular ruler or clock. Nevertheless, it does manifest itself through visits to this world, a world equipped with rulers, protractors and clocks. Any such manifestation is called a *configuration*. If we denote our Euclidean space by $\mathbb{E}^3$, a configuration κ is a map:

$$\kappa : \mathcal{B} \longrightarrow \mathbb{E}^3. \tag{A.1}$$

Technically, this map is an *embedding*, a term that we may occasionally use in this Appendix. The map assigns (smoothly) to each particle b of the body a spot $\kappa(b)$ in space, no two particles being assigned the same spot. For computational purposes, it is convenient to refer the multiplicity of possible configurations to any arbitrarily chosen fixed configuration, which will be called a *reference configuration*. In many applications, a particular reference configuration may arise naturally as the most convenient one, but ultimately the formulation should turn out to be independent of the choice of reference configuration. Expressed in other words, there should be an established rule as to how things appear to change when the reference configuration is changed, so that these apparent changes can be factored out of the physical picture. We will adopt the standard notation to the effect that quantities in the reference configuration are usually denoted with capital letters, while their counterparts in the actual *spatial configuration*, which the body happens to occupy at a certain instant of time, are denoted with lower-case letters. In particular, the coordinates in the reference configuration are denoted by X^I, ($I = 1, 2, 3$), while the spatial coordinates are denoted by x^i, ($i = 1, 2, 3$). Although not necessary, we will assume, in the spirit of simplicity, that both systems of coordinates are Cartesian.

In terms of a reference configuration $\kappa_0 : \mathcal{B} \to \mathbb{E}^3$, therefore, an arbitrary configuration $\kappa : \mathcal{B} \to \mathbb{E}^3$ can be seen as a *deformation* χ from the former to the latter. Technically, we can write the deformation as a composition embedding:

$$\chi = \kappa \circ \kappa_0^{-1} : \kappa_0(\mathcal{B}) \longrightarrow \mathbb{E}^3. \tag{A.2}$$

A deformation is expressed in coordinates by means of three (smooth) functions that provide each of the spatial coordinates x^i in terms of the three referential coordinates X^I, namely:

$$x^i = x^i(X^1, X^2, X^3), \ i = 1, 2, 3. \tag{A.3}$$

Note that we are using the same symbol for the dependent variables (x^i) as for the functions whose result they are. This device saves notation and, most of the time, does not lead to confusion.

A *motion* is a family of configurations parametrized by time (t), namely:

$$x^i = x^i(X^1, X^2, X^3; t), \ i = 1, 2, 3. \tag{A.4}$$

A.2. Observers and Frames

Just as in Classical Mechanics, the motion of a body presents itself differently to different observers. From the technical point of view, our physical space is a *Euclidean space*, a particular instance of an *affine space*, which is roughly something that becomes a vector space once an origin is chosen. This vector space is endowed with a special inner ("dot") product (consistent with the theorem of Pythagoras), which is assumed to have an intrinsic physical meaning. A *frame* consists of the choice of an origin and an orthonormal basis (three unit and mutually perpendicular vectors). The coordinates of one and the same point present themselves in two different frames, therefore, in ways that are not completely arbitrarily related, precisely because of this orthonormality. In other words, both frames must agree on the measurement of distances, and this is possible if, and only if, the coordinates x^i and $\hat{x}^i$ in the two frames are related by the formula:

$$\hat{x}^i = c^i + Q^i_j x^j, \tag{A.5}$$

where the summation convention[1] for diagonally repeated indices has been used. Here, Q^i_j are the entries of an *orthogonal matrix*, guaranteeing the preservation of the dot product, while c^i are the components (in the "hatted" frame) of the vector joining the two origins. Notice that a change of frame does not affect the reference configuration, only the spatial one.[2]

We have to talk about time. In the nonrelativistic setting, all observers agree on the measure of time, except perhaps on its origin (or "zero"). This means that two observers may disagree only to the extent that their time measurements are related by $\hat{t} = t + a$, a being a real constant. Returning now to the complete space-time notion of observer, we establish that two physical observers can differ only by a time-dependent change of frame, so that the motions recorded by two observers must be related by:

$$\hat{x}^i = c(t)^i + Q(t)^i_j x^j,$$
$$\hat{t} = t + a. \tag{A.6}$$

In closing this section, we should mention that two observers are said to be *inertially related* if $\ddot{c}^i \equiv 0$ and $\dot{Q}^i_j \equiv 0$, where a superimposed dot indicates differentiation with respect to time. These formulas correspond to the statement that the origins of the frames recede from each other at a constant velocity, while the relative orientation of the bases remains constant (no "relative angular velocity"). The equivalence relation of being inertially related splits all possible observers into equivalence classes, of which Isaac Newton affirms that one, and just one, is to be preferred to all others for the formulation of the laws of Mechanics.

A.3. Strain

It can be rightly said that an important objective of Continuum Mechanics is to predict the deformation of a body as time goes on, namely, its motion. To do this, we expect to be given such things as applied forces and initial and boundary conditions (the latter being a novelty compared with Classical Mechanics, where only initial

---

[1] See Section 2.3.
[2] In fact, the reference configuration can be seen, in principle, as just a coordinate chart of the body manifold, without any connotation of "presence" in physical space.

conditions are needed, for obvious reasons). In Classical Mechanics the prediction of the motion of a system of particles is achievable, at least in principle, by means of the correct application of universal laws of motion (Newton's laws). The situation in Continuum Mechanics is somewhat different. Universal laws will surely be at play, but there will also be a need to represent the internal constitution of the material of which the body is made, to distinguish between, say, rubber and steel. The situation is similar to a classical one in which the particles are connected by means of springs, dashpots, actuators, and so on, whose properties (stiffness, damping coefficient, etc.) have to be specified. This naive analogy should not be pushed too far, but it has a clear heuristic value.

The only reason we mention at this early, purely kinematic, stage the need for the representation of the internal constitution, is to justify what we are about to do. The experience with many materials (as well as the desire to represent nature in as simple a manner as possible) tends to show that their material response is very localized. What this means is that intuitively a point seems to "feel" only what is going on (in terms of deformations and forces) in a very small neighbourhood. If we were to think in terms of a molecular model (that is, not in the pure spirit of a continuum theory), we would rightly attribute this behavior to the fact that the interatomic potentials decay rapidly with distance, so the distant neighbours are practically irrelevant. Broad statements such as the one just made should be taken with a pinch of salt, since there are many important phenomena known (or claimed) to be nonlocal. Be that as it may, we are justified in claiming that an important kinematic quantity may turn out to be the local value of the spatial derivative of the deformation, namely, the matrix whose entries are:

$$F_I^i = \frac{\partial x^i}{\partial X^I}. \tag{A.7}$$

This matrix (which is nothing but the Jacobian matrix of the transformation from X^I to x^i) is the coordinate representation of a linear transformation (or tensor) $\mathbf{F}$ that maps vectors attached to a point at the reference configuration to vectors attached at the image point in the spatial configuration. In Continuum Mechanics this tensor is called the *deformation gradient tensor*. To interpret and relate it to our previous argument, we notice that, by the chain rule of differentiation, we may write:

$$dx^i = F_I^i \, dX^I, \tag{A.8}$$

so that, in a Leibnizian view of things, what this tensor is telling us is how a "small" segment of components dX^I is transformed by the deformation into the "small" segment with coordinates dx^i. This is the type of information that we want (if our previous argument had any value) and this is the type of information that experimentalists gather by actually attaching a short copper wire (or *strain gauge*) to the surface of a solid and then passing an electric current through it and measuring the changes in electrical resistance brought about by the change of length of the wire. But wait: The deformation gradient contains more information than just the change of length.

For example, a rigid rotation of the neighbourhood would give rise to a deformation gradient $\mathbf{F}$ given by a rotation matrix, say $\mathbf{R}$. We have already encountered such a matrix (one that preserves the dot product of vectors) and called it orthogonal. So we need a means to split the deformation gradient $\mathbf{F}$ into a part that is a mere rotation and another part that represents a true strain, that is, a change of size or shape of the neighbourhood. This decomposition is achieved by invoking the *polar decomposition* theorem of algebra. The polar decomposition theorem applies to nonsingular matrices, which is clearly the case with our matrix $\mathbf{F}$ (otherwise, $\mathbf{F}$ would collapse a small 3-dimensional region into a planar one).[3] If we assume, moreover, that the referential and spatial coordinate systems are equally oriented (both right-handed, say), the determinant of $\mathbf{F}$ must be strictly positive.

THEOREM A.1. **The polar decomposition theorem:** *Every nonsingular (square) matrix $\mathbf{F}$ is uniquely decomposable into the product $\mathbf{F} = \mathbf{RU}$ of an orthogonal matrix $\mathbf{R}$ and a symmetric positive definite matrix $\mathbf{U}$. Moreover, there exists another unique decomposition in the reverse order, $\mathbf{F} = \mathbf{VR}$, where $\mathbf{V}$ is symmetric and positive definite.*

The (elementary) proof of this theorem can be found in textbooks of linear algebra. Our interest here is to explain and interpret the result. We start by noticing that, since the determinant of $\mathbf{F}$ has been assumed to be positive, the determinant of $\mathbf{R}$ must by $+1$, thus corresponding to a pure rotation (without a subsequent reflection). As far as the positive definite matrices are concerned, in the process of proof one finds that $\mathbf{U}$ and $\mathbf{V}$ are, respectively, the square roots of $\mathbf{C} = \mathbf{F}^T\mathbf{F}$ and $\mathbf{B} = \mathbf{FF}^T$. These two tensors, themselves symmetric and positive definite, are known as the *right* and *left Cauchy-Green tensors*, respectively. A positive definite symmetric real matrix always has three positive eigenvalues and a corresponding orthogonal basis of eigenvectors (which is not unique if there are repeated eigenvalues). In this eigenvector basis, the expression of the tensor becomes a diagonal matrix. The meaning of all this is that at a point in the body, the effect of the deformation gradient is to first apply three stretches in the eigenbasis of $\mathbf{U}$, thus transforming the unit die into a brick, and then rotating the resulting brick by $\mathbf{R}$. The amount of the stretches is measured by the eigenvalues of $\mathbf{U}$. If an eigenvalue is larger than 1, then an actual stretch occurs, while a value less than 1 corresponds to a contraction. We see that $\mathbf{U}$ contains all the information regarding the actual strain of the unit die, while $\mathbf{R}$ contains all the information about its rotation. It is important to remember that the decomposition is unique, since otherwise the physical meaning would be lost. As far as the reverse (left) polar decomposition $\mathbf{F} = \mathbf{VR}$, it turns out that the eigenvalues of $\mathbf{V}$ are the same as those of $\mathbf{U}$, while the eigenvectors are rotated exactly by $\mathbf{R}$. The interpretation of this decomposition is then essentially the same as the first one, except that now we first apply the rotation and, once already in the spatial configuration, we apply the same stretches as before. The result is identical.

---

[3] Mathematically, the regularity of $\mathbf{F}$ is a consequence of having assumed that the configurations are embeddings.

REMARK A.1. **R as rotation**: The interpretation of **R** as a rotation is somewhat sloppy, since **R** is a mapping between two different vector spaces. What we are implicitly assuming is an identification between the space of the reference configuration and that of the actual configuration. This is not the best policy, nor is it necessary, but it will have to do for now.

In terms of components, the right and left Cauchy-Green tensors are given by:

$$C_{IJ} = \sum_{i=1}^{3} F_I^i F_J^i \tag{A.9}$$

and

$$B^{ij} = \sum_{I=1}^{3} F_I^i F_I^j, \tag{A.10}$$

always in Cartesian coordinates. These calculations are easy enough, but the calculation of the square roots can be cumbersome.

It is sometimes convenient to introduce a strain measure that vanishes whenever there is no strain. One such measure, known as the *Lagrangian strain tensor*, is given by:

$$\mathbf{E} = \frac{1}{2}(\mathbf{C} - \mathbf{I}), \tag{A.11}$$

where **I** is the identity tensor in the reference configuration.

BOX A.1. The Linearized Theory

The *linearized theory of strain*, also called the *infinitesimal theory* and the *small deformation theory*, is the result of assuming that, with respect to a given reference configuration, and with the identification of reference and actual spaces already alluded to in Remark A.1, the deformation gradient **F** is very close to the identity, and so are **U** and **R**. The polar decomposition can then be approximated as follows:

$$\mathbf{I} + \Delta\mathbf{F} = (\mathbf{I} + \Delta\mathbf{R})(\mathbf{I} + \Delta\mathbf{U}) \approx \mathbf{I} + \Delta\mathbf{U} + \Delta\mathbf{R}, \tag{A.12}$$

where we have neglected the term $\Delta\mathbf{R}\,\Delta\mathbf{U}$, because of the assumption of smallness of the increments $\Delta\mathbf{R}$ and $\Delta\mathbf{U}$. As a result, the polar decomposition changes from multiplicative to additive (corresponding mathematically to the passage from a Lie group to its Lie algebra), namely:

$$\Delta\mathbf{F} = \Delta\mathbf{U} + \Delta\mathbf{R}. \tag{A.13}$$

But, due to the fact that an orthogonal matrix satisfies $\mathbf{R}\mathbf{R}^T = \mathbf{I}$, a similar argument leads to the conclusion that $\Delta\mathbf{R} + \Delta\mathbf{R}^T = 0$, implying that the small rotation $\Delta\mathbf{R}$ is represented by a skew-symmetric matrix. Noting that $\Delta\mathbf{U}$ is obviously symmetric, we conclude that $\Delta\mathbf{F}$ is additively decomposed into symmetric and skew-symmetric parts. Since such a decomposition is unique, it follows that:

$$\Delta\mathbf{U} = \frac{1}{2}(\Delta\mathbf{F} + \Delta\mathbf{F}^T), \quad \Delta\mathbf{R} = \frac{1}{2}(\Delta\mathbf{F} - \Delta\mathbf{F}^T), \tag{A.14}$$

the first equation representing the small strain and the second the small rotation. We can go one step farther and assume that the two coordinate systems (referential and spatial) are actually the same, so that we can use lower-case indices for both. Since both systems are Cartesian, we can also afford the luxury of using subscripts instead of superscripts. Introducing the *displacement vector* with components:

$$u_i = x_i - X_i \tag{A.15}$$

Equation (A.14) can be written as:

$$\epsilon_{ij} = \frac{1}{2}(u_{i,j} + u_{j,i}), \quad \omega_{ij} = \frac{1}{2}(u_{i,j} - u_{j,i}), \tag{A.16}$$

where we have renamed $\Delta\mathbf{U} = \epsilon$ and $\Delta\mathbf{R} = \omega$, and where commas indicate partial derivatives in space. These formulas for strain and rotation in the realm of the geometrically linearized theory are usually all an undergraduate engineering student ever sees. In the usual notation (due to Cauchy), the off-diagonal terms of the strain are denoted with the letter γ and are usually multiplied by 2. In undergraduate courses these quantities are obtained by a careful calculation of change of size and change of shape of an infinitesimal die aligned with the coordinate system.

EXERCISE A.1. **The linearized Lagrangian strain:** Show that, upon linearization, the Lagrange strain tensor (A.11) reduces to ϵ.

A.4. Volume and Area

From the interpretation of the polar decomposition theorem (or, equivalently, from the well-known formula for change of coordinates), it follows that the ratio between the spatial volume element dv and its referential counterpart dV (occupied by the same material particles) is given by:

$$\frac{dv}{dV} = \det\mathbf{F} = \det\mathbf{U}. \tag{A.17}$$

We will also need a relation between the elements of area da and dA occupied by the same material particles in space and in the reference configuration, respectively. Let dX^I and dY^I be the components in the reference coordinates of two "small" vectors issuing from the same point, and let N_I be the components of the unit normal to the element of area they subtend. By definition of cross product, we obtain:

$$dA\, N_I = \epsilon_{IJK} dX^J dY^K, \tag{A.18}$$

where ϵ_{IJK} is the permutation symbol.[4] If dx^i and dy^i are the components of the images of dX^I and dY^I under the deformation, and n_i are the components of the

---

[4] Recall that the permutation symbol attains the value $+1$ if (i,j,k) is an even permutation of $(1,2,3)$, -1 if the permutation is odd, and 0 if there is a repeated index.

unit normal to the area they subtend, we may write:

$$n_i \, da = \epsilon_{ijk} dx^j dy^k = \epsilon_{ijk} F_J^j F_K^k dX^J dY^K$$

$$= \epsilon_{mjk} F_J^j F_K^k F_I^m (F^{-1})_i^I dX^J dY^K = \epsilon_{IJK} \det \mathbf{F} (F^{-1})_i^I dX^J dX^K$$

$$= \det \mathbf{F} (F^{-1})_i^I N_I dA, \qquad (A.19)$$

or

$$\mathbf{n} \, da = \det(\mathbf{F}) \, \mathbf{F}^{-T} \, \mathbf{N} \, dA. \qquad (A.20)$$

A.5. The Material Time Derivative

In Continuum Mechanics one works with fields over the body: scalar fields (such as temperature), vector fields (velocity), and tensor fields (strain). These fields are usually defined in the present configuration that the body happens to occupy in space, but (if the motion is known) it is often convenient to pull them back to the reference configuration. The pull-back is just a matter of composition. Let Ψ be some (scalar, say) field defined in the spatial configuration. In other words, let Ψ be given as:

$$\Psi = \Psi(x^i; t). \qquad (A.21)$$

We are abusing the notation slightly in that we let x^i stand for the three variables x^1, x^2, x^3 and, as before, we are using the same letter for the function as for its evaluation. A representation of a field in terms of spatial coordinates is often called the *Eulerian* representation. If the motion is known, we can switch to the referential (or *Lagrangian*) representation as follows:

$$\Psi = \Psi \left(x^i (X^I; t); t \right) = \tilde{\Psi} \left(X^I; t \right). \qquad (A.22)$$

Here we have been extra careful to place a tilde on top of the second function to indicate that it is certainly a different function of its arguments (X^I and t) than Ψ is of its arguments (x^i and t). In practice, even this notational distinction is dropped, hoping that the independent variables will speak for themselves.

Assume now that one needs to calculate the time derivative of this field. Will this be the time derivative of Ψ or that of $\tilde{\Psi}$? They both make sense. The first one tells us the rate of change of the quantity Ψ *at a fixed position in space*, while the second provides the rate of change of Ψ *at a fixed material particle*. To visualize this, imagine that Ψ is the temperature field. The derivative $\frac{\partial \Psi}{\partial t}$ is what a graduate student would measure and calculate while sitting with a thermometer at a laboratory chair, whereas the derivative $\frac{\partial \tilde{\Psi}}{\partial t}$ is what a thermocouple glued to the particle would allow us to determine. Because the Eulerian representation is closer to the experimental setting, but the time derivative at a fixed particle has the more intrinsic meaning, it is important to obtain a formula that permits us to calculate the latter in terms of the former. The result is called the *material time derivative*. It is the partial time derivative of $\tilde{\Psi}$, but expressed in terms of partial derivatives of Ψ. By the chain rule applied to Equation (A.22), we obtain:

$$\frac{\partial \tilde{\Psi}}{\partial t} = \frac{\partial \Psi}{\partial x^i} \frac{\partial x^i}{\partial t} + \frac{\partial \Psi}{\partial t}. \tag{A.23}$$

To avoid the cumbersome tilde, it is customary to denote the material time derivative by D/Dt, or with a superimposed dot. We note that the quantities:

$$v^i = \frac{\partial x^i(X^I;t)}{\partial t} \tag{A.24}$$

are the components of the *velocity field* $\mathbf{v}$ at the point X^I at time t. It follows that the material time derivative corrects the partial time derivative by means of a term equal to the product of the spatial gradient of the field contracted with the velocity:

$$\frac{D\Psi}{\partial t} = \frac{\partial \Psi}{\partial x^i} v^i + \frac{\partial \Psi}{\partial t}. \tag{A.25}$$

A moment's reflection reveals that this correction is exactly what one should expect. Indeed, imagine water flowing through a narrow tube, at two windows of which (separated by an axial distance Δx) two graduate students are recording the water temperature as time goes on. Imagine that each of them keeps recording a constant (but different) temperature (stationary temperature). Is the temperature of the particles passing through the windows constant? Evidently not. If the instantaneous speed of the fluid is v, the time it takes for a particle to go from the first window to the next is $\Delta t = \Delta x / v$. If the difference of the temperatures recorded by the students is $\Delta \Psi$, it is clear that the rate of change of the temperature *as would be measured by an observer moving with the particle* is $\Delta \Psi / (\Delta x / v) = (\Delta \Psi / \Delta x)\, v$, which is the 1-dimensional version of our correction term.

EXERCISE A.2. **The acceleration:** The acceleration of a particle is the second time derivative of the motion or, equivalently, the material time derivative of the velocity field. Assuming that the velocity field is known in its Eulerian representation, find the expression for the acceleration field. [Hint: apply the formula to each component.] Notice that the result contains a ("convected") term which is nonlinear in the velocity.

A.6. Change of Reference

Let a motion be referred to two different reference configurations, κ_0 and κ_1, as shown in the Figure A.1. The corresponding deformations are related by composition:

$$\kappa_0 = \kappa_1 \circ \lambda, \tag{A.26}$$

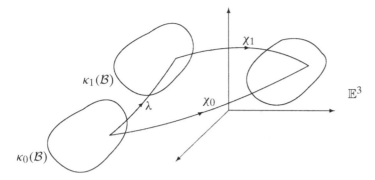

Figure A.1. Change of reference configuration

where λ is the fixed deformation of $\kappa_1(\mathcal{B})$ with respect to κ_0. By the chain rule of differentiation, we obtain for the corresponding deformation gradients the expression:

$$\mathbf{F}_0 = \mathbf{F}_1 \nabla\lambda, \tag{A.27}$$

where ∇ stands for the gradient operator. This is the desired rule for change of reference configuration. We apply it now to the material time derivatives of the deformation gradients and obtain:

$$\dot{\mathbf{F}}_0 = \dot{\mathbf{F}}_1 \nabla\lambda, \tag{A.28}$$

since λ is independent of time. It turns out, therefore, that the combination:

$$\mathbf{L} = \dot{\mathbf{F}}_0 \mathbf{F}_0^{-1} = \dot{\mathbf{F}}_1 \mathbf{F}_1^{-1} = \dot{\mathbf{F}} \mathbf{F}^{-1} \tag{A.29}$$

is independent of the reference configuration chosen. It is a purely spatial tensor called the *velocity gradient*. The name arises from the fact that, since the quantity is independent of reference configuration, we may choose the present configuration as reference, whereby the deformation gradient is instantaneously equal to $\mathbf{I}$, immediately leading (by the equality of mixed partial derivatives) to the interpretation of $\mathbf{L}$ as the spatial gradient of the velocity field. In terms of components, we have:

$$L^i_j = v^i_{,j} \quad , \tag{A.30}$$

with commas denoting partial derivatives (or covariant derivatives, if a non-Cartesian coordinate system were used). Again considering the present configuration as an instantaneous reference, and indicating corresponding quantities with the subscript t, the polar decomposition $\mathbf{F}_t = \mathbf{R}_t \mathbf{U}_t$ leads by differentiation to

$$\mathbf{L} = \dot{\mathbf{F}}_t = \dot{\mathbf{R}}_t + \dot{\mathbf{U}}_t, \tag{A.31}$$

where we have used the fact that $\mathbf{F}_t = \mathbf{R}_t = \mathbf{U}_t = \mathbf{I}$ at time t. Since $\mathbf{R}$ is always orthogonal, it follows that $\dot{\mathbf{R}}_t$ (at time t) is skew-symmetric. By the uniqueness of the decomposition into symmetric and skew-symmetric parts, we obtain that:

$$\mathbf{W} := \dot{\mathbf{R}}_t = \frac{1}{2}(\mathbf{L} - \mathbf{L}^T) \qquad (A.32)$$

and

$$\mathbf{D} := \dot{\mathbf{U}}_t = \frac{1}{2}(\mathbf{L} + \mathbf{L}^T). \qquad (A.33)$$

The tensors $\mathbf{D}$ and $\mathbf{W}$ are called, respectively, *stretching* (or rate of deformation) and *spin* (or vorticity). Compare these expressions with those of the linearized theory (Box A.1) and draw your own conclusions.

A.7. Transport Theorems

In the formulation of balance laws in Continuum Mechanics (such as the first law of thermodynamics) one is confronted with the evaluation of the rate of change of the content of an extensive physical quantity (such as energy) in the body. Let such content Φ be given by the (Riemann) integral of some density g over a spatial volume of interest v, viz.:

$$\Phi = \int_v g(x^i; t) \, dv. \qquad (A.34)$$

If we were to follow this volume as it is dragged by the motion of the body (namely, if we were to ensure that we are always enclosing the same material particles), then the rate of change of the content Φ would become a material time derivative $D\Phi/Dt$, which we intend to calculate. To achieve this aim, we first pull-back the domain of integration to the reference configuration, where the corresponding domain $V = \chi^{-1}(v)$ is independent of time. Denoting $J = \det(\mathbf{F})$, the pull-back is:

$$\Phi = \int_V g[x^i(X^I; t); t] J \, dV, \qquad (A.35)$$

by the rule of change of variables of integration. The independent variables being now X^I and t, the material time derivative boils down to a partial derivative. Moreover, since V is independent of time, we can commute the order of differentiation and integration to obtain:

$$\frac{D\Phi}{Dt} = \int_V \left[\frac{Dg}{dt} J + g \frac{DJ}{Dt} \right] dV. \qquad (A.36)$$

The derivative of the determinant (J) of a matrix ($\mathbf{F}$) depending of a parameter (t) is given by:

$$\dot{J} = J \ \text{trace} \ (\mathbf{F}^{-1}\dot{\mathbf{F}}). \qquad (A.37)$$

In view of Equations (A.29) and (A.30), we obtain:

$$\dot{J} = J \ \text{trace} \ \mathbf{L} = J \ \text{div} \ \mathbf{v}, \qquad (A.38)$$

where div stands for the spatial divergence. Plugging these results back into Equation (A.36), we obtain:

$$\frac{D\Phi}{Dt} = \int_V \left(\frac{Dg}{dt} + g \operatorname{div} \mathbf{v} \right) J \, dV, \tag{A.39}$$

which allows us to change the domain of integration back to the original v:

$$\frac{D\Phi}{Dt} = \int_v \left[\frac{Dg}{dt} + g \operatorname{div} \mathbf{v} \right] dv. \tag{A.40}$$

This result is known as *Reynolds' first transport theorem*. Similar theorems can be proven for integrals along areas and lines, but will not be needed here.

REMARK A.2. **A suspicious point**: Looking back at Reynolds' transport formula (A.40), the following point could legitimately be made. Assume that we are given a spatial region v which moves in time. This is a *control volume* which may have nothing to do with the motion of a material body. The only velocity that we can intrinsically assign to a moving volume is the normal speed at each point of its boundary ∂v, since there is no other natural way to associate points at two "successive" positions (at times t and $t + dt$) of this moving entity but to draw the normal to the boundary at a point, measure the length dn of the segment between the two surfaces and then divide by dt. Assume that a field $g(x,t)$ (such as electromagnetic energy) exists in space. Then it makes sense to ask how the content captured by the control volume changes in time. This question has nothing to do with the motion of a material, and yet we could imagine any material motion we wish (such that the particle velocities at the boundary happen to have a normal component equal to that of the surface, so that no particles escape), and then apply Reynolds' theorem! The result would obviously be correct. To solve this apparent paradox, we need only invoke our formula for the material time derivative (A.25) to rewrite (A.40) as:

$$\frac{D\Phi}{Dt} = \int_v \left[\frac{\partial g}{dt} + \frac{\partial g}{\partial x^i} v^i + g \operatorname{div} \mathbf{v} \right] dv = \int_v \left[\frac{\partial g}{dt} + \operatorname{div}(g\mathbf{v}) \right] dv. \tag{A.41}$$

By the divergence theorem, we may change the volume integral of a divergence into the total flux through the boundary, so that:

$$\frac{D\Phi}{Dt} = \int_v \frac{\partial g}{dt} \, dv + \int_{\partial v} (g\mathbf{v} \cdot \mathbf{n}) \, da, \tag{A.42}$$

where $\mathbf{n}$ is the unit normal to the element of area da of the boundary ∂v. As expected, the result depends only on the normal speed of the boundary.

A.8. The General Balance Equation

A *balance equation* in Continuum Physics is a rigorous accounting of the causes for the change in content, over a certain domain, of an extensive physical quantity. In

Continuum Mechanics, in particular, we attribute physical content only to a material body or a part thereof (as opposed to a field theory such as electromagnetism, where a vacuum can be a carrier of fields). Just as in balancing a cheque book, we want to account for the change in content of the quantity Ψ as time goes on. We will postulate that the time rate of change of the content of a physical quantity Ψ in the body is entirely due to only two causes: a *production* (or source) Π within the body and a *flux* Φ through its boundary:

$$\frac{D\Psi}{Dt} = \Pi + \Phi. \tag{A.43}$$

Naturally, both the production and the flux are rates (per unit time). This equation is the generic statement of an equation of balance.

We want, perhaps at a price, to convert this general statement into a more specific form. We can do so in either the referential ("Lagrangian") or spatial ("Eulerian") setting, and we will work in both. A configuration can be assumed to inherit the Cartesian volume element of the ambient space, so that, according to what we have been doing so far, we can represent the total content in terms of the referential volume element dV or its spatial counterpart dv as follows:

$$\Psi = \int_V G \, dV = \int_v g \, dv, \tag{A.44}$$

where V and v are corresponding material volumes in reference and space, respectively, and where G and g represent the content per unit referential and spatial volume. We are here presupposing that the content is some appropriately defined, continuous, additive set function, thereby replacing measure-theoretical issues with a mere density within the Riemann integral. The two densities are related by:

$$G = Jg. \tag{A.45}$$

Similar assumptions of smoothness will apply to the production, thus yielding:

$$\Pi = \int_V P \, dV = \int_v p \, dv, \tag{A.46}$$

with $P = Jp$. The flux term needs some more finessing. To keep the discussion to a minimum (and to avoid important philosophical questions) we will assume that a flux is defined not only through the boundary of the body, but also through any internal boundary between subbodies. Moreover, we will assume that the flux is given by an integral (over the area of interest) of some density. But, unlike the case of the volume integrals, now we are not allowed to claim that this density is a mere function of position on the surface. A simple example will clarify this situation. If you are sun-tanning in the beach, it makes a difference whether you are lying horizontally or standing up. The heat energy flows at a maximum rate when the exposed surface is perpendicular to the rays. This is as true for the outer boundary as for an internal point. When you consider an interior point of the endodermis, the flux at that point will vary depending on the orientation of the imagined surface element you consider at the point. In other words, the flux density must depend not only on position but also on some other characteristic of the boundary. The simplest assumption is that the flux depends on the boundary *only through its (local) normal*, namely, through its orientation alone (as opposed to its curvature, for example). Having made these assumptions we can write:

$$\Phi = \int_{\partial V} H(X^I, N^I; t) \, dA = \int_{\partial v} h(x^i, n^i; t) \, da. \tag{A.47}$$

To obtain the relation between the referential and spatial flux densities (H and h, respectively), we would like to invoke Equation (A.20). But there is a snag. In this equation, the dependence on the normal is explicitly linear, whereas our assumed dependence of the flux densities on the normal is so far arbitrary. It would appear that the only way out of the impasse would be to *assume* such linearity (a logical thing to do, particularly considering the example of the sun rays, whose effect on the skin is governed by a simple projection on the normal). It is a rather remarkable fact, however, that such an assumption is superfluous, since it can be *derived* from the general statement of the balance law. This result is justly called *Cauchy's theorem*, since it was Cauchy who, in a series of papers published in the 1820s, first established its validity in the context of his theory of stress. To see how this works, we will use, for example, the Lagrangian setting. The generic equation of balance so far reads:

$$\frac{D}{Dt} \int_V G \, dV = \int_V P \, dV + \int_{\partial V} H \, dA. \tag{A.48}$$

The only extra assumption we need to make (actually, already implicitly made) is that this statement of the general balance law is valid, with the same densities, for any subbody. This assumption essentially eliminates scale effects involved in phenomena such as surface tension. This being the case, the theorem follows from the *Cauchy tetrahedron argument*. By taking a tetrahedron with three faces aligned with the coordinate system and the fourth face with its vertices lying on the surface of interest, we apply the mean value theorem to each term of Equation (A.48). We let the far vertex of the tetrahedron approach the surface (along the line joining its position with the point of interest in the surface), and observe that the two volume integrals approach zero with the cube of a typical line dimension of the tetrahedron, while the surface integral does so with the square, thereby being the only survivor in the limit. But this integral consists of a sum over four triangular faces whose areas are related by the components of the unit normal to the incline. The result follows suit. It establishes that there exists a linear operator $\mathbf{H}$ (function of position alone) which, when acting on the normal $\mathbf{N}$, delivers the flux H per unit area. Observe that if the quantity being balanced is a scalar, then $\mathbf{H}$ is a vector and the linear operation is given by the dot product $H = \mathbf{H} \cdot \mathbf{N}$. If, on the other hand, the quantity being balanced is vectorial, then $\mathbf{H}$ is a tensor acting on the normal. We can, therefore, write:

$$\Phi = \int_{\partial V} \mathbf{H} \cdot \mathbf{N} \, dA, \tag{A.49}$$

and similarly for the Eulerian formulation:

$$\Phi = \int_{\partial v} \mathbf{h} \cdot \mathbf{n} \, da, \tag{A.50}$$

with an obvious notation. Invoking now Equation (A.20), we obtain the following relation between the Lagrangian and Eulerian flux densities:

$$\mathbf{H} = J \, \mathbf{F}^{-1} \, \mathbf{h}. \tag{A.51}$$

The integral statement of the generic law of balance in Lagrangian form is finally:

$$\frac{D}{Dt} \int_V G \, dV = \int_V P \, dV + \int_{\partial V} \mathbf{H} \cdot \mathbf{N} \, dA, \tag{A.52}$$

and in Eulerian form:

$$\frac{D}{Dt} \int_v g \, dv = \int_v p \, dv + \int_{\partial v} \mathbf{h} \cdot \mathbf{n} \, da. \tag{A.53}$$

These integral statements of the balance law may suffice for numerical computations (such as with the finite-volume method). Nevertheless, under suitable assumptions of smoothness, we can obtain their local (differential) forms by using the divergence theorem to convert the flux integral into a volume integral and invoking the fact that the resulting three integrals must be identically equal regardless of the domain of integration within the body. The results are, therefore,

$$\frac{\partial G}{\partial t} = P + \operatorname{Div} \mathbf{H}, \tag{A.54}$$

and

$$\frac{Dg}{Dt} + g \operatorname{div} \mathbf{v} = p + \operatorname{div} \mathbf{h}, \tag{A.55}$$

where the last equation necessitated the application of Reynolds' transport theorem. The operators Div and div stand for the referential and spatial divergence, respectively. To avoid confusion, we list the component versions of these equations:

$$\frac{\partial G}{\partial t} = P + H^I_{,I}, \tag{A.56}$$

and

$$\frac{Dg}{Dt} + g \, v^i_{,i} = p + h^i_{,i}. \tag{A.57}$$

If the quantity of departure were vectorial (whether referential or spatial), we would simply apply these equations component by component (taking advantage of the

Euclidean structure), the result being an added index (referential or spatial) to each quantity. Appropriate boundary conditions are needed, but we will not deal with them at this stage.

A.9. The Fundamental Balance Equations of Continuum Mechanics

We have devoted perhaps too much detail to the justification of the general form of an equation of balance in both the Lagrangian and the Eulerian forms. In this section we hope to reap the compensating reward when applying the general prescription just obtained to the five fundamental quantities to be balanced in a traditional Continuum Mechanics treatment. They are the following:

Conservation of mass: A balance law is said to be a *conservation law* if the production and the flux vanish identically. This is the case for the mass of a continuum.[5] Denoting by ρ_0 and ρ the referential and spatial mass densities, respectively, we obtain the Lagrangian and Eulerian differential balances as:

$$\frac{\partial \rho_0}{\partial t} = 0, \tag{A.58}$$

and

$$\frac{D\rho}{Dt} + \rho \operatorname{div} \mathbf{v} = 0. \tag{A.59}$$

The latter (Eulerian) version is known in hydrodynamics as the *continuity equation*. Expanding the material time derivative, it can also be written as:

$$\frac{\partial \rho}{\partial t} + \operatorname{div}(\rho \mathbf{v}) = 0, \tag{A.60}$$

or in other, equivalent, forms.

Balance of linear momentum: This balance is a statement of Newton's second law as applied to a deformable continuum. It is, therefore, important to recall that we must assume that our frame of reference (which we have identified with a Cartesian coordinate system) is actually *inertial*. The quantity to be balanced is the (vectorial) linear momentum, whose Lagrangian and Eulerian densities are, respectively, $\rho_0 \mathbf{v}$ and $\rho \mathbf{v}$. Note that in both cases we have a *spatial* vector to balance, whether the statement of the law is referential or spatial.[6] As for the production term, it is given by the (distributed) forces per unit volume, or *body forces*, with densities $\mathbf{B}$ and $\mathbf{b}$, respectively, in the Lagrangian and Eulerian formulations.[7] The flux terms are given by the *surface tractions* (namely, forces per unit referential or spatial area). We again emphasize that these forces, even when measured per unit referential area, are *spatial* vectors. By Cauchy's theorem, we know that the surface tractions are governed by

---

[5] In modern theories of biological growth, however, or in theories of chemically reacting mixtures (when looking at each component of the mixture), conservation of mass does not hold, and specific mass production and/or flux terms are to be included.

[6] It is possible to pull back the velocity field into a material field, but we will not pursue this avenue here.

[7] Concentrated forces could be included in a distributional (or weak) setting.

a (tensorial) flux, which we will denote by $\mathbf{T}$ and $\mathbf{t}$, respectively, for the Lagrangian and Eulerian settings. To avoid confusion, we will express the final equations in components. Plugging the various terms into the corresponding forms of the generic law of balance, and invoking the already obtained conservation of mass, we obtain the following Lagrangian and Eulerian forms of the balance of linear momentum:

$$T^{iI}{}_{,I} + B^i = \rho_0 \frac{Dv^i}{Dt}, \tag{A.61}$$

and

$$t^{ij}{}_{,j} + b^i = \rho \frac{Dv^i}{Dt}. \tag{A.62}$$

The tensorial fluxes $\mathbf{T}$ and $\mathbf{t}$ are called, respectively, the *first Piola-Kirchhoff stress* (or just the *Piola stress*) and the *Cauchy stress*. Note that, whereas the Cauchy stress is a purely spatial tensor (an automorphism of the tangent space at a point in the current configuration), the first Piola-Kirchhoff stress is a mixed tensor (a linear map between the tangent space at the reference point and its counterpart in space). What the Piola and Cauchy tensors do is to produce linearly out of vectors with components N_I and n_i, respectively, the spatial forces acting on elements of area to which they are normal. The relation between these tensors follows directly from Equation (A.51):

$$T^{iI} = J \, (F^{-1})^I_j \, t^{ij}. \tag{A.63}$$

REMARK A.3. **On the acceleration term**: Note that in the Lagrangian version (A.61), the material time derivative of the velocity (appearing on the right-hand side of the equation) reduces to a partial derivative, since the velocity field is expressed in terms of X^I and t. In the Eulerian version (A.62), on the other hand, the velocity field is expressed in terms of x^i and t, and the material time derivative includes the nonlinear convected term $\frac{\partial v^i}{\partial x^j} v^j$.

EXERCISE A.3. **Linear momentum:** Work out the details of the derivation of the differential equations of balance of linear momentum.

Balance of angular momentum: In the Newtonian Mechanics of system of particles, the law of balance of angular momentum follows from Newton's second law under the assumption that the particles of the system interact by means of forces abiding by Newton's third law ("action and reaction"). In the case of a continuum, the analog of such internal forces is the stress tensor, but the analogy is not easy to pursue rigorously. In fact, we will soon discover that in Continuum Mechanics the postulation of a law of balance of momentum leads to a *new result*, without a clear analog in discrete systems. The balance of angular momentum states that the rate of change of the total angular momentum with respect to a fixed point (the origin, say) of an inertial frame is equal to the moment of all the external forces with respect to the same point. The density of the angular momentum (namely, the moment of the momentum) is given by the vector product $\mathbf{r} \times (\rho\mathbf{v})$, where $\mathbf{r}$ is the spatial position vector. The production density is $\mathbf{r} \times \mathbf{B}$, and the flux density is $\mathbf{r} \times (\mathbf{tn})$. Expressing these cross products by means of the permutation symbol, using the general prescription of the Eulerian balance law, and invoking conservation of mass and balance of linear momentum, the somewhat surprising final result is the survival of just one term, namely:

$$\epsilon_{ijk}\, t^{jk} = 0. \tag{A.64}$$

EXERCISE A.4. **Angular momentum:** Work out the details of the derivation of the local law of balance of angular momentum to obtain the final result just stated.

By the complete skew-symmetry of the permutation symbol ϵ_{ijk}, an equivalent way to express Equation (A.64) is:

$$t^{ij} = t^{ji}, \tag{A.65}$$

namely, *the Cauchy stress is symmetric*. At this point it is appropriate to go back to the discrete analogy. There, one assumes that internal forces between the particles of the system abide by the principle of action and reaction. In the continuum case, what we have implicitly assumed is that the surface interactions are merely forces (stresses) and that there are no extra contributions of surface couples. A similar assumption was made regarding the external body forces. In other words, the only contribution to the moment equation is that arising from the moments of forces (no couple interactions). This assumption may have to be abandoned when dealing with electrically or magnetically polarizable materials. In those cases, the antisymmetric part of the stress may not vanish, but is balanced by the external body-couple.

Now that we are in possession of the Equation (A.65), we obtain the Lagrangian version by a direct use of Equation (A.63):

$$\mathbf{FT^T} = \mathbf{TF^T}. \tag{A.66}$$

Note that the first Piola-Kirchhoff tensor is not symmetric in the ordinary sense, nor could it be (since it is not an automorphism).

Balance of energy (the first law of thermodynamics): In the case of a single particle or a rigid body, a direct application of Newton's laws yields the result that the rate of change in kinetic energy K is exactly balanced by the mechanical power W_{ext} of the external forces acting on the system. The application of an external force over time along a trajectory produces (or extracts) work, and this work is entirely expended in increasing (or decreasing) the kinetic energy of the system. In a discrete nonrigid system of interacting particles, on the other hand, one can obtain, by applying Newton's equations of motion to each particle and then adding over all particles of the system, that the rate of change of the total kinetic energy of the system is equal to the power of the external forces plus the power of the internal forces. By the law of action and reaction, the power of the internal forces between two given particles is equal to the force exerted by the first on the second dot-multiplied by the relative velocity of the second with respect to the first (so that, for example, if the mutual force is one of attraction, the internal power will be positive if the particles are moving towards each other). The (positive, say) power of the external forces is now spent in part on increasing the kinetic energy and in part on overpowering the particle interactions. This purely mechanical result has an exact analog in Continuum Mechanics. If we start from the local statement of the law of balance of linear momentum (whether in Lagrangian or in Eulerian form), dot-multiply it by the velocity, integrate over the body, and apply the divergence theorem to shift a term to the boundary, we indeed obtain the result:

$$\frac{DK}{Dt} = W_{ext} + W_{int}, \tag{A.67}$$

where the kinetic energy is given by:

$$K = \int_V \frac{1}{2}\rho_0 \mathbf{v} \cdot \mathbf{v} \, dV = \int_v \frac{1}{2}\rho \mathbf{v} \cdot \mathbf{v} \, dv, \tag{A.68}$$

and the power of the external forces is:

$$W_{ext} = \int_V \mathbf{B} \cdot \mathbf{v} \, dV + \int_{\partial V} (\mathbf{TN}) \cdot \mathbf{v} \, dA = \int_v \mathbf{b} \cdot \mathbf{v} \, dv + \int_{\partial v} (\mathbf{tn}) \cdot \mathbf{v} \, da. \tag{A.69}$$

We have suggestively denoted the remaining term in (A.67) by W_{int}, to signify that this is the term that corresponds to the power of the internal forces in the discrete-system analogy. Its exact expression is:

$$W_{int} = -\int_V T^{iI} v_{i,I} \, dV = -\int_v t^{ij} v_{i,j} \, dv, \tag{A.70}$$

or, recalling that $v_{i,j}$ are the components of the velocity gradient $\mathbf{L} = \mathbf{W} + \mathbf{D}$, and that the Cauchy stress is symmetric,

$$W_{int} = -\int_v \text{trace} \, (\mathbf{tD}) \, dv, \tag{A.71}$$

where the minus sign is only indicative of the fact that, when defining the flux operator through Cauchy's tetrahedron argument, we made a choice of sign that (in the case of the stress) corresponds to the reaction (rather than the action) in the discrete analog. We will call the term W_{int} the internal power (or stress power) in a continuum.

EXERCISE A.5. **Kinetic energy:** Follow the steps indicated and derive the law of balance of kinetic energy (A.67).

We have given some attention to the law of balance of kinetic energy only to show that it is not an independent law, but just a consequence of the previous, purely mechanical, balance laws. On the other hand, our treatment has served to bring to evidence the presence and nature of the extra power term W_{int} due to the intrinsic deformability of the continuum (this term would be absent in the case of a rigid body). Continuous media, on the other hand, react to other types of energy input, particularly to what one would refer, to in everyday life as thermal energy. Indeed, it is a matter of daily experience that continuous media deform under applied heat and, conversely, deformation may lead to the emission of heat (bending a metal paper clip repeatedly until it breaks is a good experiment to reveal this common effect). There are other occurrences of nonmechanical energy sources (chemical reactions, electromagnetic fields, etc.). We will lump the nonmechanical power input (*heating*), as one might expect, into two terms: one corresponding to distributed volumetric sources (*radiation*) and the other to input across the boundaries (*conduction*):

$$W_{heat} = \int_V R\, dV + \int_{\partial V} Q\, dA = \int_v r\, dv + \int_{\partial v} q\, da. \tag{A.72}$$

The law of balance of energy (first law of thermodynamics) asserts that, in addition to the kinetic energy K, there exists another kind of energy content, called *internal energy*, U, such that the rate of change of the total energy content $K + U$ exactly balances the combined external mechanical and heating powers, namely:

$$\frac{D(K+U)}{Dt} = W_{ext} + W_{heat}. \tag{A.73}$$

Comparing this result with the balance of kinetic energy (A.67), we obtain:

$$\frac{DU}{Dt} = W_{heat} - W_{int}. \tag{A.74}$$

This equation implies roughly that an increase in internal energy can be achieved, for example, by either an input of heating power (with no strain) or working against the internal forces of the system (with no heating). To obtain the local versions of this balance law, we assume that the internal energy is given by an integral of a density u, which is usually assumed to be given per unit mass (rather than unit volume), namely:

$$U = \int_V \rho_0 u\, dV = \int_v \rho u\, dv. \tag{A.75}$$

Using the tetrahedron argument, one can show that the flux terms in the heating input in Equation (A.72) are given by referential and spatial *heat flux vectors* as $Q = -\mathbf{Q} \cdot \mathbf{N}$ and $q = -\mathbf{q} \cdot \mathbf{n}$, the minus signs being chosen so that the heat flux vectors $\mathbf{Q}$ or $\mathbf{q}$ point in the direction of the flux (if the normals are the *external* normals to the boundary of the domain of interest). The standard procedure yields the following Lagrangian and Eulerian forms of the local equations of energy balance:

$$\rho_0 \dot{u} = R - Q^I{}_{,I} + T^{il} v_{i,I}, \tag{A.76}$$

and

$$\rho \dot{u} = r - q^i{}_{,i} + t^{ij} v_{i,j} = r - \operatorname{div}\mathbf{q} + \operatorname{trace}(\mathbf{tD}). \tag{A.77}$$

Naturally, in this last (Eulerian) version, the material time derivative $\dot{u}$ includes the convected term $\nabla u \cdot \mathbf{v}$.

EXERCISE A.6. **Energy:** Provide the missing steps in the derivation of the local equation of energy balance.

Entropy inequality (the second law of thermodynamics): An important conceptual element in Continuum Mechanics is the presence of an arrow of time, that is, the natural irrevocable direction of phenomena prescribed by the second law of thermodynamics. There are different ways to deal with this delicate issue, but here we will present only the formulation based on the Clausius-Duhem inequality. There are two new elements that need to be added to the picture. The first one is a new extensive quantity, the *entropy S*, whose content will be measured in terms of the integral of a density s per unit mass:

$$S = \int_V \rho_0 s\, dV = \int_v \rho s\, dv. \tag{A.78}$$

The second element to be introduced is a new field θ, called the *absolute temperature*. It is assumed that θ is strictly positive and measurable instantaneously and locally by an appropriate instrument (such as a thermocouple). This temperature scale is consistent with the temperature appearing naturally in the theory of ideal gases. It is moreover assumed that there are two universal sources of entropy production, one volumetric and the other through the boundary. These sources are obtained, respectively, by dividing the corresponding (volume or surface) heating source by the local value of the absolute temperature. The Clausius-Duhem inequality asserts that the rate of entropy production is never less than what can be accounted for by these universal sources, namely:

$$\frac{D}{Dt} \int_V \rho_0 s\, dV \geq \int_V \frac{R}{\theta}\, dV - \int_{\partial V} \frac{\mathbf{Q} \cdot \mathbf{N}}{\theta}\, dA, \tag{A.79}$$

or, in the Eulerian version,

$$\frac{D}{Dt} \int_v \rho s\, dv \geq \int_v \frac{r}{\theta}\, dv - \int_{\partial v} \frac{\mathbf{q} \cdot \mathbf{n}}{\theta}\, da. \tag{A.80}$$

The equality corresponds to *reversibility* of a physical process, while all other processes (for which the strict inequality holds) are *irreversible*. We will later exploit this inequality to derive restrictions on the possible material responses: A material cannot exist in nature for which, for any conceivable process, the Clausius-Duhem inequality might be violated. The local forms of the inequality are obtained by the standard procedure as:

$$\rho_0 \dot{s} \geq \frac{R}{\theta} - \mathrm{Div}\left(\frac{\mathbf{Q}}{\theta}\right), \tag{A.81}$$

and

$$\rho \dot{s} \geq \frac{r}{\theta} - \mathrm{div}\left(\frac{\mathbf{q}}{\theta}\right). \tag{A.82}$$

It is often convenient to replace, as we may, the Clausius-Duhem inequality by a linear combination with the balance of energy. In particular, subtracting from the (Lagrangian) equation of energy balance (A.76) the (Lagrangian) entropy inequality

(A.81) multiplied by the absolute temperature, yields:

$$\rho_0(\dot{u} - \theta\dot{s}) \leq T^{il}v_{i,I} - \frac{1}{\theta}Q^I\theta_{,I}, \tag{A.83}$$

which has the nice feature of not involving the radiation term R. Defining now the *(Helmholtz) free energy density* per unit mass as:

$$\psi = u - \theta s, \tag{A.84}$$

we can write (A.83) in the form:

$$\rho_0\dot{\psi} + \rho_0\dot{\theta}s - T^{il}v_{i,I} + \frac{1}{\theta}Q^I\theta_{,I} \leq 0. \tag{A.85}$$

Similarly, the Eulerian form of the modified Clausius-Duhem inequality is obtained as:

$$\rho_0\dot{\psi} + \rho_0\dot{\theta}s - t^{ij}D_{ij} + \frac{1}{\theta}q^i\theta_{,i} \leq 0. \tag{A.86}$$

We have completed the formulation of all the universal laws of Continuum Mechanics, that is, those laws that apply to all material bodies, regardless of their physical constitution. They are equally valid for solids, liquids and gases of any kind. Even a cursory count of equations reveals that these laws are not sufficient to solve for all the fields involved. Nor should they be. Any science must stop at some level of description, and Continuum Mechanics stops at what might be regarded as a rather high level, whereby its description of material response loses the character of what the philosopher of science Mario Bunge would call an *interpretive explanation*. This is hardly a deficiency: It is simply a definition of the point of view already conveyed by the very name of the discipline. It is this particular point of view, namely, the consideration of matter as if it were a continuum, that has paved the way for the great successes of Continuum Mechanics, both theoretical and practical. Most of Engineering is still based on its tenets and it is likely to remain so for as long as it strives to be guided by intellectually comprehensible models rather than by computer brute force. The price to pay is that the representation of the material response is not universal but must be tailored to each material or class of materials. This tailoring is far from arbitrary. It must respect certain principles, the formulation of which is the job of the constitutive theory.

A.10. A Modicum of Constitutive Theory

If we define the *history* of a body as the pair of its motion $\chi^i = \chi^i(X^I;t)$ and the absolute temperature function $\theta = \theta(X^I;t)$ for all points of the body and for all times in the interval $(-\infty,t]$, t being the present instant, then the principles of *causality* and *determinism* assert that the history completely determines the present values of the stress tensor ($\mathbf{T}$ or $\mathbf{t}$), the heat-flux vector ($\mathbf{Q}$ or $\mathbf{q}$), the internal energy density (u) and the entropy density (s). We also assume (*principle of equipresence*) that the list of independent variables appearing in each of the functionals just mentioned is, a priori, the same in all functionals. For example, if the temperature is a determining factor for the heat flux, then it should a priori be considered determining also for the stress. The theory of materials with memory (even fading memory) is beyond the scope of this primer. Instead, we will present only a few examples of *classes of materials*, each of these classes being characterized by a dependence not on the whole history of the motion and the temperature but just on the present values of the deformation gradient, the temperature, its gradient and possibly their time derivatives. Moreover, we will assume that these materials are *local*, so that the fluxes and densities at a point depend only on the values of the independent variables (just listed) at that point. Even with these very restrictive assumptions (which, nevertheless, are general enough to encompass most material models in widespread use), we will see that the remaining two tenets of the constitutive theory, namely, the *principle of material frame indifference* and the *principle of thermodynamic consistency*, are strong enough to impose severe restrictions upon the possible constitutive laws that one might measure in the laboratory. For experimentalists (unless they have grounds to challenge the established ideas) these principles constitute a bonanza, since they drastically reduce the type and number of experiments to perform. The general validity of the principle of material frame indifference and the particular methodology to implement thermodynamic consistency that we will present have both been challenged on various grounds, which we shall not discuss herein.

A.10.1. The Principle of Material Frame Indifference and Its Applications

In Section A.2 we have discussed the effect that a change of frame (or change of observer) has upon the motion. It is clear that a change of frame will also have an effect on all observable quantities, such as deformation gradients, vorticities, temperature gradients, and so on, the exact effect depending on the intrinsic or assumed nature of the quantity at hand. The principle of material frame indifference asserts that, although the independent and dependent variables of a given constitutive equation may be affected by a change of frame, the *constitutive functions themselves are not affected*, regardless of whether or not the frames are inertially related. In plain words, what the principle is stating is that material properties such as the stiffness of a spring, the heat conductivity of a substance or the coefficient of thermal expansion can be determined in any laboratory frame.[8]

Before we can proceed to apply this important principle to particular cases, we must establish once and for all how the measurements of some of the most common physical quantities change under a change of frame, namely, a transformation of the form (A.6). The most primitive quantity is the spatial distance between two simultaneous events. By construction, the most general change of frame involves

---

[8] In fact, part of the criticism levelled against the principle of material frame indifference stems from the conceptual possibility of having materials with microscopic gyroscopic effects at the molecular level, materials which would then be sensitive to the noninertiality of the frame.

just orthogonal spatial transformation, whence it follows that *all observers agree on the distance between two simultaneous events*. A scalar quantity, the result of whose measurement is independent of the frame, is called a *frame-indifferent scalar*. On physical grounds (for example, by claiming that the length of the mercury line of a thermometer is the distance between two simultaneous events, or perhaps through a more sophisticated argument or assumption) we establish that the *absolute temperature is a frame-indifferent scalar*. Since observers agree on length, they must surely agree on volume, and they certainly should agree on the counting of particles, so that it is reasonable to assume that *mass density is a frame-indifferent scalar*. On similar grounds we will agree that *internal energy density and entropy density are frame-indifferent scalars*.

Moving now to vector quantities, we start with the oriented segment **d** joining two simultaneous events, two flashlights blinking together in the dark, as it were. Denoting by x^i and y^i the spatial coordinates of the events, applying Equation (A.6) to each point and then subtracting, we obtain:

$$\hat{y}^i - \hat{x}^i = Q_j^i \, (y^j - x^j), \tag{A.87}$$

or

$$\hat{\mathbf{d}} = \mathbf{Q}\,\mathbf{d}. \tag{A.88}$$

A vector, such as **d**, that transforms in this manner is called a *frame-indifferent vector*. The unit normal **n** to a *spatial* element of area is frame indifferent, since it can be thought of as an arrow of the type just described. The unit normal **N** to a *referential* element of area, however, is not frame indifferent, since as far as the reference configuration is concerned, a change of frame has no consequence whatsoever. There is nothing sacred about being frame indifferent. In particular, the principle of material frame indifference *does not claim that the quantities involved in a constitutive law must be frame indifferent*. Quite to the contrary, it affirms that *even though the quantities involved are in general not frame indifferent, nevertheless the constitutive functionals themselves are invariant under a change of frame*.

Next, we consider the velocity vector **v**. The observed motions are related by:

$$\hat{x}^i(X^I;\hat{t}) = c^i(t) + Q_j^i(t) \, x^i(X^I;t), \quad \hat{t} = t + a, \tag{A.89}$$

where we have repeated Equation (A.6) to emphasize the fact that the referential variables X^I remain intact. Taking the partial derivative with respect to $\hat{t}$, we obtain:

$$\hat{v}^i = \dot{c}^i + \dot{Q}_j^i x^j + Q_j^i v^j, \tag{A.90}$$

or

$$\hat{\mathbf{v}} = \dot{\mathbf{c}} + \dot{\mathbf{Q}}\mathbf{x} + \mathbf{Q}\mathbf{v}, \tag{A.91}$$

so that, as expected, the velocity is not frame indifferent, not even if the observers are inertially related (an important clue in Galilean mechanics). Taking time derivatives once more, we obtain the following rule for the change of acceleration:

$$\hat{\mathbf{a}} = \ddot{\mathbf{c}} + \ddot{\mathbf{Q}}\mathbf{x} + 2\dot{\mathbf{Q}}\mathbf{v} + \mathbf{Q}\mathbf{a}. \tag{A.92}$$

Note that the acceleration, although not frame indifferent, will appear to be so if the observers happen to be inertially related.

EXERCISE A.7. **Coriolis and centripetal terms:** Express Equation (A.92) in the standard way found in textbooks of Classical Mechanics, namely, in terms of an angular velocity vector. Identify the Coriolis and centripetal terms. [Hint: since $\mathbf{Q}$ is orthogonal, its time derivative entails a skew-symmetric matrix. In an oriented 3-dimensional Euclidean space, a skew-symmetric matrix can be replaced by a vector acting on other vectors via the cross product.]

Moving now to tensors, we are bound to define a frame-indifferent tensor as a linear transformation that takes frame-indifferent vectors into frame-indifferent vectors. It follows from this criterion that a tensor $\mathbf{A}$ is frame indifferent if it transforms according to the formula:

$$\hat{\mathbf{A}} = \mathbf{QAQ^T}. \tag{A.93}$$

EXERCISE A.8. **Transformation of tensors:** Prove the above formula for the transformation of frame-indifferent tensors under a change of frame.

Assuming that *spatial forces are frame indifferent*, it follows, according to the definition above (since spatial normals are also frame indifferent), that *the Cauchy stress is a frame-indifferent tensor*. We check now the deformation gradient. By differentiation of (A.89) with respect to X^I, we obtain:

$$\hat{\mathbf{F}} = \mathbf{QF}, \tag{A.94}$$

which shows that the deformation gradient is not frame indifferent. Its determinant J, however, is a frame-indifferent scalar. The first Piola-Kirchhoff stress tensor $\mathbf{T}$ is not frame indifferent. It transforms according to:

$$\hat{\mathbf{T}} = \mathbf{QT}. \tag{A.95}$$

EXERCISE A.9. **The velocity gradient:** Show that the velocity gradient L is not frame indifferent and that it transforms according to the formula:

$$\hat{\mathbf{L}} = \mathbf{QLQ}^T + \dot{\mathbf{Q}}Q^T. \tag{A.96}$$

[Hint: notice that to obtain $\hat{\mathbf{L}}$, one must take derivatives of $\hat{\mathbf{v}}$ with respect to $\hat{\mathbf{x}}$.]

We are now in a position to apply the principle of material frame indifference to reduce the possible forms of constitutive equations. We will consider two examples:

1. **Elasticity**: A (local) material is said to be *elastic* if the stress at a point is a function of just the present value of the deformation gradient at that point, namely:

$$\mathbf{t} = \mathbf{f}(\mathbf{F}), \tag{A.97}$$

where $\mathbf{f}$ is a tensor-valued function. Leaving aside the other constitutive functions (such as the internal energy), we will find what restrictions, if any, the principle of material frame indifference imposes on this constitutive law. According to this principle, in another frame we must have:

$$\hat{\mathbf{t}} = \mathbf{f}(\hat{\mathbf{F}}), \qquad (A.98)$$

identically for all nonsingular tensors $\mathbf{F}$. Note the conspicuous absence of a hat over the function $\mathbf{f}$, which is the whole point of the principle of frame indifference. Using Equation (A.94) and the frame-indifferent nature of the Cauchy stress, we can write (A.98) as:

$$\mathbf{Q}\,\mathbf{f}(\mathbf{F})\,\mathbf{Q}^T = \mathbf{f}(\mathbf{QF}), \qquad (A.99)$$

which is an equation that the function $\mathbf{f}$ must satisfy identically for all nonsingular $\mathbf{F}$ and all orthogonal $\mathbf{Q}$, certainly a severe restriction. To make this restriction more explicit, we use the polar decomposition theorem to write:

$$\mathbf{Q}\,\mathbf{f}(\mathbf{RU})\,\mathbf{Q}^T = \mathbf{f}(\mathbf{QRU}). \qquad (A.100)$$

Since this is an identity, we can choose (at each instant) $\mathbf{Q} = \mathbf{R}^T$ and, rearranging some terms, we obtain:

$$\mathbf{t} = \mathbf{f}(\mathbf{RU}) = \mathbf{R}\,\mathbf{f}(\mathbf{U})\,\mathbf{R}^T. \qquad (A.101)$$

What this restriction means is that the dependence of the Cauchy stress can be arbitrary as far as the strain part ($\mathbf{U}$) of the deformation gradient is concerned, but the dependence on the rotational part is canonical. Notice that, since the tensor $\mathbf{U}$ is symmetric, the number of independent variables has dropped from 9 to 6.

2. **A viscous fluid**: In elementary textbooks on fluid mechanics one sometimes finds a statement to the effect that the shearing stress in some fluids is proportional to the velocity gradient. To substantiate this conclusion, an experiment is described whereby two parallel solid plates, separated by a layer of fluid, are subjected to opposite parallel in-plane forces and the resulting relative velocity is measured. Dividing this velocity by the distance between the plates, the statement appears to be validated, at least as a first approximation, by the behavior of many fluids. Without contesting the experimental result itself, we will now show that the conclusion (even if true for that particular experiment) cannot be elevated directly to the status of a general constitutive law.

We start from an assumed law of the form:

$$\mathbf{t} = \mathbf{f}(J, \mathbf{L}), \qquad (A.102)$$

so that the stress may arise as a consequence of change of volume (compressible fluid) as well as from a nonvanishing velocity gradient. Applying the principle of frame indifference and invoking (A.96), we obtain:

$$\mathbf{t} = \mathbf{Q}^T\,\mathbf{f}(J, \mathbf{QLQ}^T + \dot{\mathbf{Q}}\mathbf{Q}^T). \qquad (A.103)$$

Choosing $\mathbf{Q} = \mathbf{I}$ and $\dot{\mathbf{Q}} \neq 0$, we obtain:

$$\mathbf{t} = \mathbf{f}(J, \mathbf{L} + \Omega), \qquad (A.104)$$

where Ω is an arbitrary skew-symmetric tensor. We may, therefore, adjust instantaneously $\Omega = -\mathbf{W}$, thereby obtaining:

$$\mathbf{t} = \mathbf{f}(J, \mathbf{D}). \tag{A.105}$$

Further restrictions arise in this case, but we will not consider them here (see, however, Exercise A.13 below). The conclusion is that the stress cannot depend on the whole velocity gradient, but only on its symmetric part (to which, for example, it may be proportional).

EXERCISE A.10. **Homogeneity of space:** Show that neither the spatial position nor the velocity can appear as independent variables in a constitutive equation.

A.10.2. The Principle of Thermodynamic Consistency and Its Applications

The second law of Thermodynamics is a restriction that Nature imposes on the observable phenomena: Certain things simply cannot happen. The point of view we will adopt to ensure that those things that should not happen never come out as a solution of the equations of Continuum Mechanics is the following: We will exclude any constitutive law for which, under any conceivable process, the Clausius-Duhem inequality might be violated, even instantaneously. In the classes of materials we have been dealing with, this statement implies that the Clausius-Duhem inequality must hold true identically for any instantaneous combination of the independent variables and their space or time derivatives. We will explore the restrictions that are obtained from the principle of thermodynamic consistency in two cases:

1. **Thermoelastic heat conductors:** In this class of materials we assume, by definition, that the constitutive variables are functions of the deformation gradient, the temperature and the temperature gradient, namely:

$$T^{il} = T^{il}(F_J^j, \theta, \theta_{,J}), \tag{A.106}$$

$$Q^I = Q^I(F_J^j, \theta, \theta_{,J}), \tag{A.107}$$

$$s = s(F_J^j, \theta, \theta_{,J}), \tag{A.108}$$

and

$$\psi = \psi(F_J^j, \theta, \theta_{,J}), \tag{A.109}$$

where, for convenience, we have substituted the free energy ψ for the internal energy u. Notice, by the way, an application of the principle of equipresence: We have assumed exactly the same list of arguments for all constitutive variables, letting thermodynamics tell us whether or not an argument should be excluded from a particular constitutive law. We plug Equations (A.106 through A.109) into (A.85), use the chain rule and group terms together to obtain the following result:

$$\left[\rho_0 \frac{\partial \psi}{\partial F_J^j} - T_j^J \right] \dot{F}_J^j + \rho_0 \left[\frac{\partial \psi}{\partial \theta} + s \right] \dot{\theta} + \rho_0 \left[\frac{\partial \psi}{\partial \theta_{,J}} \right] \dot{\theta}_{,J} + \frac{1}{\theta} Q^J \theta_{,J} \leq 0, \tag{A.110}$$

where we have used the fact that $v^j_{,J} = \dot{F}^j_J$, by the symmetry of mixed partial derivatives. The inequality we have just obtained should be valid identically for all combinations of $\dot{F}^j_J, \dot{\theta}, \theta_{,J}$, and $\dot{\theta}_{,J}$. But this inequality is *linear* in $\dot{F}^j_J, \dot{\theta}$, and $\dot{\theta}_{,J}$, since these variables do not appear anywhere except as multipliers of other expressions. This is not the case for the variable $\theta_{,J}$, since the heat-flux vector which it multiplies may depend on it, according to our constitutive assumptions (A.106–A.109). Given that a linear function cannot have a constant sign over the whole domain, we conclude that the identical satisfaction of the inequality demands the satisfaction of the following equations:

$$T^J_j = \rho_0 \frac{\partial \psi}{\partial F^j_J}, \tag{A.111}$$

$$s = -\frac{\partial \psi}{\partial \theta}, \tag{A.112}$$

$$\frac{\partial \psi}{\partial \theta_{,J}} = 0, \tag{A.113}$$

and the *residual inequality*:

$$Q^J \theta_{,J} \le 0. \tag{A.114}$$

These restrictions can be summarized as follows: The free energy density is independent of the temperature gradient (Equation (A.113)) and acts as a *potential* for the stress (Equation (A.111)) and the entropy density (Equation (A.112)), both of which are, consequently, also independent of the temperature gradient. The ten constitutive functions F^j_J and s boil down, therefore, to a single scalar function ψ. Moreover, according to the residual inequality (A.114), heat cannot flow from lower to higher temperatures, since the heat-flux vector cannot form an acute angle with the temperature gradient. If, for example, we postulate Fourier's law of conduction which establishes that the heat-flux vector is proportional to the temperature gradient, we conclude that the constant of proportionality must be negative. The coefficient of heat conduction for real materials is in fact defined as the negative of this constant, so as to be positive.

REMARK A.4. **Thermoelasticity and hyperelasticity**: If we should limit the processes that a thermoelastic material undergoes to isothermally homogeneous processes (namely, with a constant temperature in time and in space), we would arrive at the conclusion that the resulting elastic material is *hyperelastic*, namely, it is completely characterized by a stress potential, which is obtained as the restriction of the free energy density to the pre-established temperature. This argument has sometimes been adduced to claim that all elastic materials are actually hyperelastic.

EXERCISE A.11. : **Symmetry of the stress:** Show that the fact that the first Piola-Kirchhoff stress derives from a scalar potential is consistent with the symmetry of the Cauchy stress.

Naturally, the thermodynamic restrictions just derived should be combined with those arising from the principle of material frame indifference.

2. **A viscous fluid**: We now revisit the viscous fluid, whose reduced constitutive Equation (A.105) we have obtained by means of the principle of material frame indifference. To derive further restrictions dictated by the principle of thermodynamic consistence, we will add the temperature to the list of independent variables and we will supplement (A.105) with equations for the heat flux, entropy density and free-energy density. The proposed constitutive laws are, therefore, of the form:

$$t^{ij} = t^{ij}(J, \mathbf{D}, \theta), \tag{A.115}$$

$$q^{i} = q^{i}(J, \mathbf{D}, \theta), \tag{A.116}$$

$$s = s(J, \mathbf{D}, \theta), \tag{A.117}$$

and

$$\psi = \psi(J, \mathbf{D}, \theta), \tag{A.118}$$

where we are using the Eulerian formulation, for a change. Substitution of these equations into (A.86) yields

$$\left[\rho \frac{\partial \psi}{\partial J} \delta^{ij} - t^{ij}\right] D_{ij} + \rho \left[\frac{\partial \psi}{\partial \theta} + s\right] \dot{\theta} + \left[\rho \frac{\partial \psi}{\partial D_{ij}}\right] \dot{D}_{ij} + \left[\frac{1}{\theta} q^{i}\right] \theta_{,i} \leq 0, \tag{A.119}$$

where the formula for the derivative of a determinant has been used. Notice that the content within the first square brackets cannot be claimed to vanish now, since the stress may depend on $\mathbf{D}$. On the other hand, the term within the last square brackets is now zero, since we have not assumed any dependence on the temperature gradient. This is, therefore, a fluid heat insulator. To summarize, we obtain the equations:

$$\frac{\partial \psi}{\partial D_{ij}} = 0, \tag{A.120}$$

$$s = -\frac{\partial \psi}{\partial \theta}, \tag{A.121}$$

and

$$q^{i} = 0 \tag{A.122}$$

plus the residual inequality:

$$\text{trace}(\mathbf{t}_{irr} \mathbf{D}) \geq 0, \tag{A.123}$$

where we have defined the *irreversible part of the stress* as $\mathbf{t}_{irr} := \mathbf{t} - \mathbf{t}_{rev}$. The reversible part $\mathbf{t}_{rev}$ is the hydrostatic (spherical) expression:

$$\mathbf{t}_{rev} = \rho \frac{\partial \psi}{\partial J} J \mathbf{I}. \tag{A.124}$$

If, for example, the irreversible, or dissipative, part of the stress is assumed to be proportional to the rate of deformation $\mathbf{D}$, then the principle of thermodynamic consistency prescribes that the constant of proportionality, or *viscosity*, must be positive.

A.10.3. Material Symmetries

The principle of material frame indifference establishes a *universal symmetry* of all constitutive laws. Indeed, it postulates that all constitutive laws are *invariant* under the (left) action of the orthogonal group, an action which is well defined by what we have informally described as "the way different quantities transform under a change of frame." Quite apart from this universal *spatial* symmetry, constitutive laws may have additional symmetries that arise from their *material* constitution. Although some of these symmetries, in the case of solid materials, may ultimately be due to the geometric symmetries of the underlying crystal lattices, the two concepts should not be confused, for a number of reasons which is not our intention to discuss at this point. Suffice it to say that, in particular, constitutive equations may enjoy continuous symmetries, even if the underlying molecular structure is a regular lattice. From the point of view of a given constitutive law the only question that matters is the following: Which are those material transformations that, applied before each possible deformation, do not affect the response due to this deformation? Once the constitutive law is given, this is a purely mathematical question.

For the sake of brevity, we will consider the case of an elastic material, whose constitutive equation consists of a single tensor function $\mathbf{t}$ (the Cauchy stress) of a single tensor variable $\mathbf{F}$ (the deformation gradient). A somewhat simpler case would be a hyperelastic material, characterized by a single scalar function of $\mathbf{F}$. We are focusing our attention on a single point of the body, leaving for Chapter 8 the fascinating, geometrically rich, idea of comparing the responses of different points. A symmetry will, therefore, consist of a material deformation gradient[9] $\mathbf{G}$ that, when preapplied to any further deformation gradient $\mathbf{F}$, does not affect the value of $\mathbf{t}$, that is:

$$\mathbf{t}(\mathbf{FG}) = \mathbf{t}(\mathbf{F}), \tag{A.125}$$

for all nonsingular tensors $\mathbf{F}$. More pictorially, if an experimentalist has measured some material property and, about to repeat the experiment, leaves the laboratory for a moment, during which interval an evil genie decides to rotate the sample through some angle, this rotation will be a symmetry if the repetition of the experiment with the rotated sample cannot lead the experimentalist to detect or even suspect that the sample has been rotated.

It is not difficult to check that the collection of all material symmetries of a given constitutive law forms a group, called the *material symmetry group* $\mathcal{G}$ of the given law. We can say, therefore, that the constitutive law is invariant under the (right) action of its symmetry group. The symmetry group is never empty, since the identity transformation is always a symmetry.

EXERCISE A.12. **The symmetry group:** Verify that the symmetries form a group whose operation is matrix multiplication. Recall that all you need to check is associativity, existence of a unit and existence of inverses, all of which are borrowed from ordinary matrix multiplication.

---

[9] A material symmetry need not be orientation preserving, so that its determinant may be negative.

To speak of symmetries as matrices, we need to choose a reference configuration, at least locally, and the symmetry group $\mathcal{G}$ depends on the reference chosen. No information is gained or lost, however, and the various symmetry groups thus obtained are mutually *conjugate*. In principle, any nonsingular matrix could act as a symmetry of a constitutive law, so that the symmetry group could be as large as the general linear group $GL(3;\mathbb{R})$. It is usually assumed, however, that symmetries are volume preserving, which implies that all possible symmetry groups are subgroups of the *unimodular group* $\mathcal{U}$ (understood as the multiplicative group of all matrices whose determinant is equal to ± 1). An elastic material is a *solid* if its symmetry group, with respect to some reference configuration, is a subgroup of the *orthogonal group* $\mathcal{O}$. A material is called *isotropic* if its symmetry group, for some reference configuration, contains the orthogonal group. It follows that an isotropic solid is a material whose symmetry group, for some (local) reference configuration (called an *undistorted state*), is exactly the orthogonal group. A material for which the symmetry group is the whole unimodular group $\mathcal{U}$ is called a *fluid*. Naturally, in this case the group is independent of reference configuration. This notion of fluidity corresponds to the intuitive idea that a fluid (at rest) adapts itself to the shape of the receptacle without any difficulty. Similarly, the idea of solidity is that, starting from a natural stress-free state of the material, one may perhaps find rotations that leave the response unaffected, but certainly any strain ($\mathbf{U} \neq \mathbf{I}$) will be felt. A theorem in group theory allows one to establish that an isotropic material (of the type we are discussing) must be either an isotropic solid or a fluid. There is nothing in between. A material, therefore, is either a solid ($\mathcal{G} \subseteq \mathcal{O}$), a fluid ($\mathcal{G} = \mathcal{U}$), or a *crystal fluid*[10] ($\mathcal{G} \cap \mathcal{O} \neq \mathcal{O}$ and $\mathcal{G} \cap (\mathcal{U} - \mathcal{O}) \neq \emptyset$). As just remarked, crystal fluids cannot be isotropic.

To illustrate in which way the existence of a nontrivial symmetry group results in a further reduction of a constitutive law, beyond the reductions imposed by the principles of material frame indifference and of thermodynamic consistency, we will now exhibit the two classical examples of the representation of the most general constitutive law of an elastic fluid and an isotropic elastic solid.

For the elastic fluid, the symmetry condition reads:

$$\mathbf{t}(\mathbf{F}) = \mathbf{t}(\mathbf{FG}), \quad \forall \ \mathbf{F} \in GL(3;\mathbb{R}), \ \mathbf{G} \in \mathcal{U}. \tag{A.126}$$

Choosing, as we may, $\mathbf{G} = (\det \mathbf{F})^{1/3} \, (\mathbf{F}^{-1})$, we conclude that the most general constitutive equation of an elastic fluid must be of the form:

$$\mathbf{t} = \mathbf{t}(\det \mathbf{F}), \tag{A.127}$$

which completely agrees with the intuitive idea that an elastic fluid (such as an ideal gas) is, mechanically speaking, sensitive only to changes in volume.

---

[10] Not to be confused with a liquid crystal, which is a different kind of material.

As far as the isotropic elastic solid is concerned, we start by stating the symmetry condition in terms of the Cauchy stress,[11] namely:

$$\mathbf{t}(\mathbf{F}) = \mathbf{t}(\mathbf{FG}), \quad \forall \; \mathbf{F} \in GL(3;\mathbb{R}), \; \mathbf{G} \in \mathcal{O}. \tag{A.128}$$

Invoking the polar decomposition theorem A.1, we can set $\mathbf{F} = \mathbf{VR}$ and then choose $\mathbf{G} = \mathbf{R}^T$, thus obtaining a first reduction:

$$\mathbf{t} = \mathbf{t}(\mathbf{V}), \tag{A.129}$$

or, equivalently, since $\mathbf{B} = \mathbf{V}^2$,

$$\mathbf{t} = \mathbf{t}(\mathbf{B}), \tag{A.130}$$

which states that, for an isotropic elastic solid, the Cauchy stress is a function of the left Cauchy-Green tensor $\mathbf{B} = \mathbf{FF}^T$. We may now exploit the principle of material frame indifference to write:

$$\mathbf{Q}\mathbf{t}(\mathbf{B})\mathbf{Q}^T = \mathbf{t}(\mathbf{QBQ}^T), \tag{A.131}$$

for all orthogonal matrices $\mathbf{Q}$. Thus, the stress is an *isotropic function* of its matrix argument. Notice that both quantities are represented by symmetric matrices. A well-known theorem establishes, therefore, that the most general form that this function can attain is:

$$\mathbf{t} = \phi_0 \mathbf{I} + \phi_1 \mathbf{B} + \phi_2 \mathbf{B}^2, \tag{A.132}$$

where the ϕ_0, ϕ_1 and ϕ_2 are arbitrary functions of the (positive) eigenvalues of $\mathbf{B}$. This representation is, of course, valid only if the tensor $\mathbf{B}$ is measured from an undistorted state.

EXERCISE A.13. **Revisiting the viscous fluid once more:** What further reductions does the principle of material frame indifference impose on the constitutive Equation (A.105)? [Hint: use the representation theorem just mentioned.]

The representation of the general constitutive law for an isotropic elastic solid is one of the first results of the Continuum Mechanics renaissance that had its beginnings after the end of the Second World War. Having thus come full circle, it appears that this is a good point to conclude our primer.

---

[11] A simpler treatment would follow from assuming the isotropic solid to be hyperelastic, characterized, therefore, by a single scalar-valued function of $\mathbf{F}$.

Index

Printed in the United States
By Bookmasters